Finite Automata

Mark V. Lawson

A CRC Press Company
Boca Raton London New York Washington, D.C.

Library of Congress Cataloging-in-Publication Data

Lawson, Mark V.
Finite automata / Mark V. Lawson.
p. cm.
Includes bibliographical references and index.
ISBN 1-58488-255-7 (alk. paper)
1. Machine theory. I. Title.

QA267.L37 2003
511.3—dc21 2003055167

Direct all inquiries to CRC Press LLC, 2000 N.W. Corporate Blvd., Boca Raton, Florida 33431.

Visit the CRC Press Web site at www.crcpress.com

International Standard Book Number 1-58488-255-7
Library of Congress Card Number 2003055167
Printed in the United States of America 1 2 3 4 5 6 7 8 9 0
Printed on acid-free paper

For my teacher, T. Guérin-leTendre

Selves — goes itself; *myself* it speaks and spells,
Crying *Whát I dó is me: for that I came.*

Gerard Manley Hopkins

Contents

Preface

Introduction

The theory of finite automata is the mathematical theory of a simple class of algorithms that are important in computer science. Algorithms are recipes that tell us how to solve problems; the rules we learn in school for adding, subtracting, multiplying and dividing numbers are good examples of algorithms. Although algorithms have always been important in mathematics, mathematicians did not spell out precisely what they were until the early decades of the twentieth century, when they started to ask questions about the nature of mathematical proof. Perhaps motivated in part by the mechanical devices that were used to help in arithmetical calculations, mathematicians wanted to know whether there were mechanical ways of proving theorems. By mechanical, they did not intend to build real machines that could prove theorems; rather they wanted to know whether it was possible *in principle* to prove theorems mechanically. In mathematical terms, they wanted to know whether there was an algorithm for proving the theorems of mathematics. The answer to this question obviously depended on what was meant by the word 'algorithm.' Numerous definitions were suggested by different mathematicians from various perspectives and, remarkably, they all proved to be equivalent. By the end of the 1930s, mathematicians had a precise definition of what an algorithm was, and using this definition they were able to show that there were problems that could not be solved by algorithmic means. It was perhaps no accident that, as mathematicans were laying the foundations of the theory of algorithms, engineers were constructing machines that could implement algorithms as programs. Algorithms and programs are just two sides of the same coin.

Mathematicians were initially interested in the dividing line between what was and what was not algorithmic. But simply knowing that a problem could be solved by means of an algorithm was not the end of the story. Certainly, this implied you could write a program to solve the problem, but did not imply your program would run quickly or efficiently. For this reason, two approaches to classifying algorithms were developed. The first classified them according to their running times and is known as *complexity theory*, whereas the second classified them according to the types of memory used in implementing them and is known as *language theory*. In language theory, the simplest algorithms are those which can be implemented by finite automata, the subject of this book.

Finite automata were first studied in the 1950s by Stephen Kleene, and found a number of important applications in computer science: for example, in the design of computer circuits, and in the lexical analyzers of compilers. In the 1960s and 1970s, mathematicians such as Samuel Eilenberg, Marcel-Paul Schützenberger, and John Rhodes pioneered the mathematics of finite automata. More recently, other mathematicians have come to appreciate the usefulness of automata in such areas as combinatorial group theory and symbolic dynamics.

This book is intended to be an introduction to the mathematical theory of finite automata, assuming as background only a first course in discrete mathematics. The Appendix outlines the prerequisites.

Structure of the book

The book is notionally divided into two parts: Chapters 1 to 6 form the first part, and Chapters 7 to 12 the second; Chapter 7 is the bridge that links them.

PART I. This centres on Kleene's Theorem, the first major result proved about finite automata. I describe two different ways of proving this theorem with Chapter 1 common to both:

- The quickest route to proving Kleene's Theorem is the following: Sections 3.1–3.3 for constructions involving non-deterministic automata, Section 5.1 for the definition of regular expressions, and then Theorem 5.2.1 of Section 5.2, which is the proof of Kleene's Theorem itself. Chapter 2, omitting Section 2.5, can be regarded as a collection of examples.

- The route that emphasises algorithms more than proofs is the following: Sections 2.1–2.3 for practice in designing automata, Sections 3.1 and 3.2 for the accessible subset construction, Section 4.1 and Theorem 4.2.1 of Section 4.2 for ε-automata, Section 5.1 for regular expressions, and Section 5.3 for an algorithmic proof of Kleene's Theorem.

Section 2.6 on the Pumping Lemma can be read at any point after Chapter 1. Section 5.4 describes an algebraic technique for converting automata into regular expressions based on solving equations. Chapter 6, on local languages, describes a different way of converting regular expressions into automata from that used in any of the above proofs of Kleene's theorem.

PART II. This centres on the algebraic theory of recognisable languages, the main goals being Schützenberger's characterisation of star-free languages, and the Variety Theorem of Eilenberg and Schützenberger. Chapters 7 and 8 have a strong algorithmic flavour, but Chapters 9–12 are increasingly mathematical.

Chapter 7 describes how to find the smallest automaton recognising a given language. Two different techniques are given depending on whether the language is described by an automaton or by a regular expression. If the former, then Section 7.2 describes the algorithm for converting the automaton into a minimal automaton. The theory of minimal automata is developed in Sections 7.3 and 7.4. If you just want the algorithm for minimising an automaton then Sections 7.1 and 7.2 are all you need. If the language is described by a regular expression, then there is a beautiful technique, called the Method of Quotients, which will construct the minimal automaton directly from the expression. Unfortunately, there are 'issues' connected with this method, but I have relegated a discussion of these to the end of Section 7.5.

The minimal automaton is obviously important from a practical point of view, but it is also the starting point of an algebraic technique for studying recognisable languages. This is introduced in Chapter 8. Here the transition monoid of an automaton is defined and an algorithm described for computing it. At the conclusion of this chapter, semigroups and monoids are introduced.

The *point* of Chapter 8 will only become clear in Chapter 9, when we prove the algebraic counterpart of Kleene's Theorem: a language is recognisable if and only if its syntactic monoid is finite. The syntactic monoid of a recognisable language is isomorphic to the transition monoid of the minimal automaton of the language.

The main result of Chapter 9 tells us that finite monoids may be useful in studying recognisable languages. Because of this, Chapter 10 develops the theory of finite monoids we shall need. We also show how results about recognisable languages, which we previously proved using automata, can also be proved using monoids.

The real justification for studying the syntactic monoid of a recognisable language comes in Chapter 11, where we show that an important class of recognisable languages are characterised by the algebraic properties of their syntactic monoids. Specifically, we prove Schützenberger's Theorem: a language is star-free if and only if its syntactic monoid is aperiodic.

Schützenberger's Theorem opened up a whole new area of research: classifying recognisable languages by means of the algebraic properties of their syntactic monoids. In Chapter 12, we prove the Variety Theorem of Eilenberg and Schützenberger, which provides the template for proving results of this type.

Acknowledgements

In writing this book, I have used both original papers and textbooks. Individual debts are recorded in the remarks at the end of each chapter, but I would like to highlight here a few books that have been indispensable in writing my own. I learned automata theory originally from Carroll and Long [23] and Hopcroft and Ullman [65]. Their influence will be apparent in many places, particularly in the first seven chapters. The first edition of John Howie's book, now rewritten as [68], introduced me to semigroup theory. I learned about the algebraic approach to languages from Jean-Eric Pin's book [103], which is a model of concision and clarity. Anyone who works through my book will be well-placed to read Pin's. In addition, I consulted Perrin's lengthy survey article [100] and Brauer's *Handbuch* on finite automata [16] on numerous occasions. McNaughton and Papert's book [88] on star-free languages, despite being over thirty years old, is still inspiring.

On a personal level, I am especially grateful to John Fountain for reading an early version of the manuscript and for making numerous suggestions that led to considerable improvements in the exposition. I owe a further debt to John in that he taught me semigroup theory as an undergraduate, and supervised my DPhil thesis, also in semigroup theory. Victoria Gould and Peter Higgins read parts of the manuscript, contributed ideas, and corrected mistakes, as did Stuart Margolis, Benjamin Steinberg, and my algebraic Scottish colleagues in St. Andrews. At CRC Press, I would like to thank editorial assistant Jasmin Naim for being unfailingly helpful during the time I was writing the book, and Nishith Arora, Michele Berman, Suzanne Lassandro, Evelyn Meany, Jamie Sigal and the anonymous copyeditor for their assistance during the production stage. Needless to say, all errors remaining are my own.

Contact

Comments on this book are welcome. I can be contacted at:

`m.v.lawson@informatics.bangor.ac.uk`

My Website can be found at:

`http://www.informatics.bangor.ac.uk/~mvlawson/`

where I shall keep book-related information.

Chapter 1

Introduction to finite automata

The aim of this chapter is to set out the basic definitions and ideas we shall use throughout this book. In particular, we explain what we mean by a finite automaton and the language recognised by a finite automaton. Background results from discrete mathematics are outlined in the Appendix.

1.1 Alphabets and strings

Most people today are familiar with the idea of digitising information; that is, converting information from an analogue or continuous form to a discrete form. For example, it is well-known that computers deal only in 0's and 1's, but users of computers do not have to communicate with them in binary; they can interact with the computer in a great variety of ways. For example, voice recognition technology enables us to input data without using the keyboard, whereas computer graphics can present output in the form of animation. But these things are only possible because of the underlying sequences of 0's and 1's that encode this information. We begin this section therefore by examining sequences of symbols and their properties.

Information in all its forms is usually represented as sequences of symbols drawn from some fixed repertoire of symbols. More formally, any set of symbols A that is used in this way is called an *alphabet*, and any finite sequence whose components are drawn from A is called a *string over A* or simply a *string*.[1] We call the elements of an alphabet *symbols* or *letters*. The number of symbols in an alphabet A is denoted by $|A|$. The alphabets in this book will always be finite.

[1] The term *word* is often used instead of string.

Examples 1.1.1 Here are a few examples of alphabets you may have encountered.

(1) An alphabet suitable for describing the detailed workings of a computer is $\{0, 1\}$.

(2) An alphabet for representing natural numbers in base 10 is

$$\{0, 1, 2, 3, 4, 5, 6, 7, 8, 9\}.$$

(3) An alphabet suitable for writing stories in English is

$$\{a, \ldots, z, A \ldots, Z, ?, \ldots\},$$

upper and lower case letters together with punctuation symbols and a space symbol to separate different words.

(4) An alphabet for formal logic is $\{\exists, \forall, \neg, \wedge, \ldots\}$.

(5) The alphabet used in describing a programming language is called the set of *tokens* of the language. For example, in the C language, the following are all tokens:

main, printf, {, }.

(6) DNA is constructed from four main types of molecules: adenine (A), cytosine (C), guanine (G), and thymine (T). Sequences of these molecules, and so strings over the alphabet $\{A, C, G, T\}$, form the basis of genes.

□

The symbols in an alphabet do not have to be especially simple. An alphabet could consist of pictures, or each element of an alphabet could itself be a sequence of symbols. Thus the set of all Chinese characters is an alphabet in our sense although it is not an alphabet in the linguistic sense, as is the set of all words in an ordinary dictionary — a word like 'egalitarianism' would, in this context, be regarded as a single symbol. An important example of using sequences of symbols over one alphabet to represent the elements of another alphabet occurs with ASCII encoding, and also forms the basis of data-compression and error-correction codes. You might wonder why, when all information can be encoded in binary, we do not just stick with the alphabet $\{0, 1\}$. The reason is one of convenience: binary is good for computers and bad for people. That said, most of the alphabets we use in this book will just have a few elements but, again, that is just for convenience.

A string is a list and so it is formally written using brackets and commas to separate components. Thus $(0, 1, 1, 1, 0)$ is a string over the alphabet $A = \{0, 1\}$, whereas $(\text{to}, \text{be}, \text{or}, \text{not}, \text{to}, \text{be})$ is a string over the alphabet whose elements are the words in an English dictionary. The string $()$ is the empty string. However, for the remainder of this book, we shall write strings without brackets and commas and so for instance we write 01110 rather than $(0, 1, 1, 1, 0)$. The empty string needs to be recorded in some way and we denote it by ε. The set of all strings over the alphabet A is denoted by A^*, read *A star*, and the set of all strings except the empty one is denoted by A^+, read *A plus*.

If w is a string then $|w|$ denotes the total number of symbols appearing in w and is called the *length of w*. If $a \in A$ then $|w|_a$ is the total number of a's appearing in w. For example, $|\varepsilon| = 0$, and $|01101| = 5$; $|01101|_0 = 2$, and $|01101|_1 = 3$. Two strings u and v over an alphabet A are *equal* if they contain the same symbols in the same order.

Given two strings $x, y \in A^*$, we can form a new string $x \cdot y$, called the *concatenation of x and y*, by simply adjoining the symbols in y to those in x. For example, if $A = \{0, 1\}$ then both 0101 and 101010 are strings over A. The concatenation of 0101 and 101010 is denoted $0101 \cdot 101010$ and is equal to the string 0101101010. We shall usually denote the concatenation of x and y by xy rather than $x \cdot y$. If $x, y \in A^*$ then $|xy| = |x| + |y|$; when we concatenate two strings the length of the result is the sum of the lengths of the two strings. The string ε has a special property with respect to concatenation: for each string $x \in A^*$ we clearly have that $\varepsilon x = x = x\varepsilon$.

There is one point that needs to be emphasised: **the order in which strings are concatenated is important**. For example, if $A = \{a, b\}$ and $u = ab$ and $v = ba$ then $uv = abba$ and $vu = baab$ and clearly $uv \neq vu$. We have all been made painfully familiar with this fact: the spelling of the word 'concieve' is wrong, whereas the spelling 'conceive' is correct. This is because 'order matters' in spelling. In the case where A consists of only one letter, then the order in which we concatenate strings is immaterial. For example, if $A = \{a\}$ then strings in A^* are just sequences of a's, and clearly, the order in which we concatenate strings of a's is not important.

Given three strings x, y, and z, there are two distinct ways to concatenate them in this order: we can concatenate x and y first to obtain xy and then concatenate xy with z to obtain xyz, or we can concatenate y and z first to obtain yz and then concatenate x with yz to obtain xyz again. In other words, $(xy)z = x(yz)$. We say that concatenation is *associative*.[2]

If x is a string then we write x^n, when $n \geq 1$, to mean the concatenation of x with itself n-times. We define $x^0 = \varepsilon$. For example, if $x = ba$ then $(ba)^2 = baba$. The usual laws of indices hold: if $m, n \geq 0$ then $x^m x^n = x^{m+n}$.

Let $x, y, z \in A^*$. If $u = xyz$ then y is called a *factor* of u, x is called a *prefix* of u, and z is called a *suffix* of u. We call the factor y *proper* if at least one of x and z is not just the empty string. In a similar fashion we say that the prefix x (resp. suffix z) is *proper* if $x \neq u$ (resp. $z \neq u$). We say that the string u is a *substring* of the string v if $u = a_1 \ldots a_n$, where $a_i \in A$, and there

---

[2] See Chapter 9 for more on associativity. Specifically, Theorem 9.1.1 is the key property of associative operations. The proof of this result is less important than what it says.

exist strings $x_0, \ldots, x_n$ such that $v = x_0 a_1 x_1 \ldots x_{n-1} a_n x_n$. Let $x \in A^*$. We call a representation $x = u_1 \ldots u_n$, where each $u_i \in A^*$, a *factorisation of* x.

For example, consider the string $u = abab$ over the alphabet $\{a, b\}$. Then the prefixes of u are: $\varepsilon, a, ab, aba, abab$; the suffixes of u are: $\varepsilon, b, ab, bab, abab$; and the factors of u are: $\varepsilon, a, b, ab, ba, aba, bab, abab$. The strings aa, bb, abb are examples of substrings of u. Finally, $u = ab \cdot ab$ is a factorisation of u; observe that I use the $\cdot$ to emphasise the factorisation.

When discussing strings over an alphabet, it is useful to have a standard way of listing them. This can easily be done using what is known as the *tree order*[3] on A^*. Let $A = \{a_1, \ldots, a_n\}$ be an alphabet. Choose a fixed linear order for the elements of A. This is usually obvious, for example, if $A = \{0, 1\}$ then we would assume that $0 < 1$ but in principle any ordering of the elements of the alphabet may be chosen, but if a non-standard ordering is to be used then it has to be explicitly described. We now grow a tree, called the *tree over* A^*, whose root is ε and whose vertices are labelled with the elements of A^* according to the following recipe: if w is a vertex, then the vertices growing out of w are $wa_1, \ldots, wa_n$. The *tree order on* A^* is now obtained as follows: $x < y$ if and only if $|x| < |y|$, or $|x| = |y|$ and the string x occurs to the left of the string y in the tree over A^*. To make this clearer, we do a simple example. Let $A = \{0, 1\}$, where we assume $0 < 1$. The first few levels of the tree over A^* are:

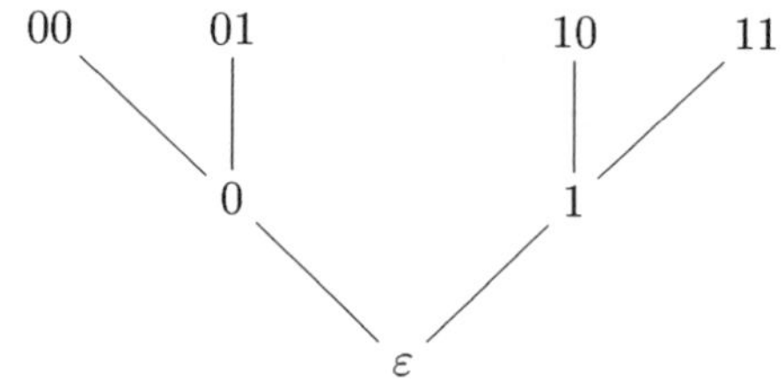

Thus the tree order for A^* begins as follows

$$\varepsilon, 0, 1, 00, 01, 10, 11, \ldots.$$

This ordering amounts to saying that a string precedes all strictly longer strings, while all the strings of the same length are listed *lexicographically*, that is to say the way they are listed in a dictionary[4] based on the ordering of the alphabet being used.

[3] Also known as the 'length-plus-lexicographic order,' which is more of a mouthful, and the 'ShortLex order.'

[4] Also known as a *lexicon*.

Exercises 1.1

1. Write down the set of prefixes, the set of suffixes, and the set of factors of the string,

aardvark,

over the alphabet $\{a, \ldots, z\}$. When writing down the set of factors, list them in order of length. Find three substrings that are not factors.

2. Let $A = \{a, b\}$ with the order $a < b$. Draw the tree over A^* up to and including all strings of length 3. Arrange these strings according to the tree order.

3. Let $A = \{a, b, c\}$. If $cc < ca < bc$ in the tree order, what is the corresponding linear order for the elements of A?

4. Let A be an alphabet. Prove that A^* is *cancellative* with respect to concatenation, meaning that if $x, y, z \in A^*$ then $xz = yz$ implies $x = y$, and $zx = zy$ implies $x = y$.

5. Let $x, y, u, v \in A^*$. Suppose that $xy = uv$. Prove the following hold:

(i) If $|x| > |u|$, then there exists a non-empty string w such that $x = uw$ and $v = wy$.

(ii) If $|x| = |u|$, then $x = u$ and $y = v$.

(iii) If $|x| < |u|$, then there exists a non-empty string w such that $u = xw$ and $y = wv$.

6. In general, if $u, v \in A^+$, then the strings uv and vu are different as we have noted. This raises the question of finding conditions under which $uv = vu$. Prove that the following two conditions are equivalent:

(i) $uv = vu$.

(ii) There exists a string z such that $u = z^p$ and $v = z^q$ for some natural numbers $p, q > 0$.

You can use Question 5 in solving this problem. Proving results about strings is often no easy matter. More combinatorial properties of strings are described in [82].

1.2 Languages

Before defining the word 'language' formally, here is a motivating example.

Example 1.2.1 Let A be the alphabet that consists of all words in an English dictionary. So A contains a very large number of elements: of the order of half a million. As we explained in Section 1.1, we can think of each English word as being a single symbol. The set A^* consists of all possible finite sequences of words. An important subset L of A^* consists of all sequences of words that form grammatically correct English sentences. Thus the sequence (to,be,or,not,to,be)$\in L$ whereas (be,be,to,to,or,not) $\notin L$. Someone who wants to understand English has to learn the rules for deciding when a string of words belongs to the set L. We can therefore think of L as being the 'English language.'[5]
□

This example motivates the following definition. For *any* alphabet A, *any* subset of A^* is called an *A-language*, or a *language over A* or simply a *language*.

Examples 1.2.2 Here are some examples of languages.

(1) In elementary arithmetic we use the alphabet,

$$A = \{0, \ldots, 9\} \cup \{+, \times, -, \div, =\} \cup \{(,)\}.$$

We can form the language L of all correct sums: thus the sequence $2 + 2 = 4$ is in L whereas the sequence $1 \div 0 = 42$ is not. Any totally meaningless string such as $\div + = 98 =$ also fails to be in L.

(2) In computer science, the set of all syntactically correct programs in a given computer language, such as Java, constitutes a language.

(3) Both $\emptyset$ and A^* are languages over A: the smallest and the largest, respectively.

□

Languages also arise from what are known as *decision problems*, which are problems whose answer is either 'yes' or 'no.' For example, we can ask whether a number is prime; the answer to this question is either a 'yes' or a 'no.' We illustrate how decision problems give rise to languages by means of an example.

Example 1.2.3 A simple graph is one with no loops and no multiple edges. A graph is said to be *connected* if any two vertices can be joined by a path. Clearly a graph is either connected or not. Thus 'Is the simple graph G connected?' is an example of a decision problem. We now show how to construct a language from this decision problem.

---

[5] In reality, membership of this set is sometimes problematic, but the languages we meet in practice will be formal and always clearly defined.

A simple graph can be represented by an adjacency matrix whose entries consist of 0's and 1's. For example, the graph G below,

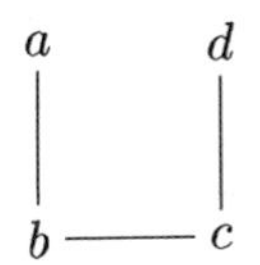

is represented by the following adjacency matrix:

$$\begin{pmatrix} 0 & 1 & 0 & 0 \\ 1 & 0 & 1 & 0 \\ 0 & 1 & 0 & 1 \\ 0 & 0 & 1 & 0 \end{pmatrix}$$

The adjacency matrix can be used to construct a binary string: just concatenate the rows of the matrix. Thus the graph G above is represented by

$$0100 \cdot 1010 \cdot 0101 \cdot 0010 = 0100101001010010,$$

which we denote by code(G). It is clear that given a binary string we can determine whether it encodes a graph and, if so, we can recapture the original graph. Thus every simple graph can be encoded by a string over the alphabet $A = \{0, 1\}$. Let

$$L = \{x \in \{0,1\}^*: x = \text{code}(G) \text{ where } G \text{ is connected}\}.$$

This is a language that corresponds to the decision problem:

'Is the simple graph G connected?'

in the sense that G answers yes if and only if code(G) $\in L$. □

This example provides evidence that languages are likely to be important in both mathematics and computer science.

Exercises 1.2

1. Is the string 0101101001011010 in the language L of Example 1.2.3?

2. Describe precise conditions for a binary string to be of the form code(G) for some simple graph G.

1.3 Language operations

In Section 1.2, we introduced languages as they will be understood in this book. We shall now define various operations on languages: that is, ways of combining languages to make new ones.

If X is any set, then $\mathsf{P}(X)$ is the set of all subsets of X, the *power set of* X. Now let A be an alphabet. A language over A is any subset of A^*, so that the set of all languages over A is just $\mathsf{P}(A^*)$. If L and M are languages over A so are $L \cap M$, $L \cup M$ and $L \setminus M$ ('relative complement'). If L is a language over A, then $L' = A^* \setminus L$ is a language called the *complement of* L. The operations of intersection, union, and complementation are called *Boolean operations* and come from set theory. Recall that '$x \in L \cup M$' means '$x \in L$ or $x \in M$ or both.' In automata theory, we usually write $L + M$ rather than $L \cup M$ when dealing with languages.

Notation If L_i is a family of languages where $1 \leq i \leq n$, then their union will be written $\sum_{i=1}^{n} L_i$.

There are two further operations on languages that are peculiar to automata theory and extremely important: the product and the Kleene star.

Let L and M be languages. Then

$$L \cdot M = \{xy\colon x \in L \text{ and } y \in M\}$$

is called the *product of* L *and* M. We usually write LM rather than $L \cdot M$. A string belongs to LM if it can be written as a string in L *followed by* a string in M. In other words, the product operation enables us to talk about the order in which symbols or strings occur.

Examples 1.3.1 Here are some examples of products of languages.

(1) $\emptyset L = \emptyset = L\emptyset$ for any language L.

(2) $\{\varepsilon\}L = L = L\{\varepsilon\}$ for any language L.

(3) Let $L = \{aa, bb\}$ and $M = \{aa, ab, ba, bb\}$. Then

$$LM = \{aaaa, aaab, aaba, aabb, bbaa, bbab, bbba, bbbb\}$$

and

$$ML = \{aaaa, aabb, abaa, abbb, baaa, babb, bbaa, bbbb\}.$$

In particular, $LM \neq ML$ in general.

□

For a language L we define $L^0 = \{\varepsilon\}$ and $L^{n+1} = L^n \cdot L$. For $n > 0$ the language L^n consists of all strings u of the form $u = x_1 \ldots x_n$ where $x_i \in L$.

The *Kleene star* of a language L, denoted L^*, is defined to be

$$L^* = L^0 + L^1 + L^2 + \ldots.$$

We also define

$$L^+ = L^1 + L^2 + \ldots.$$

Examples 1.3.2 Here are some examples of the Kleene star of languages.

(1) $\emptyset^* = \{\varepsilon\}$ and $\{\varepsilon\}^* = \{\varepsilon\}$.

(2) The language $\{a^2\}^*$ consists of the strings,

$$\varepsilon, a^2, a^4 = a^2a^2, a^6 = a^2a^2a^2, \ldots.$$

In other words, all strings over the alphabet $\{a\}$ of even length (**remember**: the empty string has even length because 0 is an even number).

(3) A string u belongs to $\{ab, ba\}^*$ if it is empty or if u can be factorised $u = x_1 \ldots x_n$ where each x_i is either ab or ba. Thus the string $abbaba$ belongs to the language because $abbaba = ab{\cdot}ba{\cdot}ba$, but the string $abaaba$ does not because $abaaba = ab \cdot aa \cdot ba$.

□

Notation We can use the Boolean operations, the product, and the Kleene star to describe languages. For example, $L = \{a, b\}^* \setminus \{a, b\}^*\{aa, bb\}\{a, b\}^*$ consists of all strings over the alphabet $\{a, b\}$ that do not contain a doubled symbol. Thus the string $ababab$ is in L whereas $abaaba$ is not. When languages are described in this way, it quickly becomes tedious to keep having to write down the brackets { and }. Consequently, from now on we shall omit them. If brackets are needed to avoid ambiguity we use (and). This notation is made rigorous in Section 5.1.

Examples 1.3.3 Here are some examples of languages over the alphabet $A = \{a, b\}$ described using our notational convention above.

(1) We can write A^* as $(a + b)^*$. To see why, observe that

$$A^* = \{a, b\}^* = (\{a\} + \{b\})^* = (a + b)^*,$$

where the last equality follows by our convention above. We have to insert brackets because $a + b^*$ is a different language. See Exercises 1.3.

(2) The language $(a + b)^3$ consists of all 8 strings of length 3 over A. This is because $(a + b)^3$ means $(a + b)(a + b)(a + b)$. A string x belongs to this language if we can write it as $x = a_1a_2a_3$ where $a_1, a_2, a_3 \in \{a, b\}$.

(3) The language $aab(a + b)^*$ consists of all strings that begin with the string aab, whereas the language $(a + b)^*aab$ consists of all strings that end in the string aab. The language $(a + b)^*aab(a + b)^*$ consists of all strings that contain the string aab as a factor.

(4) The language $(a+b)^*a(a+b)^*a(a+b)^*b(a+b)^*$ consists of all strings that contain the string aab as a substring.

(5) The language $aa(a+b)^* + bb(a+b)^*$ consists of all strings that begin with a double letter.

(6) The language $(aa+ab+ba+bb)^*$ consists of all strings of even length.

□

Exercises 1.3

1. Let $L = \{ab, ba\}$, $M = \{aa, ab\}$ and $N = \{a, b\}$. Write down the following.

(i) LM.

(ii) LN.

(iii) $LM + LN$.

(iv) $M + N$.

(v) $L(M + N)$.

(vi) $(LM)N$.

(vii) MN.

(viii) $L(MN)$.

2. Determine the set inclusions among the following languages. In each case, describe the strings belonging to the language.

(i) $a + b^*$.

(ii) $a^* + b^*$.

(iii) $(a^* + b^*)^*$.

3. Describe the following languages:

(i) a^*b^*.

(ii) $(ab)^*$.

(iii) $(a+b)(aa+ab+ba+bb)^*$.

(iv) $(a^2+b^2)(a+b)^*$.

(v) $(a+b)^*(a^2+b^2)(a+b)^*$.

(vi) $(a+b)^*(a^2+b^2)$.

(vii) $(a+b)^*a^2(a+b)^*b^2(a+b)^*$.

4. Let L be any language. Show that if $x, y \in L^*$ then $xy \in L^*$.
This is an important property of the Kleene star operation.

5. Let $L \subseteq A^*$. Verify the following:

(i) $(L^*)^* = L^*$.

(ii) $L^*L^* = L^*$.

(iii) $L^*L + \varepsilon = L^* = LL^* + \varepsilon$.

Is it always true that $LL^* = L^*$?

6. Prove that the following hold for all languages L, M, and N.

(i) $L(MN) = (LM)N$.

(ii) $L(M + N) = LM + LN$ and $(M + N)L = ML + NL$.

(iii) If $L \subseteq M$ then $NL \subseteq NM$ and $LN \subseteq MN$.

7. Let L, M, N be languages over A. Show that $L(M \cap N) \subseteq LM \cap LN$. Using $A = \{a, b\}$, show that the reverse inclusion does not hold in general by finding a counterexample.

8. Let $A = \{a, b\}$. Show that

$$(ab)^+ = (aA^* \cap A^*b) \setminus (A^*aaA^* + A^*bbA^*).$$

10. Let A be an alphabet and let $u, v \in A^*$. Prove that $uA^* \cap vA^* \neq \emptyset$ if and only if u is a prefix of v or vice versa; when this happens explicitly calculate $uA^* \cap vA^*$.

11. For which languages L is it true that $L^* = L^+$?

1.4 Finite automata: motivation

An information-processing machine transforms inputs into outputs. In general, there are two alphabets associated with such a machine: an *input alphabet* A for communicating with the machine, and an *output alphabet* B for receiving answers. For example, consider a machine that takes as input sentences in English and outputs the corresponding sentence in Russian.

There is however another way of processing strings, which will form the subject of this book. As before, there is an input alphabet A but this time each input string causes the machine to output either 'yes' or 'no.' Those input strings from A^* that cause the machine to output 'yes' are said to be *accepted* by the machine, and those strings that cause it to output 'no' are said to be *rejected.* In this way, A^* is partitioned into two subsets: the 'yes' subset we call the *language accepted by the machine*, and the 'no' subset we call the *language rejected by the machine.* A machine that operates in this way is called an *acceptor.*

Our aim is to build a mathematical model of a special class of acceptors. Before we give the formal definition in Section 1.5 we shall motivate it by thinking about real machines and then abstracting certain of their features to form the basis of our model.

To be concrete, let us think of two extremes of technology for building an acceptor and find out what they have in common. In Babbage's day the acceptor would have been constructed out of gear-wheels rather like Babbage's 'analytical engine,' the Victorian prototype of the modern computer; in our day, the acceptor would be built from electronic components. Despite their technological differences, the two different types of component involved, gear-wheels in the former and electronic components in the latter, have something in common: they can only do a limited number of things. A gear-wheel can only be in a finite number of positions, whereas many basic electronic components can only be either 'on' or 'off.' We call a specific configuration of gear-wheels or a specific configuration of on-and-off devices a *state.* For example, a clock with only an hour-hand and a minute-hand has 12×60 states that are made visible by the position of the hands. What all real devices have in common is that the total number of states is *finite.* How states are represented is essentially an engineering question.

After a machine has performed a calculation the gear-wheels or electronic components will be in some state dependent on what was being calculated. We therefore need a way of resetting the machine to an initial state; think of this as wiping the slate clean to begin a new calculation.

Every machine should do its job reliably and automatically, and so what the machine does next must be completely determined by its current state and current input, and because the state of a machine contains all the information about the configurations of all the machine's components, what a machine does next is to change state.

We can now explain how our machine will process an input string $a_1 \ldots a_n$. The machine is first re-initialised so that it is in its initial state, which we call s_0. The first letter a_1 of the string is input and this, together with the fact that the machine is in state s_0, completely determines the next state, say s_1. Next the second letter a_2 of the string is input and this, together with the fact that the machine is in state s_1, completely determines the next state, say s_2. This process continues until the last letter of the input string has been read. At this point, the machine is now ready to pass judgement on the input string. If the machine is in one of a designated set of special states called *terminal* states it deems the string to have been accepted; if not, the string is *rejected.*

To make this more concrete, here is a specific example.

Example 1.4.1 Suppose we have two coins. There are four possible ways of placing them in front of us depending on which is heads (H) and which is tails (T):

$$\text{HH}, \text{TH}, \text{HT}, \text{TT}.$$

Now consider the following two operations: 'flip the first coin,' which I shall denote by a and 'flip the second coin,' which I shall denote by b. Assume that initially the coins are laid out as HH. I am interested in all the possible ways of applying the operations a and b so that the coins are laid out as TT. The states of this system are the four ways of arranging the two coins; the initial state is HH and the terminal state is TT. The following diagram illustrates the relationships between the states and the two operations.

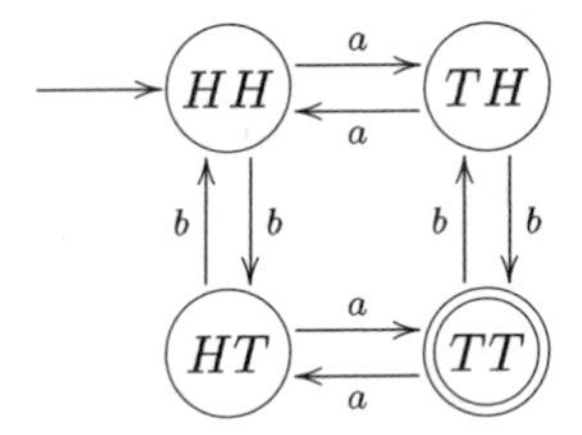

I have marked the start state with an inward-pointing arrow, and the terminal state by a double circle. If we start in the state HH and input the string *aba* Then we pass through the following states:

$$\text{HH} \xrightarrow{a} \text{TH} \xrightarrow{b} \text{TT} \xrightarrow{a} \text{HT}.$$

Thus the overall effect of starting at HH and inputting the string *aba* is to end up in the state HT. It should be clear that those sequences of a's and b's are accepted precisely when the number of a's is odd and the number of b's is odd. We can write this language more mathematically as follows:

$$\{x \in (a+b)^*\colon \mid x \mid_a \text{ and } \mid x \mid_b \text{ are odd}\}.$$

□

To summarise, our mathematical model of an acceptor will have the following features:

- A finite set representing the finite number of states of our acceptor.
- A distinguished state called the *initial state* that will be the starting state for all fresh calculations.
- Our model will have the property that the current state and the current input uniquely determine the next state.
- A distinguished set of *terminal* states.

Exercises 1.4

1. This question is similar to Example 1.4.1. Let $A = \{0, 1\}$ be the input alphabet. Consider the set A^3 of all binary strings of length 3. These will be the states. Let 000 be the initial state and 110 the terminal state. Let $a_1a_2a_3$ be the current state and let a be the input symbol. Then the next state is a_2a_3a; so we shift everything along one place to the left, the left-hand bit drops off and is lost and the right-hand bit is the input symbol. Draw a diagram, similar to the one in Example 1.4.1, showing how states are connected by inputs.

1.5 Finite automata and their languages

In Section 1.4, we laid the foundations for the following definition. A *complete deterministic finite state automaton* **A** or, more concisely, a *finite automaton* and sometimes, just for variety, a *machine* is specified by five pieces of information:

$$\mathbf{A} = (S, A, i, \delta, T),$$

where S is a finite set called the *set of states*, A is the finite *input alphabet*, i is a fixed element of S called the *initial state*, δ is a function $\delta\colon S \times A \to S$ called the *transition function*, and T is a subset of S called the set of *terminal states* (also called *final state*). The phrase 'finite state' is self-explanatory. The meanings of 'complete' and 'deterministic' will be explained below.

There are two ways of providing the five pieces of information needed to specify an automaton: 'transition diagrams' and 'transition tables.'

A *transition diagram* is a special kind of directed labelled graph: the vertices are labelled by the states S of **A**; there is an arrow labelled a from the vertex labelled s to the vertex labelled t precisely when $\delta(s, a) = t$ in **A**. That is to say, the input a causes the automaton **A** to change from state s to state t. Finally, the initial state and terminal states are distinguished in some way: we mark the initial state by an inward-pointing arrow, $\longrightarrow (i)$, and the terminal states by double circles $((t))$.[6]

Example 1.5.1 Here is a simple example of a transition diagram of a finite automaton.

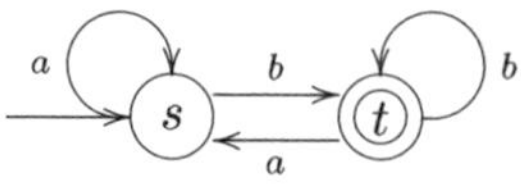

We can easily read off the five ingredients that specify an automaton from this diagram:

- The set of states is $S = \{s, t\}$.

[6]Another convention is to use outward-pointing arrows to denote terminal states, and double-headed arrows for states that are both initial and terminal.

- The input alphabet is $A = \{a, b\}$.
- The initial state is s.
- The set of terminal states is $\{t\}$.

Finally, the transition function $\delta\colon S \times A \to S$ is given by

$$\delta(s, a) = s, \quad \delta(s, b) = t, \quad \delta(t, a) = s, \text{ and } \delta(t, b) = t.$$

□

In order to avoid having too many arrows cluttering up a diagram, the following convention will be used: if the letters $a_1, \ldots, a_m$ label m transitions from the state s to the state t then we simply draw *one* arrow from s to t labelled $a_1, \ldots, a_m$ rather than m arrows labelled a_1 to a_m, respectively.

For a diagram to be the transition diagram of an automaton, two important points need to be borne in mind, both of which are consequences of the fact that $\delta\colon S \times A \to S$ is a function. First, it is impossible for two arrows to leave the same state carrying the same label. Thus a configuration such as

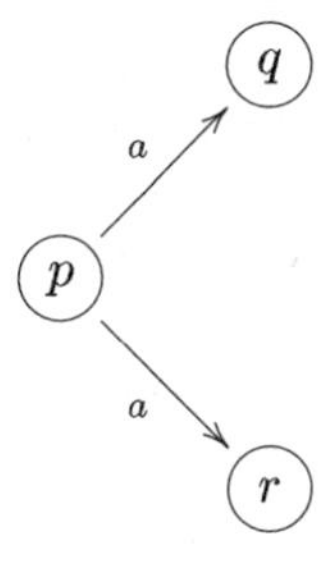

is forbidden. This is what we mean by saying that our machines are *deterministic*: the action of the machine is completely determined by its current state and current input and no choice is allowed. Second, in addition to being deterministic, there must be an arrow leaving a given state for each of the input letters; there can be no missing arrows. For this reason we say that our machines are *complete*. Incomplete automata will be defined in Section 2.2, and non-deterministic automata, which need be neither deterministic nor complete, will be defined in Section 3.2.

A *transition table* is just a way of describing the transition function δ in tabular form and making clear in some way the initial state and the set of terminal states. The table has rows labelled by the states and columns labelled by the input letters. At the intersection of row s and column a we put the element $\delta(s, a)$. The states labelling the rows are marked to indicate the initial state and the terminal states. Here is the transition table of our automaton in Example 1.5.1:

	a	b
$\rightarrow s$	s	t
$\leftarrow t$	s	t

We shall designate the initial state by an inward-pointing arrow $\rightarrow$ and the terminal states by outward-pointing arrows $\leftarrow$. If a state is both initial and terminal, then the inward and outward pointing arrows will be written as a single double-headed arrow $\leftrightarrow$.

Notation There is a piece of notation we shall frequently use. Rather than write $\delta(s,a)$ we shall write $s \cdot a$.

When you design an automaton, it really must be an automaton. This means that you have to check that the following two conditions hold:

- There is exactly one initial state.
- For each state s and each input letter $a \in A$, there is *exactly one* arrow starting at s finishing at $s \cdot a$ and labelled by a.

An automaton that satisfies these two conditions — and has a finite number of states, which is rarely an issue when designing an automaton — is said to be *well-formed.*

One thing missing from our definition of a finite automaton is how to process input strings rather than just input letters. Let $\mathbf{A} = (S, A, i, \delta, T)$ be a finite automaton, let $s \in S$ be an arbitrary state, and let $x = a_1 \ldots a_n$ be an arbitrary string. If $\mathbf{A}$ is in state s and the string x is processed then the behaviour of $\mathbf{A}$ is completely determined: for each symbol $a \in A$ and each state s' there is exactly one transition starting at s' and labelled a. Thus we pass through the states $s \cdot a_1$, $(s \cdot a_1) \cdot a_2$ and so on finishing at the state $t = (\ldots((s \cdot a_1) \cdot a_2) \ldots) \cdot a_m$. Thus there is a unique path in $\mathbf{A}$ starting at s and finishing at t and labelled by the symbols of the string $a_1 \ldots a_n$ in turn.

We can formalise this idea by introducing a new function δ^*, called the *extended transition function.* The function δ^* is the unique function from $S \times A^*$ to S satisfying the following three conditions where $a \in A$, $w \in A^*$ and $s \in S$:

(ETF1) $\delta^*(s, \varepsilon) = s$.

(ETF2) $\delta^*(s, a) = \delta(s, a)$.

(ETF3) $\delta^*(s, aw) = \delta^*(\delta(s, a), w)$.

I have claimed that there is a unique function satisfying these three conditions. This will probably seem obvious, but I have included a proof at the end of this section if you need further convincing.

Notation I shall usually write $s \cdot w$ instead of $\delta^*(s, w)$ to simplify notation.

It is important to take note of condition (ETF1): this says that the empty string has no effect on states. Thus for each state s we have that $s \cdot \varepsilon = s$.

We can now connect languages and finite automata together. Let

$$\mathbf{A} = (S, A, i, \delta, T)$$

be a complete deterministic automaton. Define the *language accepted* or *recognised* by $\mathbf{A}$, denoted $L(\mathbf{A})$, to be

$$L(\mathbf{A}) = \{w \in A^*: i \cdot w \in T\}.$$

A language is said to be *recognisable* if it is recognised by some finite automaton.

The language recognised by an automaton $\mathbf{A}$ with input alphabet A therefore consists of all strings in A^* that label paths in $\mathbf{A}$ starting at the initial state and concluding at a terminal state. There is only one string where we have to think a little to decide whether it is accepted or not. This is the empty string. Suppose first that $\varepsilon \in L(\mathbf{A})$. If i is the initial state of $\mathbf{A}$ then by definition $i \cdot \varepsilon$ is terminal because $\varepsilon \in L(\mathbf{A})$, and so i is terminal. Now suppose that the initial state i is also terminal. Because $i = i \cdot \varepsilon$, it follows from the definition that $\varepsilon \in L(\mathbf{A})$. We see that the empty string is accepted by an automaton if and only if the initial state is also terminal. This is a small point but worth remembering.

Example 1.5.2 We describe the language recognised by our machine in Example 1.5.1. We have to find all those strings in $(a + b)^*$ that label paths starting at s and finishing at t. First, any string x ending in a 'b' will be accepted. To see why let $x = x'b$ where $x' \in A^*$. If x' leads the machine to state s, then the b will lead the machine to state t; and if x' leads the machine to state t, then the b will keep it there. Second, a string x ending in 'a' will not be accepted. To see why let $x = x'a$ where $x' \in A^*$. If x' leads the machine to state s, then the a will keep it there; and if x' leads the machine to state t, then the a will send it to state s. Finally, the empty string is not accepted by this machine because the initial state is not terminal. We conclude that $L(\mathbf{A}) = A^*b$. □

To conclude this section, I give two rather technical results whose proofs are not essential in what follows. The first is the promised proof that there really is a unique extended transition function.

Proposition 1.5.3 *Let S be a finite set and A a finite alphabet. Let $\delta: S \times A \to S$ be a function. Then there is exactly one function,*

$$\Delta: S \times A^* \to S,$$

satisfying the following three conditions:

(i) $\Delta(s, \varepsilon) = s$ *where* $s \in S$.

(ii) $\Delta(s, a) = \delta(s, a)$ *where* $s \in S$ *and* $a \in A$.

(iii) $\Delta(s, aw) = \Delta(\delta(s, a), w)$ *where* $s \in S$, $a \in A$ *and* $w \in A^*$.

Proof We show first that there is at most one function satisfying these three conditions. Let Δ, Δ' be functions satisfying (i), (ii), and (iii). We prove that $\Delta(s, x) = \Delta'(s, x)$ for all $(s, x) \in S \times A^*$ by induction on the length of x. By (i) and (ii), $\Delta(s, x) = \Delta'(s, x)$ for all $s \in S$ and all strings $|x| \leq 1$. This is our base case. Our induction hypothesis is that $\Delta(s, x) = \Delta'(s, x)$ for all $s \in S$ and all strings $|x| = n$. Let y be a string of length $n + 1$. Then we can write $y = ax$ where $a \in A$ and $|x| = n$. By definition,

$$\Delta(s, y) = \Delta(\delta(s, a), x)$$

and

$$\Delta'(s, y) = \Delta'(\delta(s, a), x).$$

By the induction hypothesis $\Delta(\delta(s, a), x) = \Delta(\delta(s, a), x)$ and so $\Delta(s, y) = \Delta'(s, y)$. Thus there is at most one function satisfying conditions (i), (ii), and (iii).

The proof that there is a function satisfying these three conditions is a little more involved. First we define a sequence of functions $\Delta_1, \Delta_2, \ldots$ as follows. The function Δ_1 has domain $S \times (\varepsilon + A)$ and satisfies (i) and (ii). The function Δ_2 has domain $S \times (\varepsilon + A + A^2)$ and extends Δ_1 using (iii). The function Δ_3 has domain $S \times (\varepsilon + A + A^2 + A^3)$ and extends Δ_2 using (iii). In general, if we have defined Δ_i on domain

$$S \times (\varepsilon + A + \ldots + A^i),$$

then Δ_{i+1} has domain

$$S \times (\varepsilon + A + \ldots + A^{i+1})$$

and extends Δ_i using (iii). The sequence of functions $\Delta_1, \Delta_2, \ldots$ has the property that each function Δ_{i+1} is an extension of Δ_i. For this reason, the union of all these functions is a function we call Δ, which maps $S \times A^*$ to S. It is now easy to check that Δ satisfies (i), (ii) and (iii). □

From (ETF3), we have that $s \cdot (aw) = (s \cdot a) \cdot w$ where $a \in A$ and $w \in A^*$. This simply expresses the fact that in processing an input string we do so one symbol at a time from the front of the string. The next result generalises this.

Proposition 1.5.4 *Let* $\mathbf{A} = (S, A, i, \delta, T)$ *be an automaton. Then*

$$s \cdot (xy) = (s \cdot x) \cdot y$$

for all $x, y \in A^*$ *and* $s \in S$.

Proof We prove that for each $s \in S$ and each string $y \in A^*$ we have that

$$s \cdot (xy) = (s \cdot x) \cdot y$$

for all $x \in A^*$. We prove the result by induction on the length of x. The claim is clearly true when x is the empty string. Suppose that the claim is true for all strings x of length n and less. Let x be a string of length $n+1$. Then $x = az$ where $a \in A$ and z has length n. We have that

$$(s \cdot x) \cdot y = (s \cdot (az)) \cdot y = ((s \cdot a) \cdot z) \cdot y = (s \cdot a) \cdot (zy) = s \cdot (azy) = s \cdot (xy),$$

where

$$((s \cdot a) \cdot z) \cdot y = (s \cdot a) \cdot (zy)$$

follows from the induction hypothesis, and

$$(s \cdot a) \cdot (zy) = s \cdot (azy) = s \cdot (xy)$$

follows from the definition of the extended transition function. □

Exercises 1.5

1. For each of the following transition tables construct the corresponding transition diagram.

(i)

	a	b
$\rightarrow s_0$	s_1	s_0
s_1	s_2	s_1
$\leftarrow s_2$	s_0	s_2

(ii)

	a	b
$\leftrightarrow s_0$	s_1	s_1
s_1	s_0	s_2
$\leftarrow s_2$	s_0	s_1

(iii)

	a	b	c
$\leftrightarrow s_0$	s_1	s_0	s_2
s_1	s_0	s_3	s_0
$\leftarrow s_2$	s_3	s_2	s_0
$\leftarrow s_3$	s_1	s_0	s_1

2. Determine which of the following diagrams are finite automata and which are not, and give reasons for your answers. The alphabet in question is $A = \{a, b\}$.

(i)

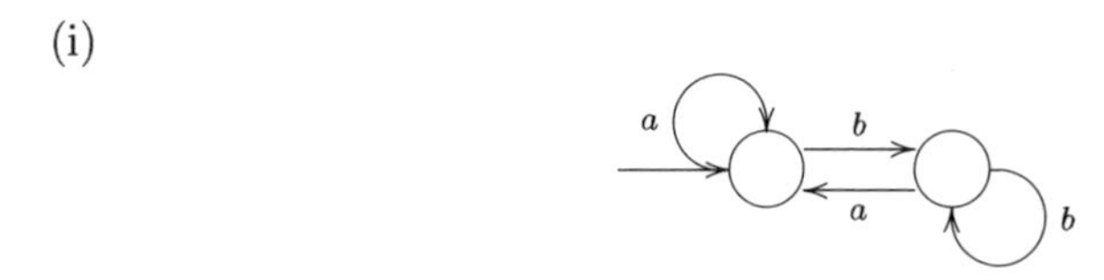

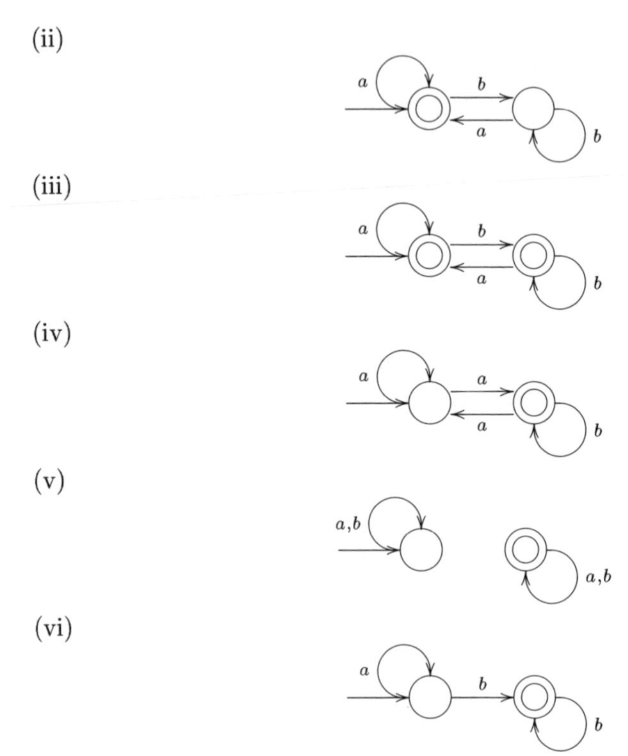

3. Let $A = \{a, b\}$ with the order $a < b$. Consider the automaton below:

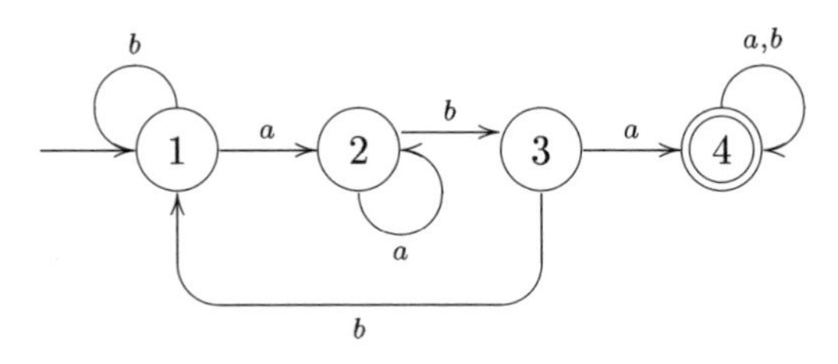

Draw up the following table: the rows should be labelled by the states, and the columns by all strings x in A^* where $0 \leq |x| \leq 3$ written in tree order. If q is a row and x is a column then the entry in the q-th row and x-th column should be the state $q \cdot x$.

4. For each of the automata below, describe the language recognised.

(i)

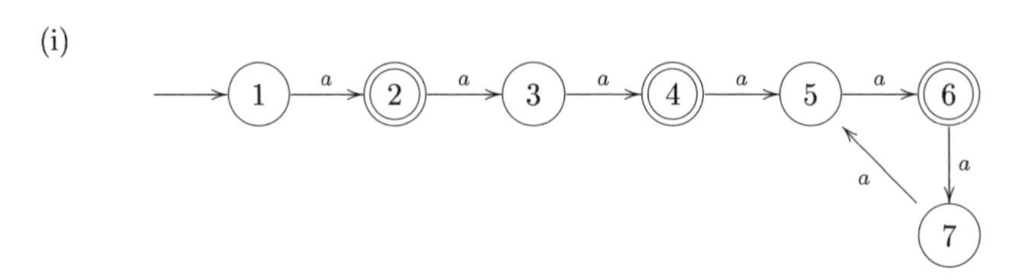

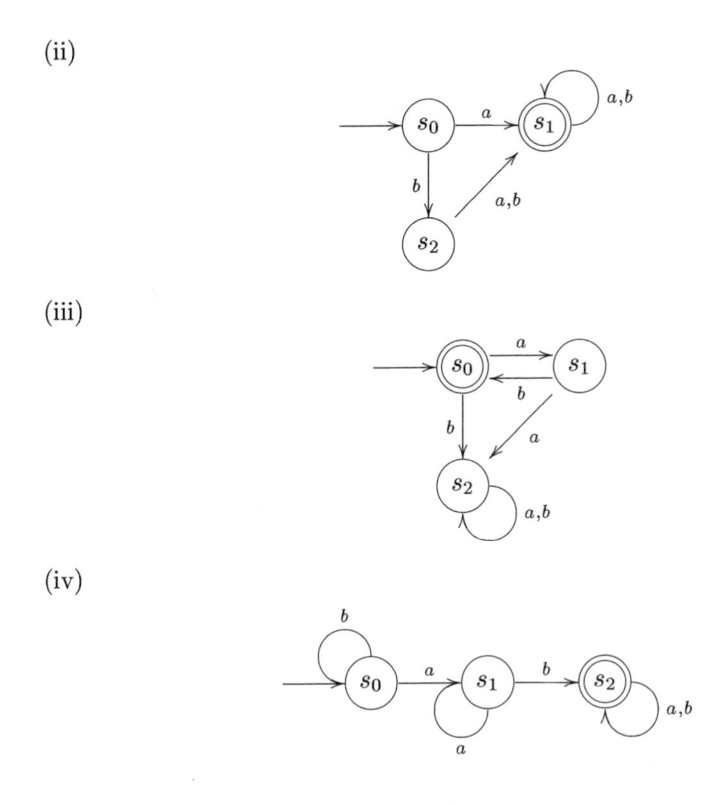

1.6 Summary of Chapter 1

- *Alphabets*: An alphabet is any finite set. The elements of an alphabet are called symbols or letters.

- *Strings*: A string is any finite sequence of symbols taken from a fixed alphabet. The empty string is denoted ε. The set of all strings taken from the alphabet A is A^*, and the set of all non-empty strings is A^+.

- *Languages*: A language over an alphabet A is any subset of A^*; this includes the two extremal subsets: the empty set $\emptyset$ and A^* itself.

- *Language operations*: There are a number of important ways of combining languages L and M to form new languages. The Boolean operations $L \cap M$, $L + M$ and L' are, respectively, intersection, union, and complementation. There are two further operations $L \cdot M$ and L^*, which are, respectively, product and Kleene star. The product $L \cdot M$ of two languages, usually written just LM, consists of all strings that can be written as a string in L followed by a string in M. The Kleene star L^* of a language consists of the empty string and all strings that can factorised as products of strings in L.

- *Finite automata*: These are special kinds of algorithms for deciding the membership of languages. They consist of a finite number of states, an input alphabet A, a distinguished initial state, a finite number of transitions labelled by the elements of A, and a finite set of terminal states. In addition, they are complete and deterministic. The language recognised or accepted by an automaton consists of all strings over A that label paths from the initial state to one of the terminal states. Completeness and determinism imply that each input string labels a unique path starting at the initial state.

1.7 Remarks on Chapter 1

I think it worth re-emphasising that although I shall view strings as sequences of symbols, such sequences can encode data that is not just text. For example, sequences of symbols can be used to encode numbers, so that in certain circumstances the language recognised by an automaton could be viewed as a set of numbers. The paper [60] is an example of this point of view. Similarly, a pair of numbers can coordinatise a point in the plane, so that we might be able to view the language recognised by an automaton as an image. How this might be done is described in [38].

The notion of a string is also capable of being generalised in a number of different ways. For example, if a string is a 'one-dimensional object,' what might its 'two-dimensional' analogue be? This is discussed in [52].

The automata studied in this book are not equipped with output. There are, however, two simple ways in which output can be generated.

A *Moore machine* [95],

$$\mathbf{A} = (Q, A, B, q_0, \delta, \lambda),$$

consists of a finite set of states Q, an input alphabet A, an output alphabet B, an initial state q_0, a transition function $\delta\colon Q \times A \to A$, and an output function $\lambda\colon Q \to B$. The transition function δ can be extended to a function $\delta^*\colon Q \times A^* \to Q$ in the usual way. We shall write $q \cdot x$ rather than $\delta^*(q, x)$. The function λ can be used to define a unique function $\lambda^*\colon Q \times A^* \to B^*$ satisfying the following two conditions:

(MoM1) $\lambda^*(q, \varepsilon) = \varepsilon$ for all $q \in Q$.

(MoM2) $\lambda^*(q, ax) = \lambda(q \cdot a)\lambda^*(q \cdot a, x)$ for all $q \in Q$, $a \in A$ and $x \in A^*$.

Such a machine determines a function $f_{\mathbf{A}}\colon A^* \to B^*$ defined by $f_{\mathbf{A}}(x) = \lambda^*(q_0, x)$. A function $f\colon A^* \to B^*$ is said to be *realised by a Moore machine* if there is a Moore machine $\mathbf{A}$ such that $f = f_{\mathbf{A}}$. Observe that the output letter generated by a Moore machine depends only on what state the machine is in.

A *Mealy machine* [90],

$$\mathbf{A} = (Q, A, B, q_0, \delta, \lambda),$$

consists of a finite set of states Q, an input alphabet A, an output alphabet B, an initial state q_0, a transition function δ: $Q \times A \to A$, and an output function λ: $Q \times A \to B$. As above, the transition function δ can be extended to a function δ^*: $Q \times A^* \to Q$, and we write $q \cdot x$ rather than $\delta^*(q, x)$. The function λ can be used to define a unique function λ^*: $Q \times A^* \to B^*$ satisfying the following two conditions:

(MeM1) $\lambda^*(q, \varepsilon) = \varepsilon$ for all $q \in Q$.

(MeM2) $\lambda^*(q, ax) = \lambda(q, a)\lambda^*(q \cdot a, x)$ for all $q \in Q$, $a \in A$ and $x \in A^*$.

Such a machine determines a function $f_{\mathbf{A}}$: $A^* \to B^*$ defined by $f_{\mathbf{A}}(x) = \lambda^*(q_0, x)$. A function f: $A^* \to B^*$ is said to be *realised by a Mealy machine* if there is a Mealy machine $\mathbf{A}$ such that $f = f_{\mathbf{A}}$. Observe that the output letter generated by a Mealy machine depends on both the state and the current input letter.

It can be proved [23] that a function is realisable by a Moore machine if and only if it is realisable by a Mealy machine. Such functions are said to be *sequential*. In the same way that acceptors lead to a theory of recognisable languages, so Moore and Mealy machines lead to a theory of sequential functions. This theory is important in circuit design [79], and is also interesting from a mathematical point of view [39, 40].

More general than either recognisable languages or sequential functions are the *rational transductions* [12]. I shall not describe these further here, but many of the ideas introduced in this book can be generalised from acceptors to other classes of rational transductions. See [25], for example, where there are also further references.

Acceptors, Moore machines and Mealy machines have certain ingredients in common that are often encountered. A *semiautomaton*,

$$\mathbf{A} = (Q, A, \delta),$$

consists of a finite set of states Q, an input alphabet A, and a transition function δ: $Q \times A \to Q$. If (Q, A, δ) is a semiautomaton, then we can construct the extended transition function $Q \times A^* \to Q$. By Proposition 1.5.4 and the definition of the extended transition function, the following two conditions hold:

(A1) $q \cdot \varepsilon = q$ for each $q \in Q$.

(A2) $q \cdot (xy) = (q \cdot x) \cdot y$ for each $q \in Q$ and $x, y \in A^*$.

We say that A^* *acts on* Q *(on the right)*. Thus semiautomata are the same things as 'actions of A^* on finite sets.'

Three papers laid the foundations of finite automata theory: Turing's 1936 paper [135], McCulloch and Pitts' paper [85] of 1943, and Kleene's paper [74] developing his RAND report of 1951.

In 1936, Turing introduced what are now known as Turing machines, briefly described in the Remarks on Chapter 2, as models of general information-processing machines. This was several years before computers, as now understood, were invented. Turing introduced his machines to solve a long-standing problem in logic, universally known by its German name 'Entscheidungsproblem,' meaning 'decison problem.' The particular decision problem in question concerned first-order logic: was there an algorithm that would accept a formula in first-order logic and output 'yes' if it was a theorem and 'no' otherwise. Using his machines, Turing proved that there was no such algorithm. Until the mid-1970s, Turing's name was largely known only within the confines of mathematics and computer science. However, it was at this time that information about the activities of British scientists during World War II started to be published. As a result, Turing's involvement in the decoding of German ciphers at Bletchley Park became more widely known. The work of Bletchley Park is described in [124]. Turing's life and tragically early death at the age of 41 put a human face to a highly creative mathematician. Andrew Hodges' book [63] is a detailed account of Turing's life and work.

Turing's paper was rooted in mathematical logic, McCulloch and Pitts' on the other hand grew out of research into the structure of the brain and the recognition by neurophysiologists of the central role played by brain neurons — brain cells — in processing information. McCulloch and Pitts constructed a mathematical model of such neurons and showed how networks of these cells could be used to process information. Although the details of their model are almost certainly wrong, they showed how 'something like a brain' could do 'brain-like' things. A description of their work can be found in [6] and also [93]. Brains themselves are discussed in [31].

In 1951, Stephen Kleene was asked by the RAND Corporation to analyse McCulloch and Pitt's paper. The report he produced contributed to his paper [74], which was not formally published until 1956. It was in this paper that Kleene described the languages, or as he termed them 'events,' which could be recognised by McCulloch and Pitts neural-network models. This was the basis for what is now known as 'Kleene's Theorem' and is the subject of Chapter 5.

Turing's paper on the one hand, and those by McCulloch and Pitts, and Kleene on the other, represent what we now recognise as two different approaches to automata: the abstract and the structural. In the abstract approach, the states are not analysed any further, they are simply a given, whereas in the structural approach states are represented by specific data types. For example, a state could correspond to a particular allocation of bits to the components of a computer memory. During the 1950s, papers on automata theory appeared using both approaches, however mathematicians came to favour the abstract approach and engineers the structural. These two aspects to the theory of finite automata are brought together in [92]

The foundational work of the 1930s, 1940s and early 1950s, led to a decade in which the basic concepts and theorems of automata theory were discovered as a result of individual and collaborative effort amongst mathematicians, linguists, and electrical engineers including: Huffman [69], Schützenberger [114], Mealy [90], Chomsky [27], Moore [95], Medvedev [91], Myhill [97], and Nerode [98]. The definition of finite automaton given in this chapter is due to Rabin and Scott [109], whose paper, appearing at the end of the 1950s, served as a useful reference on automata theory in subsequent work.

If you want to know more about the history of finite automata, the essay by Perrin [101] is interesting, and there are surveys of important papers in Brauer [16]. The collections of papers that appear in [118] and [96] convey something of the flavour of this early work. References to work on automata theory in the former Soviet bloc can be found in [50] and [55] as well as Brauer [16].

The theory of finite automata is an established part of theoretical computer science, and so any book dealing with this subject will contain accounts of finite automata to a greater or lesser extent. Textbooks that contain chapters on finite automata, at approximately the same level as this one, are [23], [34], [66], [75], [80], [113], and [123].

The books by Büchi [22], Conway [35], Eilenberg[7] [39, 40], and Minsky [93] are classics: those by Büchi and Minsky are eminently readable and Conway's is inventive and thought-provoking. Eilenberg's two volumes provided the key texts for the algebraic theory of automata, which we develop in Chapters 9 to 12.

Two other techniques for handling regular languages which are not discussed in this book are based on logic, see Straubing [129] and Thomas [130], and formal power series, see Berstel and Reutenauer [14].

Applications of finite automata and their languages cover an enormous range. The book by Petzold [102] is an elementary introduction to circuit design and [79] a more advanced one; Aho, Sethi, and Ullman [1] explain how finite automata form one of the ingredients in designing compilers; Friedl [47] describes the thousand-and-one uses of regular expressions to professional programmers — such expressions are equivalent to finite automata as we shall prove in Chapter 5; searching for patterns in texts can be carried out efficiently using automata [37]; the collection of papers to be found in [112] demonstrates the usefulness of finite automata in natural language processing; Lind and Marcus [81] show how finite automata, under the alias of 'sofic system,' can be used in encoding information, a further useful introduction to these ideas is [10]; von Haeseler [58] uses finite automata to generate sequences of numbers; Sims [122] uses finite automata to describe some algorithms in group theory; Epstein et al [43] explain how finite automata form an important tool in combinatorial group theory and geometry; Thurston [132] interweaves groups, tilings, dynamical systems, and finite automata; Grigorchuk et al [57] actually build groups from automata; finally, Pin [103] develops the algebraic theory of recognisable languages within finite semigroup theory.

[7]It is important to add that Volume B contains important contributions by Bret Tilson.

Chapter 2

Recognisable languages

In Chapter 1, we introduced finite automata and the languages they recognise. In this chapter we do two things. First, we look at ways of constructing certain kinds of automata. Second, we describe a technique for showing that some languages are *not* recognisable.

2.1 Designing automata

Designing an automaton $\mathbf{A}$ to recognise a language L is more an art than a science. However, it is possible to lay down some guidelines. In this section, I will describe the general points that need to be born in mind, and in Sections 2.2 to 2.5, I will describe some specific techniques for particular kinds of languages.

Automata can be regarded as simple programming languages, and so the methods used to write programs can be adopted to help design automata. Whenever you write a program to solve a problem, it is good practice to begin by formulating the algorithm that the program should implement. By the same token, before trying to construct an automaton to recognise a language, you should first formulate an algorithm that accomplishes the task of recognising the language. There is however an important constraint: your algorithm must only involve using a fixed amount of memory. One way of ensuring this is to imagine how you would set about recognising the strings belonging to a language for *extremely large* inputs. When you have done this, you can then implement your algorithm by means of an automaton.

Once you have a design, $\mathbf{A}$, it is easy to check that it is well-formed — this is equivalent to checking that a program is syntactically correct — but the crucial point now is to verify that your automaton really recognises the language L in question. This involves checking that two conditions hold:

(1) Each string accepted by $\mathbf{A}$ is in L.

(2) Each string in L is accepted by $\mathbf{A}$.

I have emphasised that two conditions must hold, because it is a very common mistake to check only (1). If both of these conditions are satisfied then you have solved the problem. But if either one of them is violated then you have to go back and modify your machine and try again. Unfortunately, it is easier to show that your automaton *does not* work than it is to show it *does*. To show that your machine is wrong it is enough to find just one string x that is in the language L but not accepted by the machine $\mathbf{A}$, or is accepted by the machine $\mathbf{A}$ but is not in the language L. The difficulty is that A^* contains infinitely many strings and your machine has to deliver the correct response to each of them.

The minimum you should do to show that your well-formed automaton solves the problem is to try out some test strings on it, making sure to include both strings belonging to the language and those that do not. However even if your automaton passes these tests with flying colours it could still be wrong. There are two further strategies that you can use. First, if you find you have made a mistake then try to use the nature of the mistake to help you see how to correct the design. Second, try to be clear about the function of each state in your machine: each state should be charged with detecting some feature of that part of the input string that has so far been read. Some of the ideas that go into constructing automata are illustrated by means of examples in the following sections. At the same time I shall describe some useful techniques for constructing automata. Although I often use alphabets with just two letters, this is just for convenience, similar examples can be constructed over other alphabets.

Exercises 2.1

1. Let $\mathbf{A}$ be the automaton:

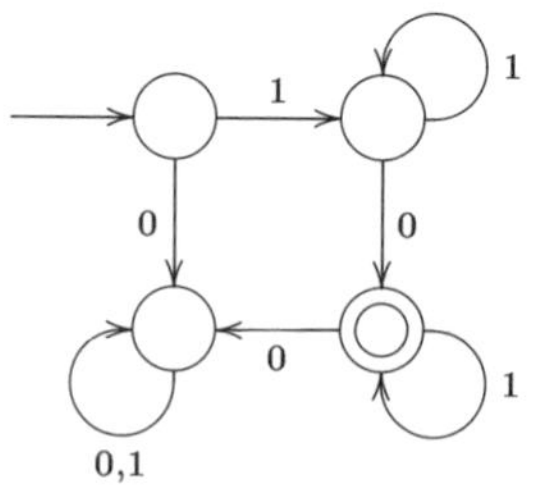

and let $L = 1^+0$. Show that $L \subseteq L(\mathbf{A})$, but that the reverse inclusion does not hold. Describe $L(\mathbf{A})$.

2. Let **A** be the automaton:

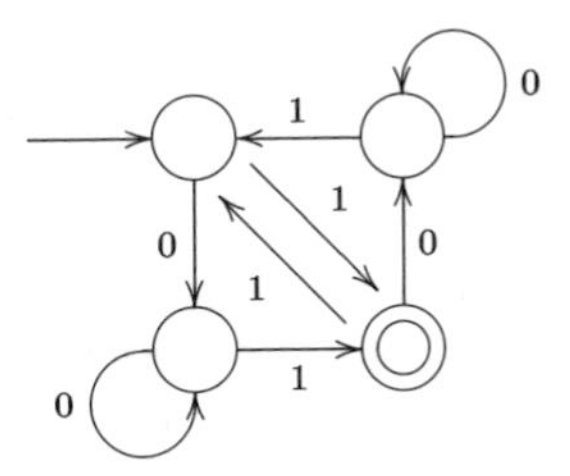

Show that every string in $L(\mathbf{A})$ has an odd number of 1's, but that not every string with an odd number of 1's is accepted by **A**. Describe $L(\mathbf{A})$.

3. Let **A** be the automaton:

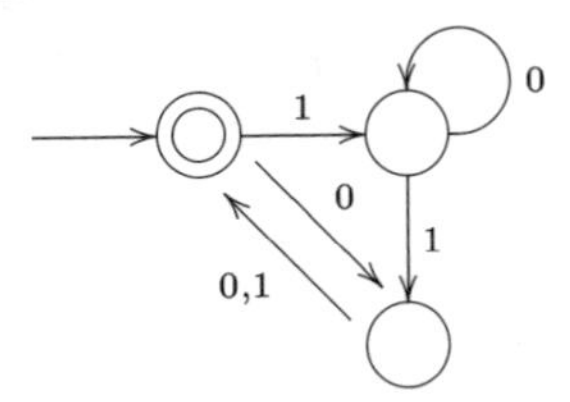

Let L be the language consisting of all strings over $\{0,1\}$ containing an odd number of 1's. Show that neither $L \subseteq L(\mathbf{A})$ nor $L(\mathbf{A}) \subseteq L$. Describe $L(\mathbf{A})$.

2.2 Incomplete automata

A useful design technique is illustrated by the following example.

Example 2.2.1 Construct an automaton to recognise the language $L = \{abab\}$. We first construct a 'spine' as follows:

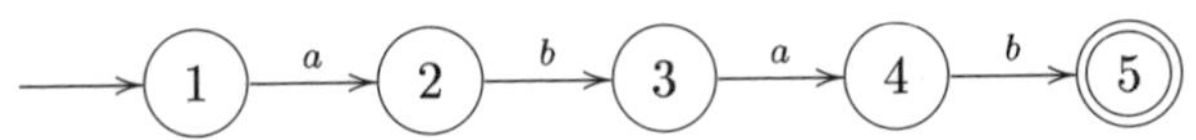

This diagram has a path labelled *abab* from the initial state to the terminal state and so we have ensured that the string *abab* will be recognised. However, it fails to be an automaton because there are missing transitions. It is tempting to put loops on the states and label the loops with the missing symbols, but this is exactly the wrong thing to do (why?). Instead, we add a new state, called a 'sink state,' to which we can harmlessly lead all unwanted transitions. In this way, we obtain the following automaton.

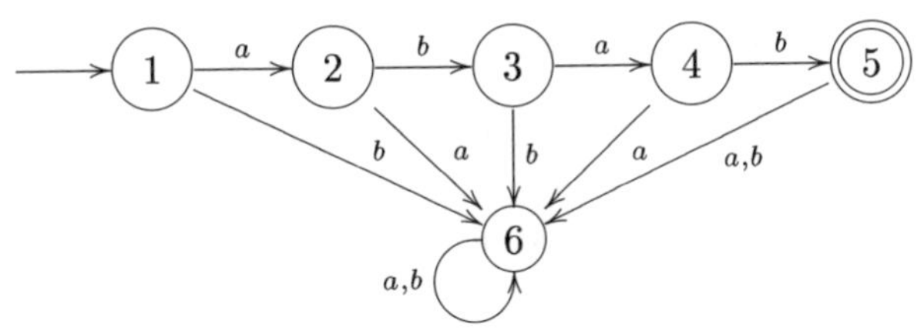

□

The idea behind the above example can be generalised. The 'spine' we constructed above is an example of an incomplete automaton; this is just like an automaton except that there are missing transitions. More formally, we define them as follows. An *incomplete automaton* is defined in exactly the same way as a complete deterministic automaton except that the transition *function* δ is replaced by a *partial function.* This means that $\delta(s, a)$ is not defined for some $(s, a) \in S \times A$; in such cases, we say that the machine *fails.*

Let $\mathbf{A} = (S, A, i, \delta, T)$ be an incomplete automaton. To define the language accepted by this machine we proceed as in the complete case, and with a similar justification. The extended transition function δ^* is defined as before, except this time δ^* is a partial function from $S \times A^*$ to S. More precisely, δ^* is defined as follows. Let $a \in A$, $w \in A^*$ and $s \in S$:

(ETF1) $\delta^*(s, \varepsilon) = s$.

(ETF2) $\delta^*(s, a) = \delta(s, a)$.

(ETF3) $\delta^*(s, aw) = \begin{cases} \delta^*(\delta(s, a), w) & \text{if } \delta(s, a) \text{ is defined} \\ \text{not defined} & \text{else.} \end{cases}$

We now define $L(\mathbf{A})$ to consist of all those strings $x \in A^*$ such that $\delta^*(i, x)$ is defined and is a terminal state.

In a complete automaton, there is *exactly one* path in the machine starting at the initial state and labelled by a given string. In an incomplete automaton, on the other hand, there is *at most one* path starting at the initial state and labelled by a given string. The language recognised by an incomplete automaton still consists of all strings over the input alphabet that label paths beginning at the initial state and ending at a terminal state.

It is easy to convert an incomplete machine into a complete machine that recognises the same language.

Proposition 2.2.2 *For each incomplete automaton* $\mathbf{A}$ *there is a complete automaton* $\mathbf{A}^c$ *such that* $L(\mathbf{A}^c) = L(\mathbf{A})$.

Proof Let $\mathbf{A} = (S, A, i, \delta, T)$. Define $\mathbf{A}^c$ as follows. Let ∞ be any symbol not in S. Then $S \cup \{\infty\}$ will be the set of states of $\mathbf{A}^c$; its initial state is i and its set of terminal states is T. The transition function of γ of $\mathbf{A}^c$ is defined as follows. For each $a \in A$ and $s \in S \cup \{\infty\}$

$$\gamma(s, a) = \begin{cases} \delta(s, a) & \text{if } \delta(s, a) \text{ is defined} \\ \infty & \text{if } s \neq \infty \text{ and } \delta(s, a) \text{ is not defined} \\ \infty & \text{else.} \end{cases}$$

Because the machine $\mathbf{A}$ is sitting inside $\mathbf{A}^c$, it is immediate that $L(\mathbf{A}) \subseteq L(\mathbf{A}^c)$. To prove the reverse inclusion, observe that any string that is accepted by $\mathbf{A}^c$

cannot pass through the state ∞ at any point. Thus the string is essentially being processed by **A**. □

We say that a state s in an automaton is a *sink state* if $s \cdot a = s$ for each $a \in A$ in the input alphabet. Thus the state ∞ in our construction above is a sink state, and the process of converting an incomplete machine into a complete machine is called *completion (by adjoining a sink state)*. The automaton $\mathbf{A}^c$ is called the *completion* of **A**.

It is sometimes easier to design an incomplete automaton to recognise a language and than to complete it by adjoining a sink state then to try to design the automaton all in one go. We can apply this idea to show that any *finite* language is recognisable. We illustrate this result by means of an example.

Example 2.2.3 Consider the finite language $\{b, aa, ba\}$. The starting point for our automaton is the part of the tree over $\{a, b\}$, which contains all strings of length 2 and smaller:

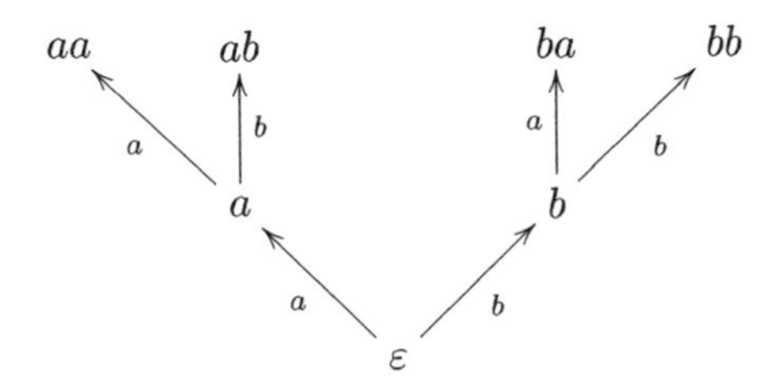

Notice that I have used labelled arrows rather than edges. This is used to build an incomplete automaton that recognises $\{b, aa, ba\}$: the vertices of the tree become the states, the initial state is the vertex labelled ε, and the terminal states are the vertices labelled with the strings in the language.

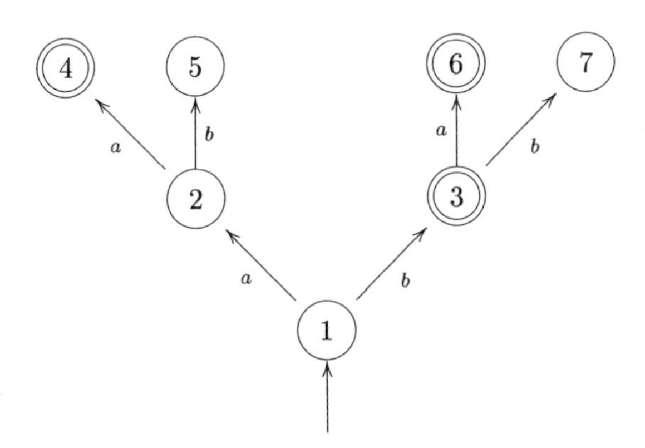

This incomplete automaton is then completed by the addition of a sink state. We thus obtain the following automaton that recognises $\{b, aa, ba\}$.

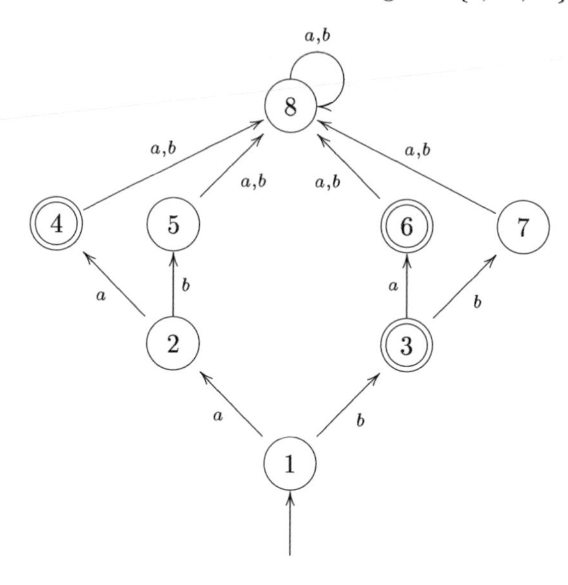

□

Our example of an automaton that recognises the language $\{b, aa, ba\}$ raises another point. Another (incomplete) machine that recognises this language is

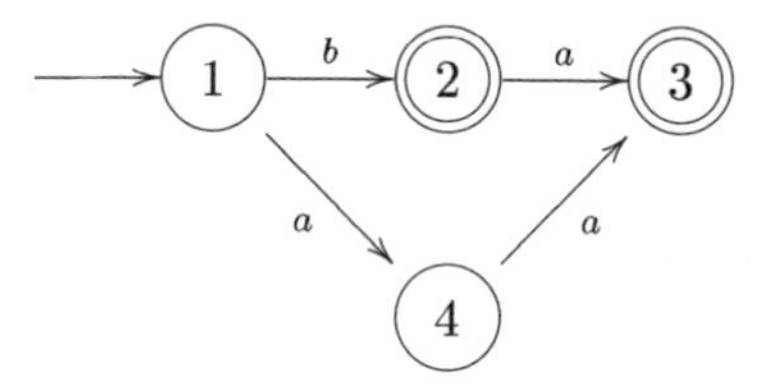

Thus by adjoining a sink state, we need only 5 states to recognise $\{b, aa, ba\}$ instead of the 8 in our example above. The question of finding the smallest number of states to recognise a given language is one that we shall pursue in Chapter 7.

The proof of the following is now left as an exercise.

Proposition 2.2.4 *Every finite language is recognisable.* □

Exercises 2.2

1. Construct an automaton to recognise the language

$$L = \{\varepsilon, ab, a^2b^2, a^3b^3\}.$$

2. Write out a full proof of Proposition 2.2.4.

2.3 Automata that count

Counting is one of the simplest ways of describing languages. For example, we might want to describe a language by restricting the lengths of the strings that can appear, or by requiring that a particular letter or pattern appears a certain number of times. We shall also see that there are limits to what automata can do in the realm of counting. We begin with a simple example.

Example 2.3.1 Construct an automaton to recognise the language

$$L = \{x \in (a+b)^*\colon |\,x\,| \text{ is even }\}.$$

The first step in constructing an automaton is to ensure that you understand what the language is. In this case, $x \in L$ precisely if $|\,x\,| = 0, 2, 4, \ldots$. The empty string ε is accepted since $|\,\varepsilon\,| = 0$, and so the initial state will also have to be terminal. In this case, we shall only need two states: one state remembers that we have read an even number of symbols and another that remembers that we have read an odd number of symbols. We therefore obtain the following automaton.

If, instead, we wanted to construct an automaton that recognised the language of strings over $\{a, b\}$ of *odd* length, then we would simply modify the above machine by making state 0 non-terminal and state 1 terminal. □

To generalise the above example, I shall need some terminology. The set of integers, that is the set of positive and negative whole numbers, is denoted $\mathbb{Z}$. The word 'number' will almost always mean 'integer' from now on. If $a, b \in \mathbb{Z}$ we say that a *divides* b, or that b is *divisible by* a, or that b is a *multiple* of a, if $b = aq$ for some $q \in \mathbb{Z}$; this is written mathematically as $a \mid b$. If $a, b \in \mathbb{Z}$ and $a > 0$ then we can write $b = aq + r$ where $0 \leq r < a$. The number q is called the *quotient* and r is called the *remainder*. The quotient and remainder are uniquely determined by a and b meaning that if $b = aq' + r'$ where $0 \leq r' < a$ then $q = q'$ and $r = r'$. This result is called the 'Remainder Theorem' and is one of the basic properties of the integers.

Using this terminology, let us look again at odd and even numbers. If we divide a number by 2, then there are exactly two possible remainders: 0 or 1. A number that has no remainder when divided by 2 is just an even number and a number that leaves the remainder 1 when divided by 2 is just an odd number. It is an accident of history that English, and many other languages, happen to have single words that mean 'leaves no remainder when divided by 2' and 'leaves remainder 1 when divided by 2.'

Now let us look at what happens when we divide a number by 3. This time there are three possible cases: 'leaves no remainder when divided by 3,' 'leaves remainder 1 when divided by 3,' and 'leaves remainder 2 when divided by 3.' In this case, there are no single words in English that we can use to substitute for each of these phrases, but this does not matter.

Let $n \geq 2$ be an integer, and let a and b be arbitrary integers. We say that a is *congruent to b modulo n*, written as

$$a \equiv b \pmod{n},$$

if $n \mid (a - b)$. An equivalent way of phrasing this definition is to say that a and b have the same remainder when divided by n. Put $\mathbb{Z}_n = \{0, 1, \ldots, n-1\}$, the set of possible remainders when a number is divided by n.

Using this notation, we see that a is even precisely when $a \equiv 0 \pmod{2}$ and is odd when $a \equiv 1 \pmod{2}$. If $a \equiv b \pmod{2}$ then we say they have the *same parity*: they are either both odd or both even. If a number $a \equiv 0 \pmod{n}$ then it is divisible by n.

Now that we have this terminology in place, we can generalise Example 2.3.1.

Example 2.3.2 Construct an automaton recognising the language

$$L = \{x \in (a+b)^* \colon |x| \equiv 1 \pmod{4}\}.$$

In this case, a string x is in L if its length is $1, 5, 9, 17, \ldots$. In other words, it has length one more than a multiple of 4. Notice that we are not interested in the exact length of the string. It follows that we must reject strings that have lengths $4q$, $4q+2$, or $4q+3$ for some q; we do not need to worry about strings of length $4q+4$ because that is itself a multiple of 4. In other words, there are only four possibilities, and these four possibilities will be represented by four states in our machine. I will label them $0, 1, 2$, and 3, where the state r means 'the length of the string read so far is $4q+r$ for some q.' The automaton that recognises L is therefore as follows:

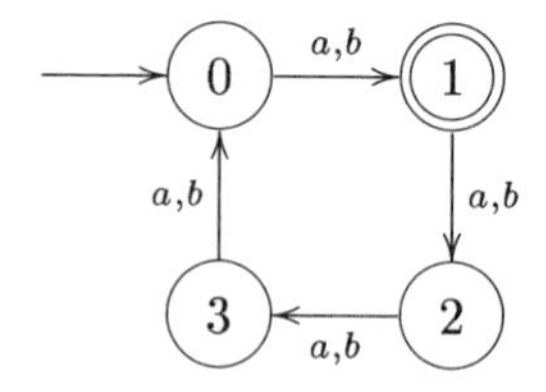

□

It should now be clear that we can easily construct automata to recognise any language L of the form,

$$L = \{x \in (a+b)^* \colon |\, x \,| \equiv r \pmod{n}\},$$

for any $n \geq 2$ and $0 \leq r < n$. We now turn to a different kind of counting.

Example 2.3.3 Construct an automaton which recognises the language

$$L = \{x \in (a+b)^*: |x| < 3\}.$$

Here we are required to determine length up to some number; this is called 'threshold counting.' We have to accept the empty string, all strings of length 1, and all strings of length 2; we reject all other strings. We are therefore led to the following automaton:

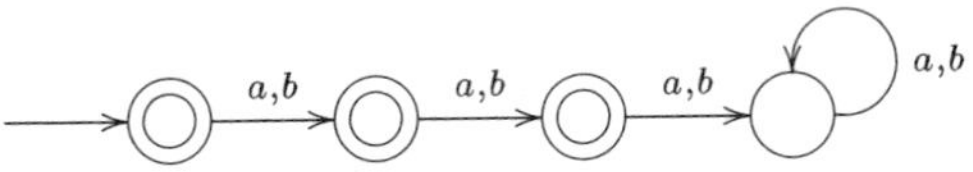

a,b a,b a,b a,b

□

In the above example, there is nothing sacrosanct about the number 3. Furthermore, we can easily modify our machine to deal with similar but different conditions on $|x|$ such as $|x| \leq 3$ or $|x| = 3$ or $|x| \geq 3$ or where $|x| > 3$.

Examples 2.3.1, 2.3.2, and 2.3.3 are typical of the way that counting is handled by automata: we can determine length modulo a fixed number, and we can determine length relative to some fixed number.

Our next result shows that there are limits to what we can count using finite automata.

Proposition 2.3.4 *The language*

$$L = \{a^n b^n: n \in \mathbb{N}\}$$

is not recognisable.

Proof When we say an automaton **A** recognises a language L we mean that it recognises *precisely* the strings in L and no others.

We shall argue by contradiction. Suppose $\mathbf{A} = (S, A, s_0, \delta, T)$ is a finite automaton such that $L = L(\mathbf{A})$. Let

$$q_n = s_0 \cdot a^n \text{ and } t_n = q_n \cdot b^n,$$

where $n \geq 0$. Thus q_n is the name we give the state that we reach when starting in the initial state s_0 and inputting the string a^n; and t_n is the name we give the state that we reach when starting in q_n and inputting b^n. Then $s_0 \cdot (a^n b^n) = t_n$ and so t_n is a terminal state because $a^n b^n \in L$. We claim that if $i \neq j$ then $q_i \neq q_j$. Suppose to the contrary that $q_i = q_j$ for some $i \neq j$. Then

$$s_0 \cdot (a^i b^j) = q_i \cdot b^j = q_j \cdot b^j = t_j.$$

But this would imply $a^i b^j \in L$ and we know $i \neq j$. Since this cannot happen, we must have $i \neq j$ implies $q_i \neq q_j$ and so **A** has infinitely many states. This is a contradiction. □

The problem with the language $\{a^n b^n : n \in \mathbb{N}\}$ is that we have to compare the number of a's with the number of b's and there can be an arbitrary number of both. Notice that we can construct an automaton that recognises a^*b^*:

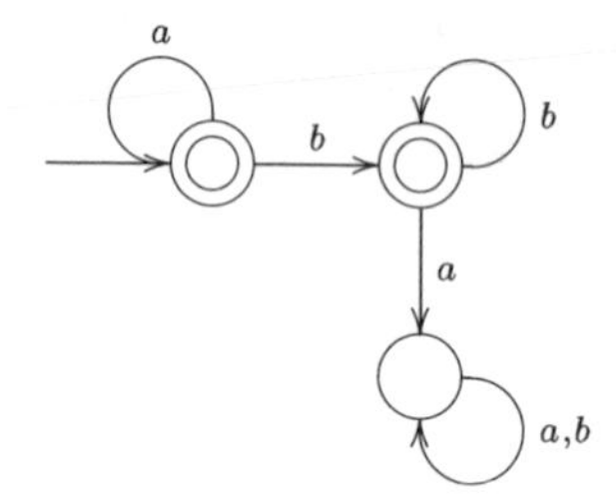

Thus an automaton can check that all the a's precede all the b's.

The above result raises the following important question: how can we distinguish between recognisable and non-recognisable languages? We shall investigate this question in Section 2.6.

Exercises 2.3

1. Let $A = \{a, b\}$. Construct finite automata for the following languages.

(i) All strings x in A^* such that $|x| \equiv 0 \pmod 3$.

(ii) All strings x in A^* such that $|x| \equiv 1 \pmod 3$.

(iii) All strings x in A^* such that $|x| \equiv 2 \pmod 3$.

(iv) All strings x in A^* such that $|x| \equiv 1$ or $2 \pmod 3$.

2. Construct a finite automaton to recognise the language

$$L = \{x \in (a+b)^* : |x|_a \equiv 1 \pmod 5\}.$$

3. Let $A = \{0, 1\}$. Construct finite automata to recognise the following languages.

(i) All strings x in A^* where $|x| < 4$.

(ii) All strings x in A^* where $|x| \leq 4$.

(iii) All strings x in A^* where $|x| = 4$.

(iv) All strings x in A^* where $|x| \geq 4$.

(v) All strings x in A^* where $|x| > 4$.

(vi) All strings x in A^* where $|x| \neq 4$.

(vii) All strings x in A^* where $2 \leq |x| \leq 4$.

4. Construct a finite automaton to recognise the language

$$\{x \in (a+b)^*: |x|_a \leq 4\}.$$

5. Show that the language $\{a^i b^j : i \equiv j \pmod 2\}$ is recognisable.

6. Let $A = \{a, b, c\}$. Construct a finite automaton recognising those strings in A^*, where the string abc occurs an odd number of times.

2.4 Automata that locate patterns

In this section, we shall show that the languages xA^*, A^*xA^*, and A^*x are all recognisable where A is any alphabet and x is any non-empty string. We begin with the simplest case: we show that the languages xA^* are recognisable.

Proposition 2.4.1 *Let A be an alphabet and let $x \in A^+$ be a string of length n. The language xA^* can be recognised by an automaton with $n+2$ states.*

Proof Because $x \in xA^*$ the string x itself must be accepted by any prospective automaton. So if $x = a_1 \ldots a_n$ where each $a_i \in A$, then we must have the following states:

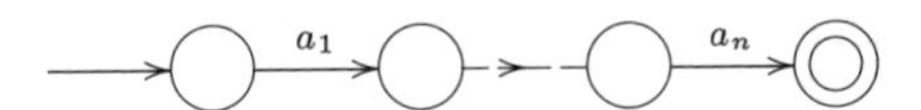

If we now put a loop on the last state labelled with the elements of A, we shall then have an incomplete automaton recognising xA^*. It is now a simple matter to complete this automaton to obtain one recognising the same language. □

We now turn to the problem of showing that languages of the form A^*xA^* are recognisable. This is not quite as straightforward and so we begin with an example to illustrate the ideas involved.

Example 2.4.2 Construct an automaton that recognises the language $L = (a+b)^*aba(a+b)^*$. In other words, all strings that contain aba as a factor. The first point to note is that aba should itself be accepted. So we can immediately write down the spine of the machine:

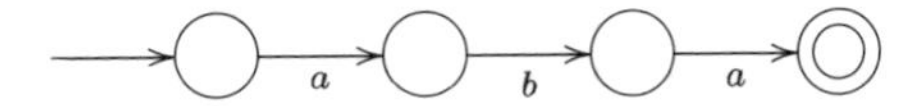

Once we have ascertained that an input string contains the factor aba we do not care what follows. So we can certainly write

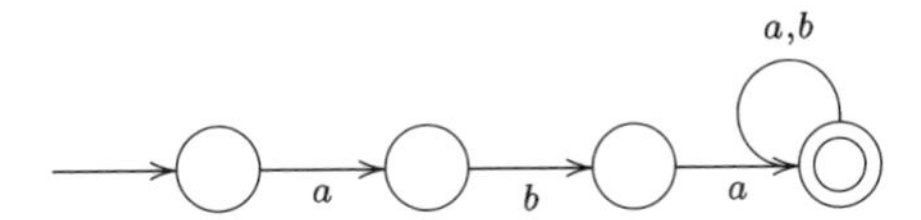

To find out what to do next, put yourself in the position of having to detect the string aba in a very long input string. As a finite automaton you can only read one letter at a time, so imagine that you are constrained to view the input string one letter at a time through a peephole. If you are reading a b, then you are not interested, but as soon as you read an a you are: you make a mental note 'I have just read an a.' If you read a b next, then you get even more interested: you make a mental note 'I have just read ab;' if instead you read an a, then you simply stay in the 'I have just read an a' state. The next step is the crucial one: if you read an a, then you have located the string aba, you do not care what letters you read next; if on the other hand you read a b, then it takes you back to the 'uninterested' state. We see now that the four states on our spine correspond to: 'uninterested,' 'just read an a,' 'just read ab' and 'success!' The automaton we require is therefore the following one:

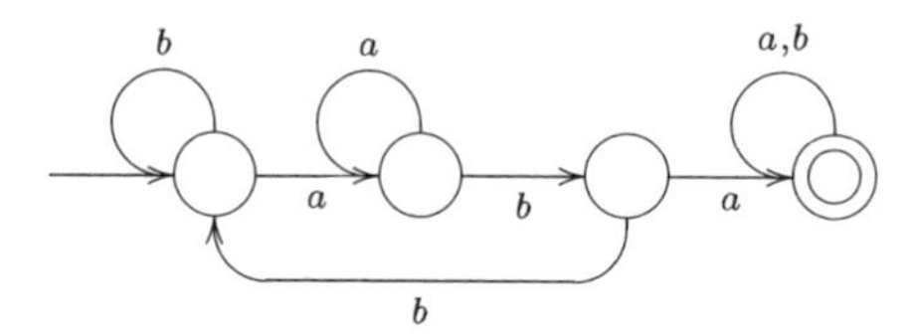

□

Our example above can be used to formulate a general principle for constructing automata that recognise languages of the form A^*xA^* where $x \in A^+$.

Proposition 2.4.3 *Let A be an alphabet and $x \in A^+$ a string of length n. The language A^*xA^* can be recognised by an automaton with $n+1$ states.*

Proof The first step is to construct the spine as we did in our example. If $x = a_1 \ldots a_n$, then this spine will have $n+1$ states: the first one is initial, the last is terminal, and the transitions are labelled in turn a_i for $i = 1, \ldots, n$; the last state also carries a loop labelled A.

Because we read an input string from left to right, each of these $n+1$ states is really storing which *prefix* of x we have read in the input: from the first state representing ε to the last representing x itself. To work out where to put the missing transitions, suppose that we are in the state corresponding to the prefix y of x, where $y = a_1 \ldots a_i$ and that the next letter of the input string we read is a. There are two cases to consider.

(Case 1): suppose that $a = a_{i+1}$, that is ya is also a prefix of x. Then we simply move to the next state to the right along the spine.

(Case 2): suppose that $a \neq a_{i+1}$. It is tempting to think that we have to go back to the initial state, but this is not necessarily so. The string ya is not a prefix of x; however we can always find a *suffix* of ya that is a *prefix* of x; we do not exclude the possibility that this suffix could be ε. Choose the *longest* suffix z of ya that is a prefix of x. The transition we require is then

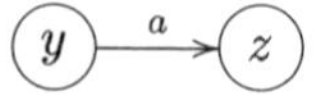

More generally, for a fixed string x and arbitrary string u we denote by $\sigma_x(u)$ the longest suffix of u that is a prefix of x. Thus $z = \sigma_x(ya)$.

Notice that (Case 1) is really included in the rule stated in (Case 2) because if ya is a prefix of x then $\sigma_x(ya) = ya$. □

We illustrate the design technique contained in the above result by the following example.

Example 2.4.4 Construct an automaton to recognise the language $A^*ababbA^*$. We construct two tables: the transition table and an auxiliary table that will help us to complete the transition table. We begin by entering in the transition table the transitions on the spine:

	a	b
ε	a	1
a	2	ab
ab	aba	3
aba	4	$abab$
$abab$	5	$ababb$
$ababb$	$ababb$	$ababb$

I have numbered the transitions we still have to find. The auxiliary table below gives the calculations involved in finding them.

	u	proper suffixes of u	$\sigma_{ababb}(u)$
1	b	ε	ε
2	aa	ε, a	a
3	abb	ε, b, bb	ε
4	$abaa$	ε, a, aa, baa	a
5	$ababa$	$\varepsilon, a, ba, aba, baba$	aba

The first column, labelled u, is the concatenation of the prefix labelling the state and the input letter: $\varepsilon \cdot b$, $a \cdot a$, $ab \cdot b$, $aba \cdot a$, and $abab \cdot a$, respectively. The last column, labelled $\sigma_{ababb}(u)$, is the 'longest suffix of the string in the first column, which is a prefix of $ababb$'; we use the middle column to determine this string. We can now complete the transition table using the auxiliary table:

	a	b
$\rightarrow \varepsilon$	a	ε
a	a	ab
ab	aba	ε
aba	a	$abab$
$abab$	aba	$ababb$
$\leftarrow ababb$	$ababb$	$ababb$

It is now easy to draw the transition table of the required automaton. □

Finally, we show that languages of the form A^*x are recognisable.

Proposition 2.4.5 *Let A be an alphabet and $x \in A^+$ a string of length n. The language A^*x can be recognised by an automaton with $n+1$ states*

Proof The construction is similar to the one contained in Proposition 2.4.3. The first step is to construct the spine. If $x = a_1 \ldots a_n$, then this spine will have $n+1$ states: the first one is initial, the last is terminal, and the transitions are labelled in turn a_i for $i = 1, \ldots, n$. In this case, the last state does *not* carry a loop labelled A. We now carry out exactly the same procedure as in Proposition 2.4.3, except this time we apply it also to the terminal state. □

Exercises 2.4

1. Let $A = \{a, b\}$. Construct finite automata to recognise the following languages.

(i) All strings in A^* that begin with ab.

(ii) All strings in A^* that contain ab.

(iii) All strings in A^* that end in ab.

2. Let $A = \{0, 1\}$. Construct automata to recognise the following languages.

(i) All strings that begin with 01 or 10.

(ii) All strings that start with 0 and contain exactly one 1.

(iii) All strings of length at least 2 whose final two symbols are the same.

3. Let $A = \{a, b\}$. Construct an automaton to recognise the language

$$A^*aaA^*bbA^*.$$

4. Let $A = \{a, b, c\}$. Construct an automaton to recognise all strings that begin or end with a double letter.

2.5 Boolean operations

In describing languages we frequenty use words such as 'and' 'or' and 'not.' For example, we might describe a language over the alphabet $\{a, b, c\}$ to consist of all strings that satisfy the following condition: they begin with *either* an a *or* a c *and* do *not* contain ccc as a factor. In this section, we describe algorithms for constructing automata where the description of the languages involves Boolean operations.

Example 2.5.1 Consider the language

$$L = \{x \in (a+b)^*: |x| \equiv 1 \pmod 4\},$$

of Example 2.3.2. We showed in Section 2.3 how to construct an automaton $\mathbf{A}$ to recognise L. Consider now the language $L' = A^* \setminus L$. We could try to build a machine from scratch that recognises L' but we would much prefer to find some way of adapting the machine $\mathbf{A}$ we already have to do the job. The strings in L' are those $x \in (a+b)^*$ such that $|x| \equiv 0$ or $|x| \equiv 2$ or $|x| \equiv 3 \pmod 4$. It follows that the machine $\mathbf{A}'$ recognising L' is

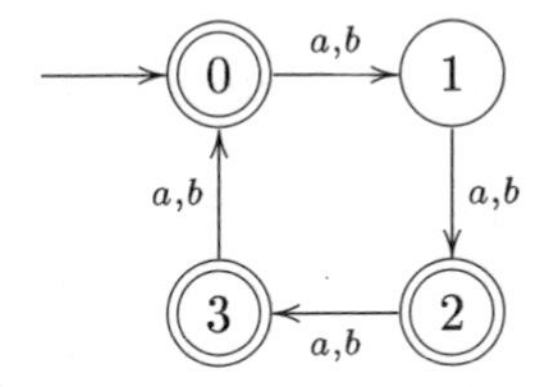

We can see that this was obtained from $\mathbf{A}$ by interchanging terminal and non-terminal states. □

The above example turns out to be typical.

Proposition 2.5.2 *If L is recognised by $\mathbf{A} = (S, A, i, \delta, T)$ then L' is recognised by $\mathbf{A}' = (S, A, i, \delta, T')$ where $T' = S \setminus T$.*

Proof The automaton $\mathbf{A}'$ is exactly the same as the automaton $\mathbf{A}$ *except* that the terminal states of $\mathbf{A}'$ are the non-terminal states of $\mathbf{A}$. We claim that $L(\mathbf{A}') = L'$. To see this we argue as follows. By definition $x \in L(\mathbf{A}')$ if and only if $i \cdot x \in S \setminus T$, which is equivalent to $i \cdot x \notin T$, which means precisely that $x \notin L(\mathbf{A})$. This proves the claim. □

Example 2.5.3 A language $L \subseteq A^*$ is said to be *cofinite* if L' is finite. We proved in Proposition 2.2.4 that every finite language is recognisable. It follows by Proposition 2.5.2 that every cofinite language is recognisable. This example is hopefully an antidote to the mistaken view that people sometimes get when first introduced to finite automata: **the languages recognised by automata need not be finite!** □

The following example motivates our next construction using Boolean operations.

Example 2.5.4 Consider the language

$$N = \{x \in (a+b)^*: x \in (a+b)^* aba (a+b)^* \text{ and } |x| \equiv 1 \pmod 4\}.$$

If we define

$$L = \{x \in (a+b)^*\colon |x| \equiv 1 \pmod 4\} \text{ and } M = (a+b)^*aba(a+b)^*,$$

then $N = L \cap M$. Automata that recognise L and M, respectively, are

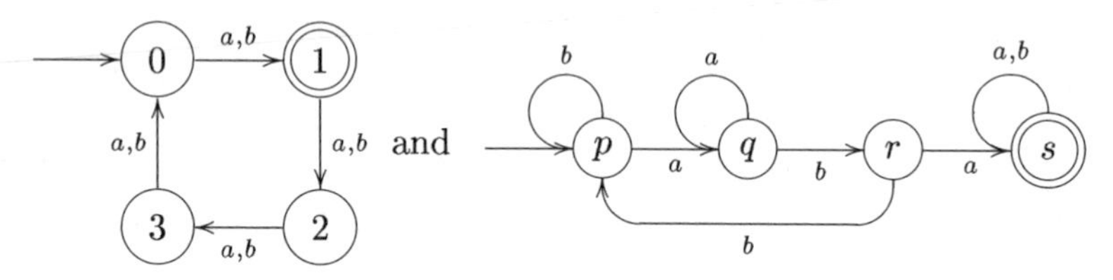

We would like to combine these two automata to build an automaton recognising $N = L \cap M$. To discover how to do this, we need only reflect on how we would decide if a string x is in N: we would run it on the left-hand machine and on the right-hand machine, and we would accept it if and only if when it had been read, both left- and right-hand machines were in a terminal state. To do this, we could run x first on one machine and then on the other, but we could also run it on both machines at the same time. Thus x is input to the left-hand machine in state 0, and a copy on the right-hand machine in state p. The subsequent states of both machines can be recorded by an ordered pair (l, r) where l is the current state of the left-hand machine and r is the current state of the right-hand machine. For example, $abba$ causes the two machines to run through the following pairs of states:

$$(0,p), (1,q), (2,r), (3,p), (0,q).$$

The string $abba$ is not accepted because although 0 is a terminal state in the left-hand machine, q is not a terminal state in the right-hand machine. □

The above example illustrates the idea behind the following result.

Proposition 2.5.5 *If L and M are recognisable languages over A then so is $L \cap M$.*

Proof Let $L = L(\mathbf{A})$ and $M = L(\mathbf{B})$ where $\mathbf{A} = (S, A, s_0, \delta, F)$ and $\mathbf{B} = (T, A, t_0, \gamma, G)$. Put

$$\mathbf{A} \times \mathbf{B} = (S \times T, A, (s_0, t_0), \delta \times \gamma, F \times G),$$

where

$$(\delta \times \gamma)((s,t), a) = (\delta(s,a), \gamma(t,a));$$

we write

$$(\delta \times \gamma)((s,t), a) = (s,t) \cdot a = (s \cdot a, t \cdot a),$$

as usual. It is easy to check that if x is a string, then the extended transition function has the form

$$(s,t) \cdot x = (s \cdot x, t \cdot x).$$

We claim that $L(\mathbf{A} \times \mathbf{B}) = L \cap M$. By definition $x \in L(\mathbf{A} \times \mathbf{B})$ if and only if $(s_0, t_0) \cdot x \in F \times G$. But this is equivalent to $s_0 \cdot x \in F$ and $t_0 \cdot x \in G$, which says precisely that $x \in L(\mathbf{A}) \cap L(\mathbf{B}) = L \cap M$, and so the claim is proved. □

We have dealt with complementation and intersection, so it is natural to finish off with union. The idea is similar to Proposition 2.5.5.

Proposition 2.5.6 *If L and M are recognisable languages over A then so is $L + M$.*

Proof Let $L = L(\mathbf{A})$ and $M = L(\mathbf{B})$, where $\mathbf{A} = (S, A, s_0, \delta, F)$ and $\mathbf{B} = (T, A, t_0, \gamma, G)$. Put

$$\mathbf{A} \sqcup \mathbf{B} = (S \times T, A, (s_0, t_0), \delta \times \gamma, (F \times T) + (S \times G)).$$

We claim that $L(\mathbf{A} \sqcup \mathbf{B}) = L + M$. By definition $x \in L(\mathbf{A} \sqcup \mathbf{B})$ if and only if $(s_0, t_0) \cdot x \in (F \times T) + (S \times G)$. This is equivalent to $s_0 \cdot x \in F$ *or* $t_0 \cdot x \in G$, because $s_0 \cdot x \in S$ *and* $t_0 \cdot x \in T$ always hold. Hence $x \in L(\mathbf{A}) + L(\mathbf{B}) = L + M$, and the claim is proved. □

There is an alternative proof of Proposition 2.5.6, which relies only on Propositions 2.5.2 and 2.5.5 together with a little set theory. See Exercises 2.5.

Observe that the only difference between automata constructed in Proposition 2.5.5 and Proposition 2.5.6 lies in the definition of the terminal states: to recognise the *intersection* of two languages the terminal states are those ordered pairs (s, t) where s *and* t are terminal; to recognise the *union* of two languages the terminal states are those ordered pairs (s, t) where s *or* t is terminal.

Propositions 2.5.2, 2.5.5, and 2.5.6 can be applied in at least two different ways. First, they can be used to construct an automaton recognising a language described in terms of Boolean operations; second, they can be used to prove that a language described in terms of Boolean operations is recognisable without actually constructing a specific machine. We give examples of each of these two applications.

Example 2.5.7 Let $A = \{a, b\}$. We wish to construct an automaton to recognise the language

$$L = \{x \in A^*\colon |\, x \,|_a \text{ is even and } |\, x \,|_b \text{ is odd}\}.$$

This language is the intersection of

$$M = \{x \in A^*\colon |\, x \,|_a \text{ is even}\} \text{ and } N = \{x \in A^*\colon |\, x \,|_b \text{ is odd}\}.$$

It is easy to construct automata that recognise these languages separately; the machine $\mathbf{A}$ below recognises M:

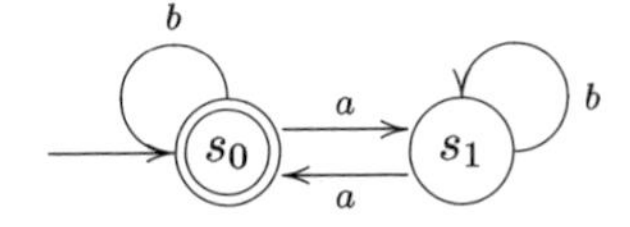

and the machine **B** below recognises N:

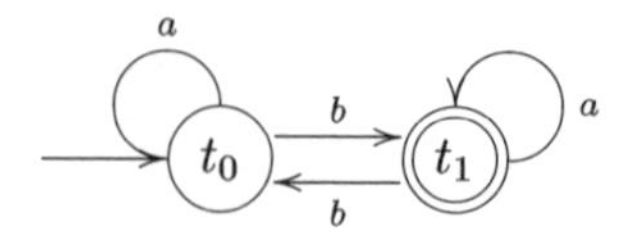

To construct the machine $\mathbf{A} \times \mathbf{B}$ (and similar comments apply to the construction of $\mathbf{A} \sqcup \mathbf{B}$) we proceed as follows. The set of states of $\mathbf{A} \times \mathbf{B}$ is the set $S \times T$, where S is the set of states of $\mathbf{A}$ and T is the set of states of $\mathbf{B}$. In this case,

$$S \times T = \{(s_0, t_0), (s_0, t_1), (s_1, t_0), (s_1, t_1)\}.$$

We draw and label these four states. Mark the initial state, which is (s_0, t_0). Mark the set of terminal states, which in this case is just (s_0, t_1); it is only at this point that the constructions of $\mathbf{A} \times \mathbf{B}$ and $\mathbf{A} \sqcup \mathbf{B}$ differ. It remains now to insert all the transitions. For each $a \in A$ and each pair $(s, t) \in S \times T$, calculate $(s, t) \cdot a$ which by definition is just $(s \cdot a, t \cdot a)$. For example,

$$(s_0, t_0) \cdot a = (s_0 \cdot a, t_0 \cdot a) = (s_1, t_0)$$

and

$$(s_0, t_0) \cdot b = (s_0 \cdot b, t_0 \cdot b) = (s_0, t_1).$$

We therefore draw an arrow labelled a from the state labelled (s_0, t_0) to the state labelled (s_1, t_0), and an arrow labelled b from the state labelled (s_0, t_0) to the state labelled (s_0, t_1). Continuing in this way, the machine $\mathbf{A} \times \mathbf{B}$, that recognises the language $L = M \cap N$, has the following form:

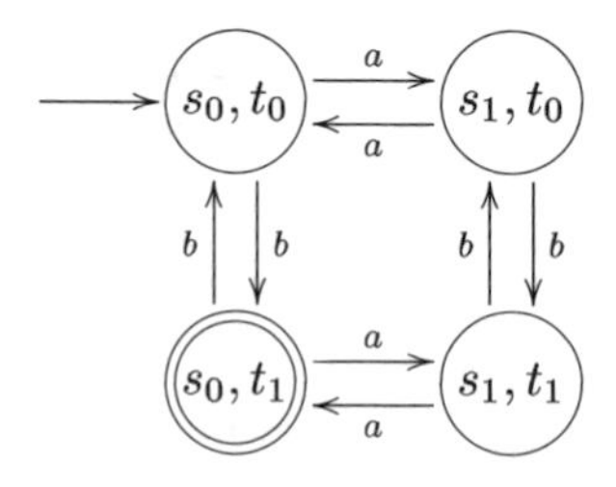

□

Example 2.5.8 We show how Propositions 2.5.2, 2.5.5, and 2.5.6 can be used to prove a language is recognisable without actually constructing a recognising automaton. We prove that the language,

$$P = (aA^* \cap A^*b) \setminus A^*ccA^*,$$

over the alphabet $A = \{a, b, c\}$ is recognisable. Put $L = aA^*$, $M = A^*b$, and $N = A^*ccA^*$. By Section 2.4, these languages are recognisable. Since N is recognisable, N' is recognisable by Proposition 2.5.2. Since L and M are recognisable, $L \cap M$ is recognisable by Proposition 2.5.5. Finally, $L \cap M$ and N' are both recognisable and so $P = (L \cap M) \cap N' = (L \cap M) \setminus N$ is recognisable by Proposition 2.5.5. □

Exercises 2.5

1. Construct separate automata to recognise each of the languages below:

$$L = \{w \in \{0,1\}^*: |\, w\,|_0 \equiv 1 \pmod 3\}$$

and

$$M = \{w \in \{0,1\}^*: |\, w\,|_1 \equiv 2 \pmod 3\}.$$

Use Proposition 2.5.5 to construct an automaton that recognises $L \cap M$.

2. Let

$$L = \{x \in a^*: |x| \equiv 0 \pmod 3\} \text{ and } M = \{x \in a^*: |x| \equiv 0 \pmod 5\}.$$

Construct automata that recognise L and M, respectively. Use Proposition 2.5.6 to construct an automaton that recognises $L + M$.

3. Prove that if L and M are recognisable languages over A, then so is $L \setminus M$.

4. Show how the constructions of $\mathbf{A}'$ and $\mathbf{A} \times \mathbf{B}$ combined with one of de Morgan's laws enables $\mathbf{A} \sqcup \mathbf{B}$ to be constructed.

5. Show that if $L_1, \ldots, L_n$ are each recognisable, then so too are $L_1 + \ldots + L_n$ and $L_1 \cap \ldots \cap L_n$.

6. Let A be an alphabet. A language L over A is said to be *definite* if it can be written in the form $L = X + A^*Y$, where X and Y are finite languages. Prove that every definite language is recognisable.

You can find out more about definite languages in [18] and [99].

2.6 The Pumping Lemma

In this section, we shall show that there is a property shared by all recognisable languages: 'the pumping property.' Consequently, if a language does not have this property it cannot be recognisable. There are, however, non-recognisable languages that have the pumping property as well, so it is important to remember that we can only use it to show that a language is *not* recognisable.

We begin with the key result referred to as the 'Pumping Lemma' although we shall label it as a theorem because of its significance.

Theorem 2.6.1 (The Pumping Lemma) *Let L be a recognisable language over the alphabet A. Then there is a number n such that the following holds: for each $w \in L$ such that $|w| \geq n$ we can find strings x,y and z such that the following four properties hold:*

(i) $w = xyz$.

(ii) $|xy| \leq n$.

(iii) $|y| \geq 1$.

(iv) *For all $i \geq 0$ we have that $xy^iz \in L$* ('pumping property').

Proof If L is finite let n be greater than the length of the longest string in L. Then the result holds vacuously.

The more interesting case is where L is infinite. Since L is recognisable, there is an automaton $\mathbf{A}$ such that $L = L(\mathbf{A})$. Let $\mathbf{A}$ have n states. By assumption there are strings in L of length at least n. Let $w = a_1 \ldots a_m \in L$ where $m \geq n$. Consider the sequence of states:

$$s_0,\, s_1 = s_0 \cdot a_1,\, s_2 = s_0 \cdot (a_1 a_2), \ldots,\, s_m = s_0 \cdot (a_1 \ldots a_m).$$

Since $m \geq n$ it follows that $m + 1 > n$. But $s_0, s_1, \ldots, s_n$ is a list of states of length $n + 1$ so there must be some repetition of states in this list; let s_i be the first repeated state in this list and s_j the first instance of the repeat of s_i. Then $0 \leq i < j \leq n$.

We have the following schematic diagram of the path in $\mathbf{A}$ labelled by the string x:

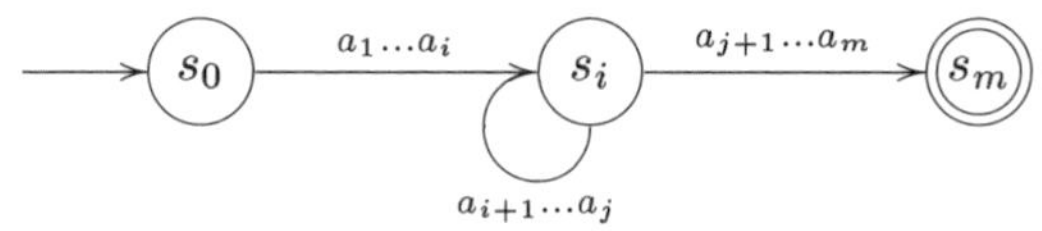

Put $x = a_1 \ldots a_i$, $y = a_{i+1} \ldots a_j$, and $z = a_{j+1} \ldots a_m$. It is clear that conditions (i), (ii), and (iii) hold by construction, and that (iv) follows as we may circle the loop at s_i any number of times (including none) before proceeding on to the terminal state. □

We now give some examples to show how the Pumping Lemma can be used to prove that a language is not recognisable.

Example 2.6.2 We shall prove that $L = \{a^n b^n \colon n \geq 0\}$ is not recognisable using the Pumping Lemma.

Suppose that L, clearly an infinite language, is recognisable. Then by the Pumping Lemma, there is a number n such that for each string $w \in L$ with $|w| \geq n$ the four consequences of the Pumping Lemma hold. The specific string $a^n b^n \in L$ and $|a^n b^n| \geq n$. Thus there exist strings $x, y, z \in A^*$ such that

(i) $a^n b^n = xyz$.

(ii) $|xy| \leq n$.

(iii) y is not empty.

(iv) $xy^i z \in L$ for all $i \geq 0$.

Observe that property (ii) tells us that xy consists entirely of a's, and that property (iii) tells us that y consists of at least one a. It follows that z consists of all the b's and possibly some a's.

Consider the string $xz = xy^0 z$. By the pumping property, $xz \in L$. However, by our observations above, $|xz|_b = n$, but $|xz|_a = n - |y|_a < n$. It follows that $xz \notin L$.

We have therefore arrived at a contradiction. It follows that our original assumption that L was recognisable is false. Thus L is not recognisable. □

The next example shows in effect that the set of primes is not recognisable.

Example 2.6.3 We shall prove that $L = \{a^p \colon p \text{ prime}\}$ is not recognisable using the Pumping Lemma: that is to say no finite automaton can be used to decide primality of the integers.

By a theorem of Euclid, L is an infinite set. Suppose that L is recognisable. Then there is a number n such that for each string $w \in L$ with $|w| \geq n$ the four consequences of the pumping lemma hold.

Let p be a prime chosen so that $p > n$. The string $a^p \in L$ and $|a^p| \geq n$. Thus there exist strings $x, y, z \in A^*$ such that

(i) $a^p = xyz$.

(ii) $|xy| \leq n$.

(iii) y is not empty.

(iv) $xy^i z \in L$ for all $i \geq 0$.

By the pumping property, $xy^{p+1}z \in L$. We now calculate the length of this string:

$$|xy^{p+1}z| = |x| + |y^{p+1}| + |z| = |x| + (p+1)|y| + |z| = |xyz| + p|y|.$$

By assumption, $|xyz| = p$. By consequence (iii), y is not empty and so $y = a^m$ for some $m \geq 1$. Thus $|xy^{p+1}z| = p(1+m)$. But $p(1+m)$ is not a prime because $m \geq 1$.

We have therefore arrived at a contradiction. It follows that our original assumption that L was recognisable is false. Thus L is not recognisable. □

Sometimes the Pumping Lemma can be used in conjunction with the results in Section 2.5 to show that a language is not recognisable.

Example 2.6.4 Prove that the language

$$L = \{x \in (a+b)^*\colon |\, x\,|_a = |\, x\,|_b\}$$

is not recognisable. Suppose that L were recognisable. At the conclusion of Section 2.3, we proved that the language a^*b^* is recognisable. By Proposition 2.5.5, the language $L \cap a^*b^*$ is therefore recognisable. However, this is just the language $\{a^n b^n\colon n \geq 0\}$, which we proved to be non-recognisable using the Pumping Lemma in Example 2.6.2. This is a contradiction, and so L is not recognisable. □

Exercises 2.6

1. Use the Pumping Lemma to prove that each of the following languages is not recognisable.

(i) $L = \{a^n b^{2n}\colon n \geq 0\}$.

(ii) $L = \{a^m\colon \text{where } m \text{ is a square}\}$.

(iii) $L = \{x \in (a+b)^*\colon |\, x\,|_a < |\, x\,|_b\}$.

(iv) $L = \{a^{2^n}\colon n \geq 0\}$.

2. Let $\mathbf{A}$ be an automaton with n states recognising the language L, which is finite. Show that for each string $x \in L$ we have that $|\, x\,| < n$.

3. Let $L \subseteq b^*$ be any non-recognisable language. Put $M = a^+L + b^*$. Show that M satisfies the conclusions of the Pumping Lemma. In Question 3 of Exercises 4.2, you will be asked to prove that M is not recognisable.

This example shows that the Pumping Lemma can only be used to show that a language is not recognisable, not that it is recognisable. There are, however, more advanced versions of the Pumping Lemma that do deliver necessary and sufficient conditions. See [136] for more information.

2.7 Summary of Chapter 2

- *Incomplete automata*: An automaton is incomplete if there are missing transitions. An incomplete automaton $\mathbf{A}$ can easily be converted into a complete automaton $\mathbf{A}^c$ recognising the same language by simply adding a sink state: this is a state to which missing transitions are connected but from which there is no escape.

- *Automata that count*: By arranging states in a circle it is possible to count modulo n. Automata can also be constructed to count relative to a threshold by arranging the states in a line.

- *Automata recognising patterns*: There are algorithms for constructing automata to recognise the languages xA^*, A^*xA^*, and A^*x where x is any non-empty string.

- *Recognising Boolean combinations of languages*: If $L = L(\mathbf{A})$ and $M = L(\mathbf{B})$, then there are algorithms for combining $\mathbf{A}$ and $\mathbf{B}$ to recognise $L + M$ and $L \cap M$. This is also an algorithm to convert $\mathbf{A}$ into an automaton recognising L'.

- *The Pumping Lemma*: The Pumping Lemma describes a property shared by all recognisable languages. It is sometimes possible to prove that a language is not recognisable by showing that it does not have this property. In particular, the language $\{a^n b^n \colon n \geq 0\}$ is one example of a non-recognisable language.

2.8 Remarks on Chapter 2

This chapter can be regarded as a collection of examples illustrating a number of different kinds of automata and languages.

Incomplete automata can provide a more concise way for describing a language than complete automata. They are also interesting in their own right. For example, an important class of incomplete automata are those in which each input letter induces a partial injection on the set of states. Such automata are said to be *reversible*. They provide a useful tool in studying finitely generated subgroups of free groups. The papers [104] and [84] are good starting points for this important work. The languages recognised by reversible automata are discussed by Angluin from the point of view of language learning in [4].

Finite languages are the simplest kinds of recognisable languages; nevertheless interesting questions can still be posed about them. For example, the remark following Example 2.2.3 illustrates the problem of finding the smallest automaton recognising a given finite language, where by smallest I mean one having the smallest number of states. The algorithm intimated in Example 2.2.3 will not in general yield such an automaton. The general question of 'minimising automata' is described in Chapter 7. However in the case of finite languages a more direct approach is possible. Using a description of the finite language by means of a regular expression, Section 9 of Chapter 1 of the book [113] determines the smallest number of states needed to recognise a given finite language together with the associated number of terminal states. In the case of a two-letter alphabet A, the subsets of A^n for a fixed n have been investigated from the point of view of minimal automata in [24].

The two different ways of counting using automata, described in Section 2.3, were highlighted on page 5 of [88]. In Chapter 11, we classify those languages in which counting, in the sense of modulo n counting, need not be used in recognising them. The languages of Section 2.4 turn out to be particular instances of such languages, and they are discussed further in the notes at the end of Chapter 11.

The automata of Section 2.4 are also simple cases of pattern-matching algorithms. Propositions 2.4.3 and 2.4.5, which are just variants of the same idea, are automata-theoretic manifestations of the Knuth-Morris-Pratt algorithm. This is proved on page 875 of the voluminous [36]. This algorithm is also discussed in [2]. Further examples of pattern-matching algorithms and their applications can be found in [126]. Automata-theoretic algorithms for locating substrings of a string together with applications are discussed in [8].

The Pumping Lemma was first proved in [9].

I would like to place automata in a slightly wider context so that we can better appreciate why they should be interesting. To do this, I will introduce a more concrete model of an automaton. We regard an automaton **A** as a black box equipped with a read-head and two lights: one labelled 'accept' and one labelled 'reject.' There is also a reset-button, which re-initialises the automaton. The input to the automaton is written on a paper tape that is right infinite: it has a beginning but no end. This tape is divided into squares. The symbols of the input are written into the successive squares on the tape starting from the leftmost square. So if our input string is $abab$, then the tape will look like this:

a	b	a	b				$\cdots$

To process the input string, the read-head of **A** is placed over the leftmost square and the reset button of **A** is pushed. From now on **A** runs completely automatically: it reads the contents of the current square, changes its internal state as a result and moves one square to the right. When **A** reads a blank square, then it knows that it has read a complete input string; if it is in a terminal state, then the 'accept' light is activated, whereas if it is in a non-terminal state, the 'reject' light is activated, after which **A** goes into a rest state and does nothing until the reset button is pushed. It is clear, I think, that it would not be too difficult to build such a machine from any finite automaton and, conversely, any such machine could be modelled by a finite automaton.

My reason for describing finite automata in such concrete terms is that two limitations in their design are now obvious: the read-head is restricted to move always one square to the right, in addition it is unable to overwrite the contents of a square. Automata in which the tape-head can move left as well as right are called two-way automata. Despite appearances, they are in fact no more powerful than ordinary automata. Proofs of this result can be found in [109] and [120]. However, if we allow the two-way automata also to overprint the contents of a square, then the situation changes radically. Such automata are called *Turing machines* after Alan Turing who first studied them in the 1930s. Turing machines can be used as language acceptors, just like finite automata; indeed, it is clear that every recognisable language is Turing recognisable. However, it is not hard to design a Turing machine to recognise the language $L = \{a^i b^i : i \geq 0\}$, which we proved not to be recognisable by a finite automaton. Thus Turing machines are strictly more powerful than finite automata. But just how powerful are they? It is clear that anything a Turing machine can compute must be algorithmic since the basic actions a Turing machine can carry out are clearly algorithmic. Thus it is certainly true that Turing machines are special kinds of algorithms for language acceptors. But more is true: it is believed that *every* language L for which there is an algorithm that outputs 'yes' precisely when a string is in L can be implemented by means of a suitable Turing machine. This belief is usually termed the *Church-Turing Thesis.*

There is considerable evidence for the Church-Turing Thesis and as a result it is usual to define algorithmic problems to be those solvable by a Turing machine. Because of the Church-Turing Thesis, wherever algorithms play a role in mathematics Turing machines can be used to implement them. Since finite automata are really nothing other than special kinds of Turing machines, it becomes a little less surprising that finite automata should arise in different branches of mathematics.

We conclude that finite automata are special kinds of algorithms for solving some decision problems, and although they are limited in what they can do, this very limitation means that they can be handled more easily than more general kinds of algorithms.

Chapter 3

Non-deterministic automata

In Chapter 2, we looked at various ways of constructing an automaton to recognise a given language. However, we did not get very far. The reason was that automata as we have defined them are quite 'rigid': from each state we must have exactly one transition labelled by each element of the input alphabet. This is a strong constraint and restricts our freedom considerably when designing an automaton. To make progress, we need a tool that is easy to use and can help us design automata. This is the role played by the non-deterministic automata we introduce in this chapter. A non-deterministic automaton is exactly like an automaton except that we allow multiple initial states and we impose no restrictions on transitions as long as they are labelled by symbols in the input alphabet. Using such automata, it is often easy, as we shall see, to construct a non-deterministic automaton recognising a given language. However, this would be useless unless we had some way of converting a non-deterministic automaton into a deterministic one recognising the same language. We describe an algorithm that does exactly this. We can therefore add non-deterministic automata to our toolbox for building automata.

Non-deterministic automata also enable us to establish a link with what are known as 'grammars': these are devices for *generating* strings belonging to a language. We prove that the languages recognised by finite automata are precisely the languages generated by what are called 'right linear grammars.'

3.1 Accessible automata

There are many different automata that can be constructed to recognise a given language. All things being equal, we would like an automaton with the smallest number of states. In Chapter 7, we will investigate this problem in detail. For the time being, we shall look at one technique that may make an automaton smaller without changing the language it recognises, and that will play an important role in our algorithm for converting a non-deterministic automaton into a deterministic automaton in Section 3.2.

Let $\mathbf{A} = (S, A, i, \delta, T)$ be a finite automaton. We say that a state $s \in S$ is *accessible* if there is a string $x \in A^*$ such that $i \cdot x = s$. A state that is not accessible is said to be *inaccessible*. An automaton is said to be *accessible* if every state is accessible. In an accessible automaton, each state can be reached from the initial state by means of some input string. Observe that the initial state itself is always accessible because $i \cdot \varepsilon = i$. It is clear that the inaccessible states of an automaton can play no role in accepting strings, consequently we expect that they could be removed without the language being changed. This turns out to be the case as we now show.

Let $\mathbf{A} = (S, A, i, \delta, T)$ be a finite automaton. Define a new machine,

$$\mathbf{A}^a = (S^a, A, i^a, \delta^a, T^a),$$

as follows:

- S^a is the set of accessible states in S.
- $i^a = i$.
- $T^a = T \cap S^a$, the set of accessible terminal states.
- δ^a has domain $S^a \times A$, codomain S^a but otherwise behaves like δ.

The way $\mathbf{A}^a$ is constructed from $\mathbf{A}$ can be put very simply: erase all inaccessible states from $\mathbf{A}$ and all transitions that either start or end at an inaccessible state.

Proposition 3.1.1 *Let $\mathbf{A} = (S, A, i, \delta, T)$ be a finite automaton. Then $\mathbf{A}^a$ is an accessible automaton and $L(\mathbf{A}^a) = L(\mathbf{A})$.*

Proof It is clear that $\mathbf{A}^a$ is a well-defined accessible automaton. It is also obvious that $L(\mathbf{A}^a) \subseteq L(\mathbf{A})$. To show that $L(\mathbf{A}^a) = L(\mathbf{A})$, it only remains to prove that $L(\mathbf{A}) \subseteq L(\mathbf{A}^a)$. Let $x \in L(\mathbf{A})$. Then $i \cdot x \in T$, and every state in the path labelled by x from i to $i \cdot x$ is accessible. Thus this path also lies in $\mathbf{A}^a$. It follows that $x \in L(\mathbf{A}^a)$, as required. □

The automaton $\mathbf{A}^a$ is called the *accessible part* of $\mathbf{A}$. When the number of states is small, it is easy to construct $\mathbf{A}^a$.

Example 3.1.2 Let $\mathbf{A}$ be the automaton below:

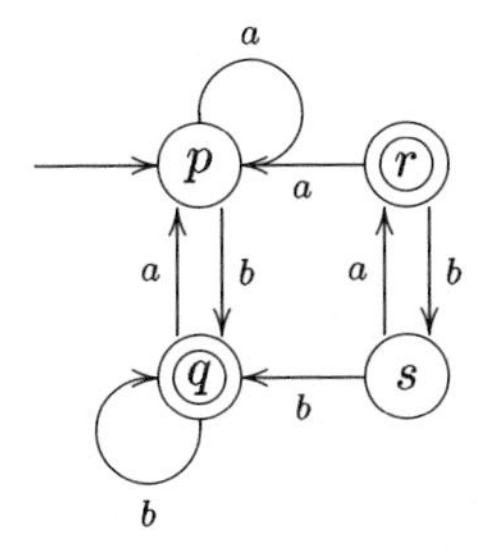

It is clear that p and q are both accessible since $p = p \cdot \varepsilon$ and $q = p \cdot b$, and that neither r nor s are accessible. Thus in this case, $\mathbf{A}^a$ is the following:

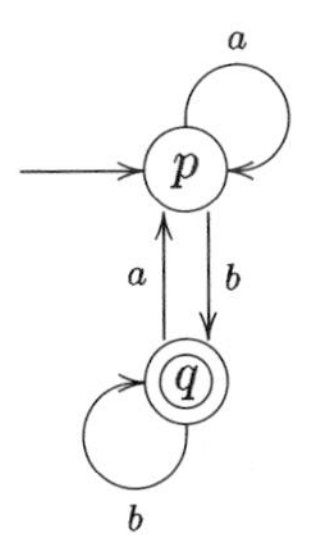

This machine is obtained by erasing the non-accessible states r and s and all transitions to and from these two states. □

However, when there are many states, the construction of $\mathbf{A}^a$ is not quite so straightforward. The following lemma lays the foundations for an algorithm for constructing $\mathbf{A}^a$. It says that if a state is accessible, then it can be reached by a string whose length is strictly less than the number of states in the machine.

Lemma 3.1.3 *Let $\mathbf{A} = (S, A, s_0, \delta, T)$ be a finite automaton with set of states S. If s is an accessible state, then there exists a string $x \in A^*$ such that $|x| < |S|$ and $s_0 \cdot x = s$.*

Proof Let $s \in S$ be an accessible state. By definition there exists $x \in A^*$ such that $s_0 \cdot x = s$. Let $x \in A^*$ be a string of *smallest possible length* such that $s_0 \cdot x = s$. We would like to show that $|x| < |S|$, so for the sake of argument, assume instead that $|x| \geq |S|$. Let $x = x_1 \ldots x_n$ where $x_i \in A$ and $n = |x| \geq |S|$. Consider the sequence of states,

$$s_0,\ s_1 = s_0 \cdot x_1,\ s_2 = s_0 \cdot (x_1 x_2), \ldots,\ s_n = s_0 \cdot (x_1 \ldots x_n).$$

Since $n \geq |S|$ it follows that $n + 1 > |S|$. But $s_0, s_1, \ldots, s_n$ is a list of states of length $n + 1$ so there must be some repetition of states in this list. Let $i \neq j$

be subscripts such that $s_i = s_j$. We have the following schematic diagram of the path in **A** labelled by the string x:

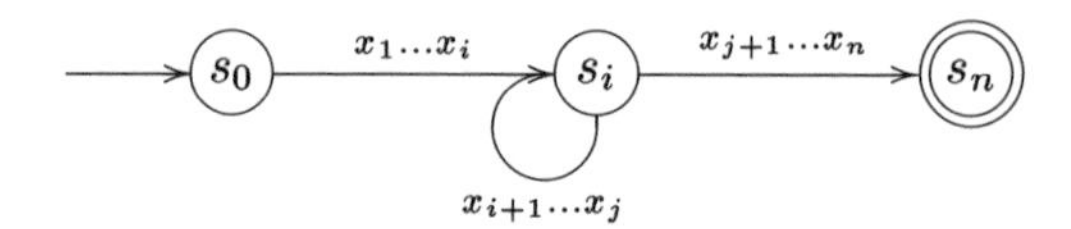

Put $x' = x_1 \ldots x_i x_{j+1} \ldots x_n$; in other words, cut out the factor of x which labels the loop. Then $|x'| < |x|$ and $s_0 \cdot x' = s$, which contradicts our choice of x. Consequently, we must have $|x| < |S|$. □

The above result implies that we can find S^a in the following way. Let $n = |S|$ and denote the initial state by s_0. If $X \subseteq S$ and $L \subseteq A^*$ then define

$$X \cdot L = \{x \cdot a \colon x \in X \text{ and } a \in L\}.$$

The set of strings over A of length at most $n-1$ is just

$$\sum_{i=0}^{n-1} A^i = A^0 + \ldots + A^{n-1}.$$

Thus Lemma 3.1.3 can be expressed in the following way:

$$S^a = s_0 \cdot \left(\sum_{i=0}^{n-1} A^i \right) = \sum_{i=0}^{n-1} s_0 \cdot A^i.$$

To calculate the terms in this union, we need only calculate in turn the sets,

$$S_0 = \{s_0\}, \quad S_1 = S_0 \cdot A, \quad S_2 = S_1 \cdot A, \ldots, S_{n-1} = S_{n-2} \cdot A,$$

because $S_j = s_0 \cdot A^j$.

These calculations can be very easily put into the form of a sequence of rooted trees. By the 'distance' between two vertices in a tree we mean the length of the shortest path joining the two vertices. The 'height' of the tree is the length of the longest path from root to leaf with no repeated vertices. The root of the tree is labelled s_0. For each $a \in A$ construct an arrow from s_0 to $s_0 \cdot a$. In general, if s is the label of a vertex, then draw arrows from s to $s \cdot a$ for each $a \in A$. The vertices at the distance j from the root are precisely the elements of $s_0 \cdot A^j$. Thus the process will terminate when the tree has height $n-1$. The vertices of this tree are precisely the accessible states of the automaton. The automaton $\mathbf{A}^a$ can now be constructed from **A** by erasing all non-accessible vertices and the transitions that go to or from them.

The drawback of this algorithm is that if the automaton has n states, then all of the tree to height $n-1$ has to be drawn. However such a tree contains repeated information: a state can appear more than once and, where it is repeated, no *new* information will be constructed from it.

The following construction omits the calculation of $s \cdot A$ whenever s is a repeat, which means that the whole tree is not constructed; in addition, it also enables us to detect when all accessible states have been found without having to count.

Algorithm 3.1.4 (Transition tree of an automaton) Let $\mathbf{A}$ be an automaton. The *transition tree* of $\mathbf{A}$ is constructed inductively in the following way. We assume that a linear ordering of A is specified at the outset so we can refer meaningfully to 'the elements of A in turn':

(1) The root of the tree is s_0 and we put $T_0 = \{s_0\}$.

(2) Assume that T_i has been constructed; vertices in T_i will have been labelled either 'closed' or 'non-closed.' The meaning of these two terms will be made clear below. We now show how to construct T_{i+1}.

(3) For each non-closed leaf s in T_i and for each $a \in A$ in turn construct an arrow from s to $s \cdot a$ labelled by a; if, in addition, $s \cdot a$ is a repeat of any state that has already been constructed, then we say it is *closed* and mark it with a $\times$.

(4) The algorithm terminates when all leaves are closed.

□

We have to prove that the algorithm above is what we want.

Proposition 3.1.5 *Let* $|S| = n$*. Then there exists an* $m \leq n$ *such that every leaf in* T_m *is closed, and* S^a *is just the set of vertices in* T_m*.*

Proof Let $s \in S^a$ and let $x \in A^*$ be the smallest string in the tree order such that $s_0 \cdot x = s$. Then s first appears as a vertex in the tree $T_{|x|}$. By Lemma 3.1.3, the latest an accessible state can appear for the first time is in T_{n-1}. Thus at worst all states in T_n are closed. □

The transition tree not only tells us the accessible states of $\mathbf{A}$ but can also be used to construct $\mathbf{A}^a$ as follows: erase the $\times$'s and glue leaves to interior vertices with the same label. The diagram that results is the transition diagram of $\mathbf{A}^a$. All of this is best illustrated by means of an example.

Example 3.1.6 Consider the automaton $\mathbf{A}$ pictured below:

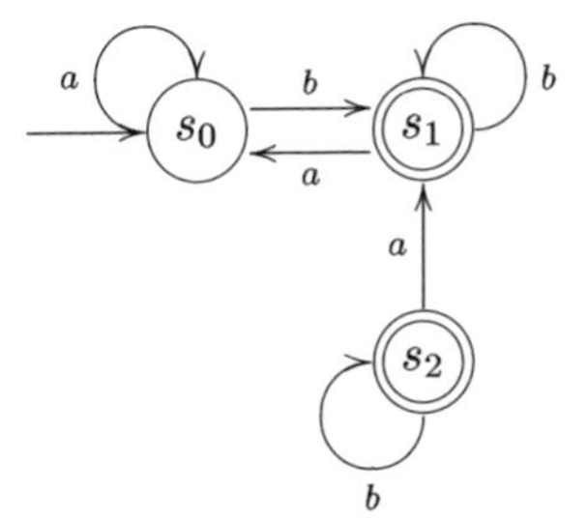

We shall step through the algorithm for constructing the transition tree of $\mathbf{A}$ and so construct $\mathbf{A}^a$. Of course, it is easy to construct $\mathbf{A}^a$ directly in this case, but it is the algorithm we wish to illustrate.

The tree T_0 is just

s_0

The tree T_1 is

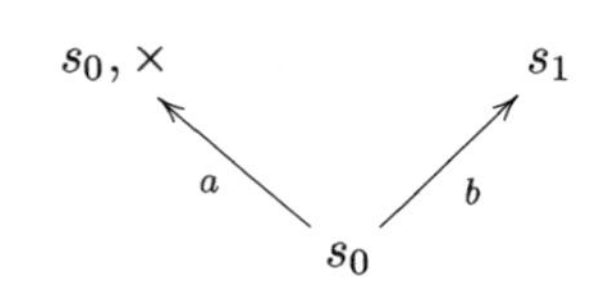

The tree T_2 is

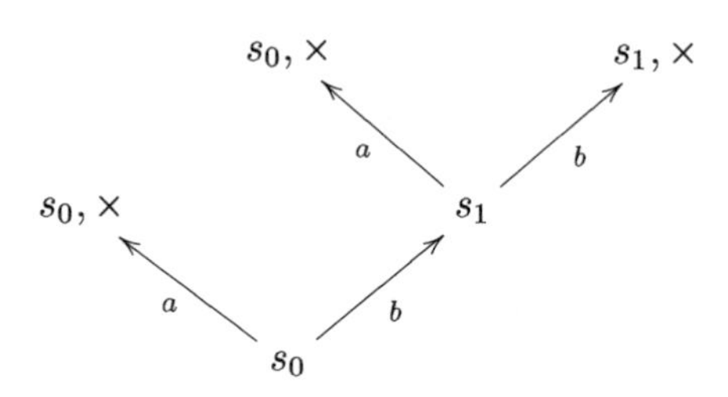

T_2 is the transition tree because every leaf is closed. This tree can be transformed into an automaton as follows. Erase all $\times$'s, and mark initial and terminal states. This gives us the following:

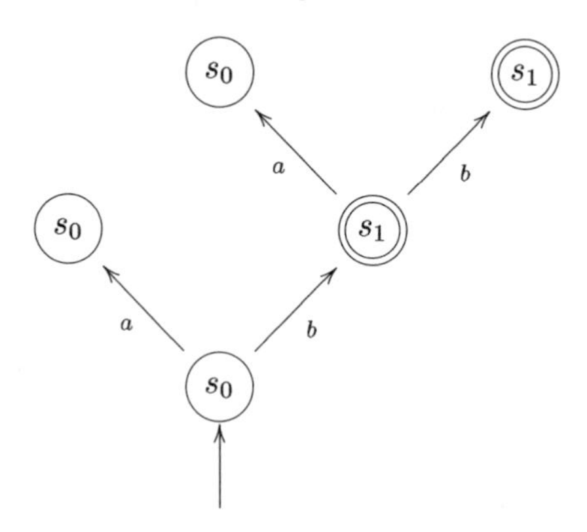

Now glue the leaves to the interior vertices labelling the same state. We obtain the following automaton:

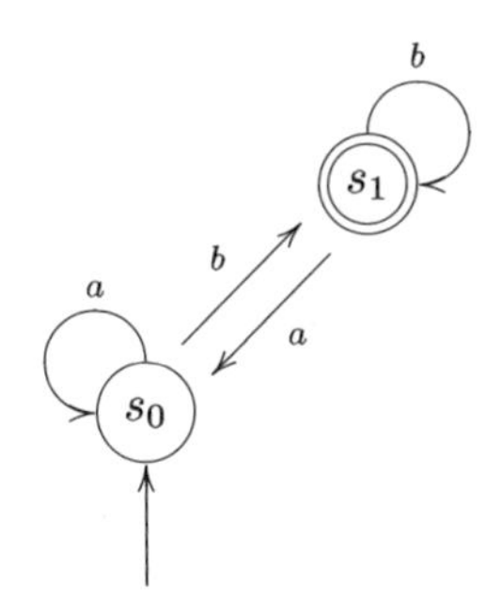

This is the automaton $\mathbf{A}^a$. □

Exercises 3.1

1. Construct transition trees for the automata below.

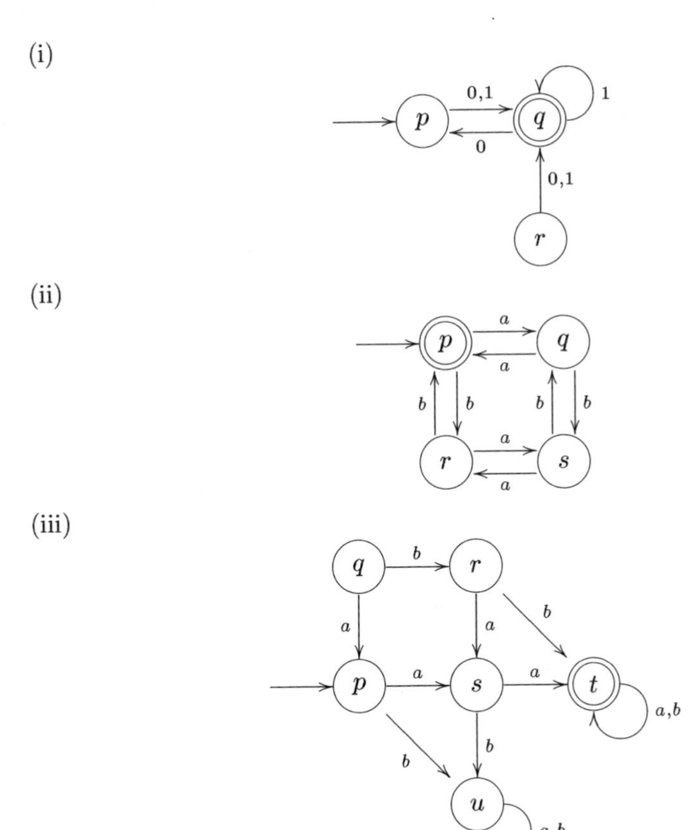

3.2 Non-deterministic automata

Deterministic automata are intended to be models of real machines. The non-deterministic automata we introduce in this section should be regarded as tools helpful in designing deterministic automata rather than as models of real-life machines. To motivate the definition of non-deterministic automata, we shall consider a simple problem.

Let A be an alphabet. If $x = a_1 \ldots a_n$ where $a_i \in A$, then the *reverse of* x, denoted $\text{rev}(x)$, is the string $a_n \ldots a_1$. We define $\text{rev}(\varepsilon) = \varepsilon$. Clearly, $\text{rev}(\text{rev}(x)) = x$, and $\text{rev}(xy) = \text{rev}(y)\text{rev}(x)$. If L is a language then the *reverse of* L, denoted by $\text{rev}(L)$, is the language

$$\text{rev}(L) = \{\text{rev}(x) \colon x \in L\}.$$

Consider now the following question: if L is recognisable, then is $\text{rev}(L)$ recognisable? To see what might be involved in answering this problem, we consider an example.

Example 3.2.1 Let $L = (aa + bb)(a + b)^*$, the language of all strings of a's and b's that begin with a double letter. This language is recognised by the following automaton:

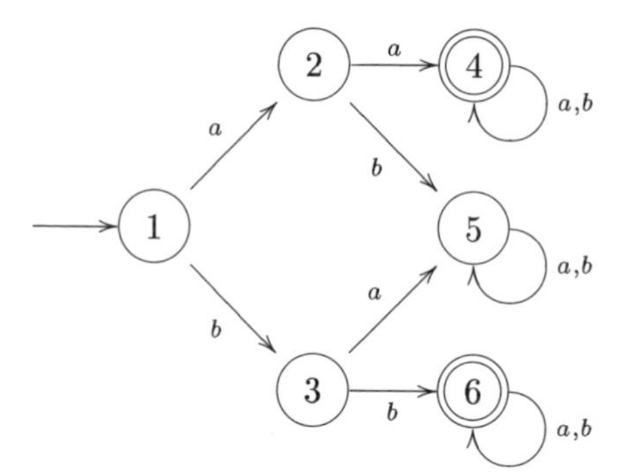

In this case, $\text{rev}(L)$ is the language of all strings of a's and b's that *end* with a double letter. In order to construct an automaton to recognise this language, it is tempting to modify the above automaton in the following way: reverse all the transitions, and interchange initial and terminal states, like this:

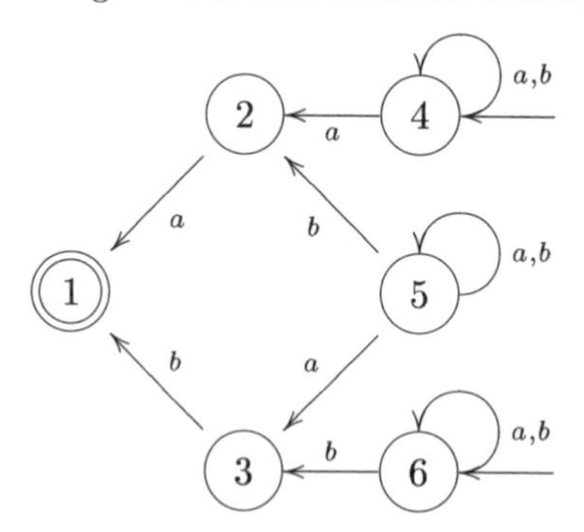

This diagram violates the rules for the transition diagram of a complete, deterministic finite-state automaton in two fundamental ways: there is more than one initial state, and there are forbidden configurations. However, if we put to one side these fatal problems, we do notice an interesting property of this diagram: the strings that label paths in the diagram, which begin at one of the initial states and conclude at the terminal state, form precisely the language rev(L): those strings that *end* in a double letter. □

This diagram is in fact an example of a non-deterministic automaton. After giving the formal definition below, we shall prove that every non-deterministic automaton can be converted into a deterministic automaton recognising the same language. An immediate application of this result can be obtained by generalising the example above; we will therefore be able to prove that the reverse of a recognisable language is also recognisable.

Recall that if X is a set, then $\mathsf{P}(X)$ is the power set of X, the set of all subsets of X. The set $\mathsf{P}(X)$ contains both $\emptyset$ and X as elements. A *non-deterministic automaton* **A** is determined by five pieces of information:

$$\mathbf{A} = (S, A, I, \delta, T),$$

where S is a finite set of states, A is the input alphabet, I is a set of initial states, $\delta: S \times A \to \mathsf{P}(S)$ is the transition function, and T is a set of terminal states.

In addition to allowing any number of initial states, the key feature of this definition is that $\delta(s, a)$ is now a subset of S (possibly empty!). We can draw transition diagrams and transition tables just as we did for deterministic machines. The transition table of the machine we constructed in Example 3.2.1 is as follows:

	a	b
$\leftarrow 1$	$\emptyset$	$\emptyset$
2	$\{1\}$	$\emptyset$
3	$\emptyset$	$\{1\}$
$\rightarrow 4$	$\{2,4\}$	$\{4\}$
5	$\{3,5\}$	$\{2,5\}$
$\rightarrow 6$	$\{6\}$	$\{3,6\}$

It now remains to define the analogue of the 'extended transition function.' For each string $x \in A$ and state $s \in S$ we want to know *the set of all possible states* that can be reached by paths starting at s and labelled by x. The formal definition is as follows and, as in the deterministic case, the function being defined exists and is unique. The function δ^* is the unique function from $S \times A^*$ to $\mathsf{P}(S)$ satisfying the following three conditions where $a \in A$, $w \in A^*$ and $s \in S$:

(ETF1) $\delta^*(s, \varepsilon) = \{s\}$.

(ETF2) $\delta^*(s,a) = \delta(s,a)$.

(ETF3) $\delta^*(s,aw) = \sum_{q\in\delta(s,a)} \delta^*(q,w)$.

Condition (ETF3) needs a little explaining. Suppose that $\delta(s,a) = \{q_1,\ldots,q_n\}$. Then condition (ETF3) means

$$\delta^*(q_1,w) + \ldots + \delta^*(q_n,w).$$

Notation We shall usually write $s \cdot x$ rather than $\delta^*(s,x)$, but it is important to remember that $s \cdot x$ is a *set* in this case.

Let us check that this definition captures what we intended.

Lemma 3.2.2 *Let $x = a_1 \ldots a_n \in A^*$. Then $t \in s \cdot x$ if and only if there exist states $q_1,\ldots,q_n = t$ such that $q_1 \in s \cdot a_1$, and $q_i \in q_{i-1} \cdot a_i$ for $i = 2,\ldots,n$.*

Proof The condition simply says that $t \in s \cdot x$ if and only if the string x labels some path in $\mathbf{A}$ starting at s and ending at t. Observe that $t \in \delta^*(s,ax)$ if and only if $t \in \delta^*(q,x)$ for some state $q \in \delta(s,a)$. By applying this observation repeatedly, starting with $ax = a_1 \ldots a_n$, we obtain the desired result. □

The language $L(\mathbf{A})$ is defined to be

$$L(\mathbf{A}) = \{w \in A^* \colon \left(\sum_{q\in I} q \cdot w\right) \cap T \neq \emptyset\}.$$

That is, the language recognised by a non-deterministic automaton consists of all strings that label paths starting at one of the initial states and ending at one of the terminal states.

It might be thought that because there is a degree of choice available, non-deterministic automata might be more powerful than deterministic automata, meaning that non-deterministic automata might be able to recognise languages that deterministic automata could not. In fact, this is not so. To prove this, we shall make use of the following construction. Let $\mathbf{A} = (S, A, I, \delta, T)$ be a non-deterministic automaton. We construct a deterministic automaton $\mathbf{A}^d = (S^d, A, i^d, \Delta, T^d)$ as follows:

- $S^d = \mathsf{P}(S)$; the set of states is labelled by the subsets of S.
- $i^d = I$; the initial state is labelled by the subset consisting of all the initial states.
- $T^d = \{Q \in \mathsf{P}(S) \colon Q \cap T \neq \emptyset\}$; the terminal states are labelled by the subsets that contain at least one terminal state.

- For $a \in A$ and $Q \in \mathsf{P}(S)$ define

$$\Delta(Q,a) = \sum_{q \in Q} q \cdot a;$$

this means that the subset $\Delta(Q,a)$ consists of all states in S that can be reached from states in Q by following a transition labelled by a.

It is clear that $\mathbf{A}^d$ is a complete, deterministic, finite automaton.

Theorem 3.2.3 (Subset construction) *Let* $\mathbf{A}$ *be a non-deterministic automaton. Then* $\mathbf{A}^d$ *is a deterministic automaton such that* $L(\mathbf{A}^d) = L(\mathbf{A})$.

Proof The main plank of the proof will be to relate the extended transition function Δ^* in the deterministic machine $\mathbf{A}^d$ to the extended transition function δ^* in the non-deterministic machine $\mathbf{A}$. We shall prove that for any $Q \subseteq S$ and $x \in A^*$ we have that

$$\Delta^*(Q,x) \quad = \quad \sum_{q \in Q} \delta^*(q,x). \tag{3.1}$$

This is most naturally proved by induction on the length of the string x.

For the base step, we prove the theorem holds when $x = \varepsilon$. By the definition of Δ, we have that $\Delta^*(Q,\varepsilon) = Q$, whereas by the definition of δ^* we have that $\sum_{q \in Q} \delta^*(q,\varepsilon) = \sum_{q \in Q} \{q\} = Q$.

For the induction hypothesis, assume that (3.1) holds for all strings $x \in A^*$ satisfying $|x| = n$. Consider now the string y where $|y| = n+1$. We can write $y = ax$ where $a \in A$ and $x \in A^*$ and $|x| = n$. From the definition of Δ^* we have

$$\Delta^*(Q,y) = \Delta^*(Q,ax) = \Delta^*(\Delta(Q,a),x).$$

Put $Q' = \Delta(Q,a)$. Then

$$\Delta^*(Q,y) = \Delta^*(Q',x) = \sum_{q' \in Q'} \delta^*(q',x)$$

by the induction hypothesis. From the definitions of Q' and Δ we have that

$$\sum_{q' \in Q'} \delta^*(q',x) = \sum_{q \in Q} \left(\sum_{q' \in \delta(q,a)} \delta^*(q',x) \right).$$

By the definition of δ^* we have that

$$\sum_{q \in Q} \left(\sum_{q' \in \delta(q,a)} \delta^*(q',x) \right) = \sum_{q \in Q} \delta^*(q,ax) = \sum_{q \in Q} \delta^*(q,y).$$

This proves the claim.

We can now easily prove that $L(\mathbf{A}) = L(\mathbf{A}^d)$. By definition,

$$x \in L(\mathbf{A}) \Leftrightarrow \left(\sum_{q \in I} \delta^*(q, x)\right) \cap T \neq \emptyset.$$

From the definition of the terminal states in $\mathbf{A}^d$ this is equivalent to

$$\sum_{q \in I} \delta^*(q, x) \in T^d.$$

We now use equation (3.1) to obtain

$$\Delta^*(I, x) \in T^d.$$

This is of course equivalent to $x \in L(\mathbf{A}^d)$. □

The automaton $\mathbf{A}^d$ is called the *determinised* version of $\mathbf{A}$.

Notation Let $\mathbf{A}$ be a non-deterministic automaton with input alphabet A. Let Q be a set of states and $a \in A$. We denote by $Q \cdot a$ the set of all states that can be reached by starting in Q and following transitions labelled only by a. In other words, $\Delta(Q, a)$ in the above theorem. We define $Q \cdot A = \sum_{a \in A} Q \cdot a$.

Example 3.2.4 Consider the following non-deterministic automaton $\mathbf{A}$:

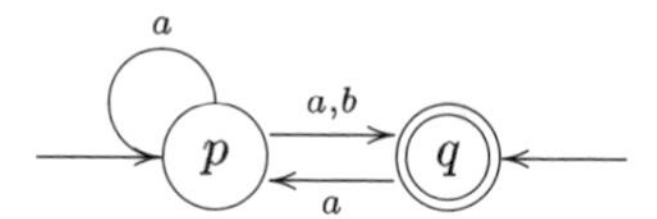

The corresponding deterministic automaton constructed according to the subset construction has 4 states labelled $\emptyset$, $\{p\}$, $\{q\}$, and $\{p, q\}$; the initial state is the state labelled $\{p, q\}$ because this is the set of initial states of $\mathbf{A}$; the terminal states are $\{q\}$ and $\{p, q\}$ because these are the only subsets of $\{p, q\}$ that contain terminal states of $\mathbf{A}$; the transitions are calculated according to the definition of Δ. Thus $\mathbf{A}^d$ is the automaton:

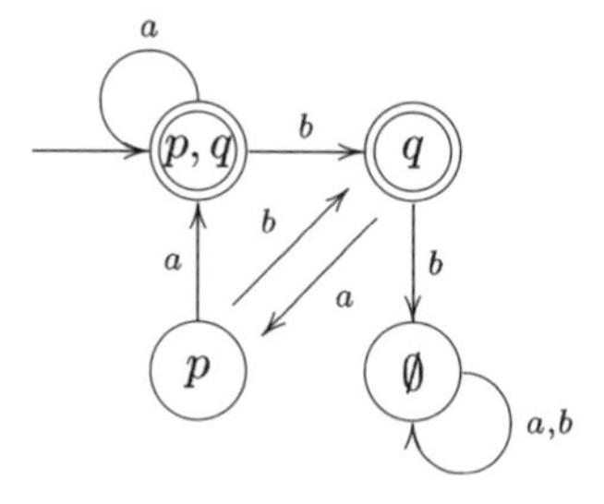

Observe that the state labelled by the empty set is a sink state, as it always must be. □

The obvious drawback of the subset construction is the huge increase in the number of states in passing from $\mathbf{A}$ to $\mathbf{A}^d$. Indeed, if $\mathbf{A}$ has n states, then $\mathbf{A}^d$ will have 2^n states. This is sometimes unavoidable as we ask you to show in Exercises 3.2, but often the machine $\mathbf{A}^d$ will contain many inaccessible states. There is an easy way of avoiding this: construct the transition tree of $\mathbf{A}^d$ directly from $\mathbf{A}$ and so construct $(\mathbf{A}^d)^a = \mathbf{A}^{da}$. This is done as follows.

Algorithm 3.2.5 (Accessible subset construction) The input to this algorithm is a non-deterministic automaton $\mathbf{A} = (S, A, I, \delta, T)$ and the output is $\mathbf{A}^{da}$, an accessible deterministic automaton recognising $L(\mathbf{A})$. The procedure is to construct the transition tree of $\mathbf{A}^d$ directly from $\mathbf{A}$. The root of the tree is the set I. Apply the algorithm for the transition tree by constructing as vertices $Q \cdot a$ for each non-closed vertex Q and input letter a. □

We show how this algorithm works by applying it to the non-deterministic automaton constructed in Example 3.2.1.

Example 3.2.6 The root of the tree is labelled $\{4, 6\}$, the *set* of initial states of the non-deterministic automaton. The next step in the algorithm yields the tree:

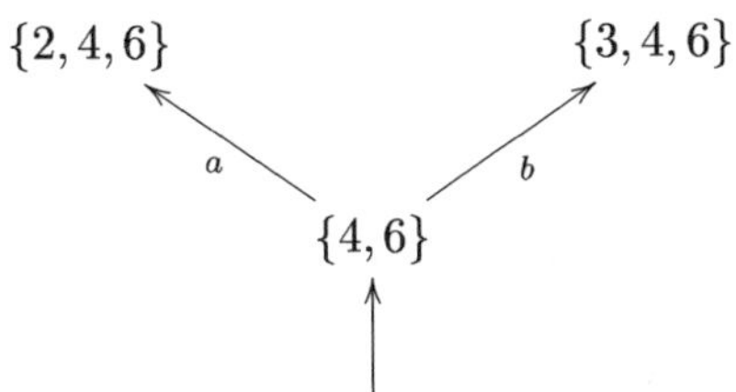

Continuing with the algorithm, we obtain the transition tree of $(\mathbf{A}^d)^a = \mathbf{A}^{da}$:

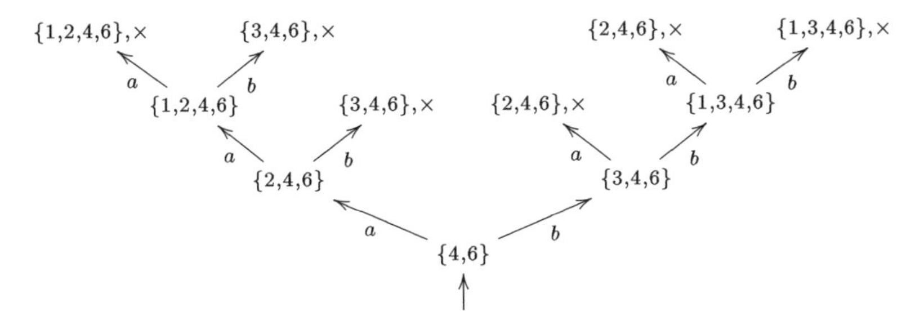

Finally, we obtain the automaton $\mathbf{A}^{da}$ pictured below:

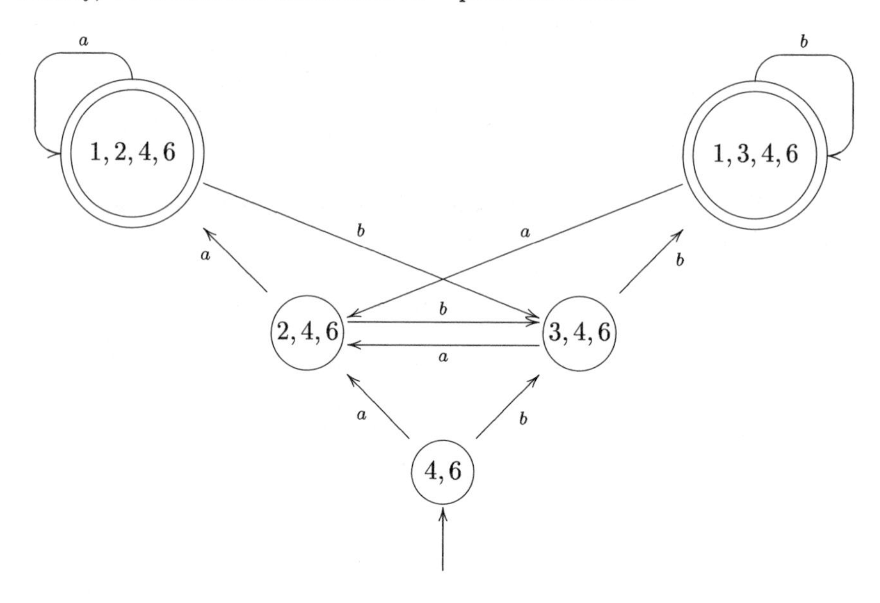

□

Exercises 3.2

1. Apply the accessible subset construction to the non-deterministic automata below. Hence find deterministic automata which recognise the same language in each case.

(i)

(ii)

(iii)

(iv)

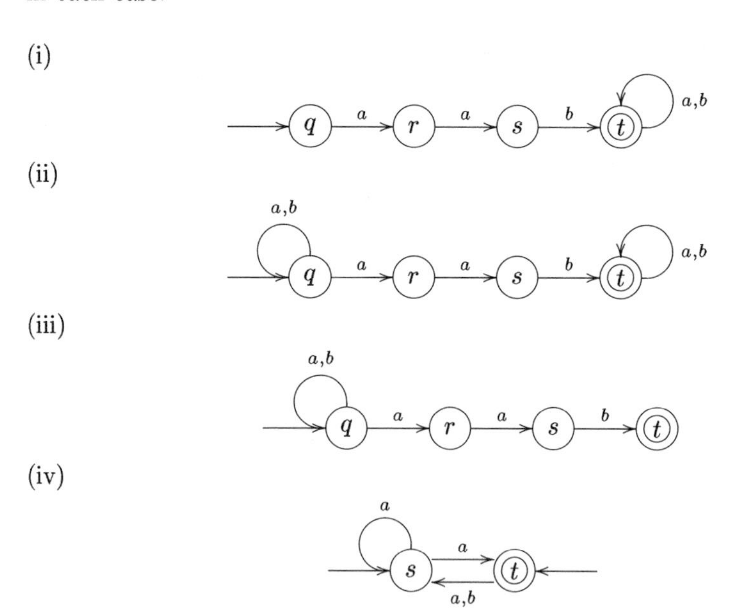

2. Find a non-deterministic automaton with 4 states that recognises the language $(0+1)^*1(0+1)^2$. Use the accessible subset construction to find a deterministic automaton that recognises the same language.

3. Let $n \geq 1$. Show that the language $(0+1)^*1(0+1)^{n-1}$ can be recognised by a non-deterministic automaton with $n+1$ states. Show that any deterministic automaton that recognises this language must have at least 2^n states.

This example shows that an exponential increase in the number of states in passing from a non-deterministic automaton to a corresponding deterministic automaton is sometimes unavoidable.

3.3 Applications

Non-deterministic automata make designing certain kinds of automata easy: we may often simply write down a non-deterministic automaton and then apply the accessible subset construction. It is however worth pointing out that the automata obtained by applying the accessible subset construction will often have some rather obvious redundancies and can be simplified further. The general procedure for doing this will be described in Chapter 7.

In this section, we look at some applications of non-deterministic automata. Our first result generalises Example 3.2.1.

Proposition 3.3.1 *Let L be recognisable. Then* $\text{rev}(L)$*, the reverse of L, is recognisable.*

Proof Let $L = L(\mathbf{A})$, where $\mathbf{A} = (S, A, I, \delta, T)$ is a non-deterministic automaton. Define another non-deterministic automaton $\mathbf{A}^{\text{rev}}$ as follows:

$$\mathbf{A}^{\text{rev}} = (S, A, T, \gamma, I),$$

where γ is defined by $s \in \gamma(t, a)$ if and only if $t \in \delta(s, a)$; in other words, we reverse the arrows of $\mathbf{A}$ and relabel the initial states as terminal and vice versa. It is now straightforward to check that $x \in L(\mathbf{A}^{\text{rev}})$ if and only if $\text{rev}(x) \in L(\mathbf{A})$. Thus $L(\mathbf{A}^{\text{rev}}) = \text{rev}(L)$. □

The automaton $\mathbf{A}^{\text{rev}}$ is called the *reverse* of $\mathbf{A}$.

Non-deterministic automata provide a simple alternative proof to Proposition 2.5.6.

Proposition 3.3.2 *Let L and M be recognisable. Then $L + M$ is recognisable.*

Proof Let $\mathbf{A}$ and $\mathbf{B}$ be non-deterministic automata, recognising L and M, respectively. Lay them side by side; the result is a non-deterministic automaton recognising $L + M$. □

In Section 2.4, we considered the recognisability of languages defined in terms of the presence of patterns of various kinds. The following result shows how easy the proofs become in this case when we use non-deterministic automata.

Proposition 3.3.3 *Let A be an alphabet and let $w = a_1 \ldots a_n$ be a non-empty string over A. Each of the languages wA^*, A^*wA^* and A^*w is recognisable.*

Proof Each language can be recognised by the respective non-deterministic automaton below, which I have drawn schematically:

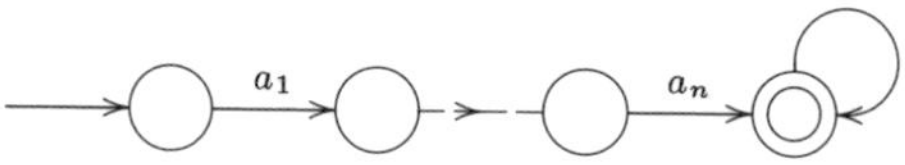

A
a_1 a_n

and

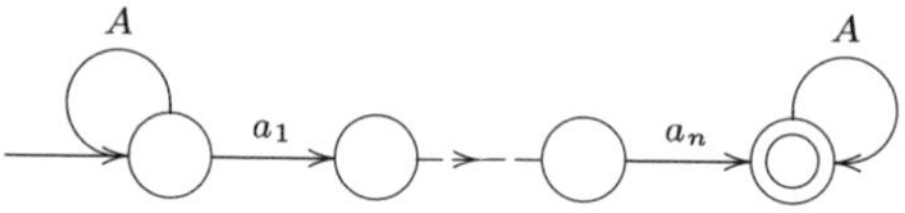

A A
a_1 a_n

and

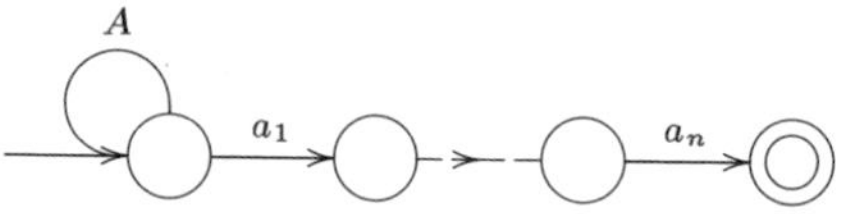

A
a_1 a_n

Each of these can be converted into an equivalent deterministic automaton using the accessible subset construction. □

In Chapter 1 we introduced the product of two languages and the Kleene star of a language.

Proposition 3.3.4 *If L and M are recognisable, then LM is recognisable.*

Proof Let $\mathbf{A} = (S, A, S_0, \delta, F)$ and $\mathbf{B} = (Q, A, Q_0, \gamma, G)$ be non-deterministic automata, recognising L and M, respectively. We shall combine these two automata into a new, non-deterministic automaton $\mathbf{C}$ such that $L(\mathbf{C}) = LM$. It is no loss in generality to assume that $S \cap Q = \emptyset$. We define $\mathbf{C} = (R, A, R_0, \xi, H)$ as follows:

- $R = S \cup Q$; the set of states of the new machine is the union of the sets of states of the two original machines.
- $R_0 = S_0$; the set of initial states of the new machine is just the set of initial states of $\mathbf{A}$.

- H: the set of terminal states is $F \cup G$ if one of the initial states of **B** is also terminal, otherwise the set of terminal states is just G, the set of terminal states of **B**.
- Transition function ξ: all the transitions in **A** are carried forward to **C** as are all the transitions in **B**; we also introduce new transitions from the states in F to the states in $Q_0 \cdot A$. We do this as follows: for each $s \in F$ and for each transition $q_0 \xrightarrow{a} q$ in **B** where $q_0 \in Q_0$ add the transition $s \xrightarrow{a} q$ to ξ in **C**. A picture of this may be helpful.

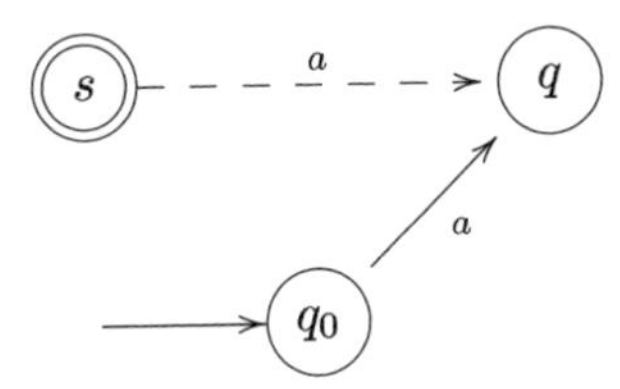

The automaton **C** is a well-defined, non-deterministic automaton. We show that $L(\mathbf{C}) = LM$.

First we prove that $L(\mathbf{C}) \subseteq LM$. Let $x \in L(\mathbf{C})$. There are two possibilities. If $\varepsilon \in M$, then either x labels a path in **C** from a state in S_0 to a state in F, or x labels a path in **C** from a state in S_0 to a state in G, whereas if $\varepsilon \notin M$ then only the second case can occur. In the first case, $x \in L(\mathbf{A})$ and so $x = x\varepsilon \in LM$. In the second case, x must traverse one of the new transitions. Thus $x = uav$, where $s_0 \xrightarrow{u} s$, where $s \in F$, $s \xrightarrow{a} q$, and $q \xrightarrow{v} q'$, where $q' \in G$. By construction, $q_0 \xrightarrow{a} q$. It follows that $av \in L(\mathbf{B})$. Thus $x \in LM$. Hence $L(\mathbf{C}) \subseteq LM$.

To prove the reverse inclusion, let $x \in LM$. There are two possibilities: if $\varepsilon \in M$, then either $x \in L$ or $x = uw$ where $u \in L$, $w \in M$ and $w \neq \varepsilon$, whereas if $\varepsilon \notin M$, then only the second case can occur. In the first case, $x \in L(\mathbf{C})$. In the second case, we can write $x = uw$ where w is, by assumption, non-empty. Thus $w = av$ for some $a \in A$ and $v \in A^*$. We can now essentially reverse our argument from the first part of the proof above to get $x \in L(\mathbf{C})$. □

The new transitions in **C** can be described as follows: for each state in F copy the transitions out of the states in Q_0 in **B**.

Example 3.3.5 Let

$$L = \{x \in (a+b)^*\colon |x|_b \text{ is odd}\}$$

and let

$$M = \{x \in (a+b)^*\colon |x|_a \text{ is even}\}.$$

An automaton **A** that recognises L is

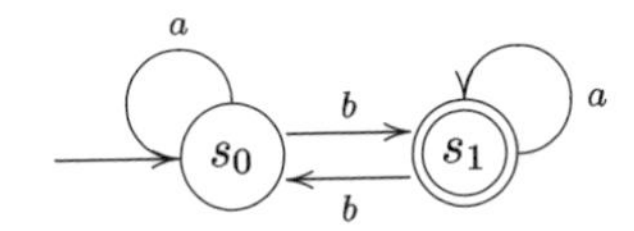

and an automaton **B** that recognises M is

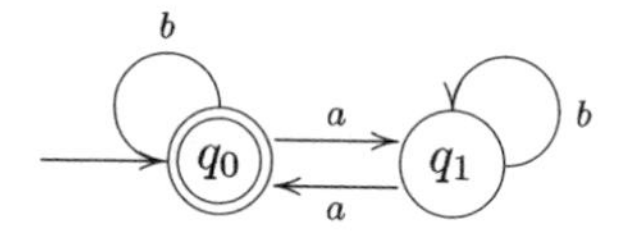

Using the construction in Proposition 3.3.4, an automaton **C** that recognises LM is

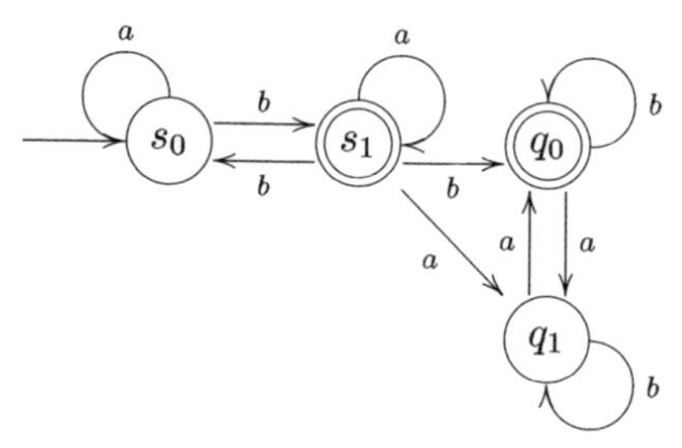

□

Proposition 3.3.6 *If L is recognisable, then L^* is recognisable.*

Proof Let $\mathbf{A} = (S, A, S_0, \delta, F)$ be a non-deterministic automaton such that $L(\mathbf{A}) = L$. We construct a new non-deterministic automaton,

$$B = (Q, A, Q_0, \gamma, G),$$

as follows:

- The set of states Q is just the set S together with a new state $\diamond$.
- The set of initial states Q_0 is just the set of initial states S_0 together with $\diamond$.
- The set of terminal states G is the set of terminal states F together with $\diamond$.
- The transition function γ: all the transitions in **A** are carried forward; we also introduce new transitions from states in F to states in $S_0 \cdot A$. We do this as follows: for each state $s \in F$ and each transition $s' \xrightarrow{a} s''$ where $s' \in S_0$ add the transition $s \xrightarrow{a} s''$ to γ in **B**. A picture of this may be helpful.

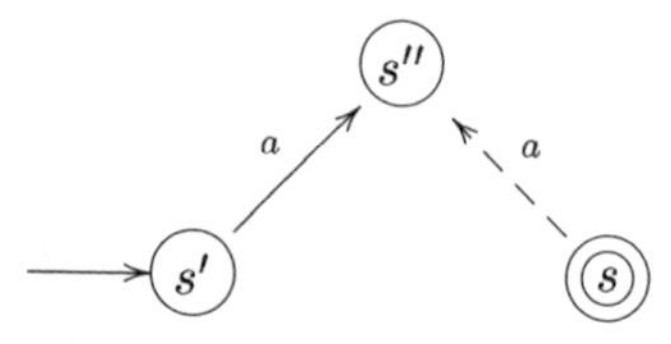

It is clear that **B** is a well-defined, non-deterministic automaton. We prove that $L(\mathbf{B}) = L^*$. The only role of the state $\diamond$ is to recognise ε. So in what follows we can limit ourselves to non-empty strings.

Let $x \in L(\mathbf{B})$. Then there is a path in **B**, which I shall call 'successful,' starting at a state in S_0 and ending at a state in F labelled by x. If x uses none of the new transitions, then $x \in L$. Otherwise we can factor $x = u_1 a_1 u_2 a_2 \ldots u_n$, where the symbols $a_1, a_2 \ldots$ are the labels of the new transitions occurring in the successful path. From the definition of the new transitions, this implies that each of the strings $u_1, (a_1 u_2), \ldots, (a_{n-1} u_n)$ is accepted by $L(\mathbf{A})$ and so belongs to L. It follows that $x \in L^*$.

To prove the reverse inclusion, let $x \in L^*$ be a non-empty string. We can factorise $x = v_1 \ldots v_n$ where each $v_i \in L$ and is non-empty. If $n = 1$, then clearly $x \in L(\mathbf{B})$. Assume $n \geq 2$. We can write each $v_i = a_i u_i$ for $i \geq 2$. We can now essentially reverse our argument from the first part of the proof to get $x \in L(\mathbf{B})$. □

Example 3.3.7 Let $L = \{ab, ba\}$. An incomplete automaton **A** recognising L is

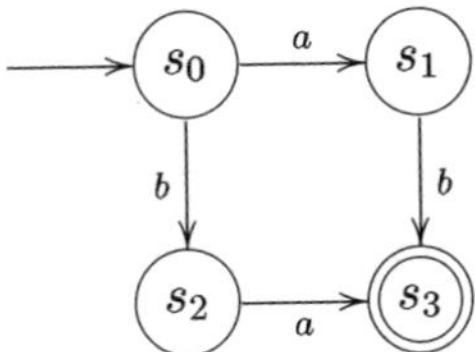

Using the construction of Proposition 3.3.6, an automaton **B** recognising L^* is

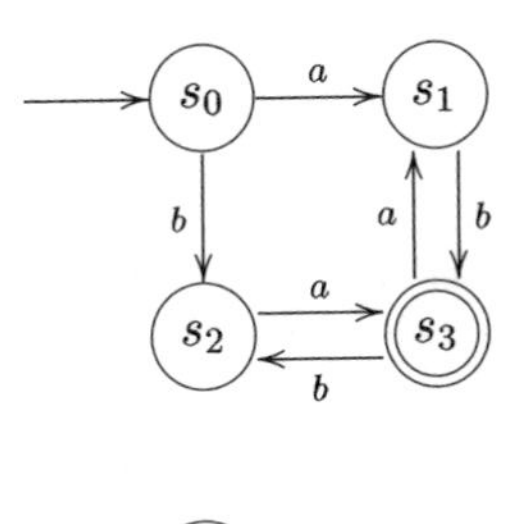

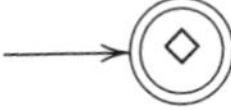

□

Exercises 3.3

1. Construct non-deterministic automata recognising the following languages over the alphabet $A = \{a, b\}$.

(i) $(a^2 + ab + b^2)(a + b)^*$.

(ii) $(a+b)^*(a^2+ab+b^2)$.

(iii) $(a+b)^*(aaa+bbb)(a+b)^*$.

(iv) $(a^2+ba+b^2+ba^2+b^2a)^*$.

(v) $((b^2)^*(a^2b+ba))^*$.

2. Construct an incomplete automaton $\mathbf{A} = (S, A, i, \delta, T)$ such that the automaton $\mathbf{B} = (S, A, i, \delta, S \setminus T)$ does not recognise $L(\mathbf{A})'$, the complement of $L(\mathbf{A})$.

It is only possible to prove that the complement of a recognisable language is recognisable using complete deterministic automata.

3.4 Trim automata

We defined the idea of an accessible automaton for complete deterministic automata in Section 3.1. We now extend the definition to arbitrary non-deterministic automata.

Notation Let

$$\mathbf{A} = (S, A, I, \delta, T)$$

be a non-deterministic automaton. If $Q \subseteq S$ and $x \in A^*$ then we define

$$Q \cdot x = \sum_{q \in Q} \delta^*(q, x),$$

the set of all states that can be reached starting in Q and following paths labelled x. If $L \subseteq A^*$ then we define

$$Q \cdot L = \sum_{x \in L} Q \cdot x,$$

the set of all states that can be reached starting in Q and following paths labelled by elements of L. Using this notation, the language $L(\mathbf{A})$ recognised by $\mathbf{A}$ is simply

$$L(\mathbf{A}) = I \cdot A^* \cap T \neq \emptyset.$$

Let $\mathbf{A}$ be a non-deterministic automaton with state set S and input alphabet A. We say that a state $s \in S$ is *accessible* if there is some string $x \in A^*$, which labels a path from one of the initial states to s. An automaton $\mathbf{A}$ is *accessible* if every state is accessible. The set of accessible states is $I \cdot A^*$. It follows that any states of $\mathbf{A}$ that are not in $I \cdot A^*$ can play no role in accepting or rejecting a string and can therefore be removed. Put

$$S^a = I \cdot A^* \text{ and } T^a = S^a \cap T,$$

and let δ^a be the restriction of δ to $S^a \times A$. Put $\mathbf{A}^a = (S^a, A, I, \delta^a, T^a)$. In other words, the automaton $\mathbf{A}^a$ is constructed by erasing all non-accessible states and any transitions to or from them. It is clear that $L(\mathbf{A}) = L(\mathbf{A}^a)$. We call $\mathbf{A}^a$ the *accessible part* of $\mathbf{A}$. We already have an algorithm that computes the set S^a; this is the 'accessible subset construction' of Section 3.1. Apply this algorithm to $\mathbf{A}$, and the states that occur as labels of the vertices in this algorithm are precisely the accessible states.

We now define a notion dual[1] to that of accessibility. Let $\mathbf{A}$ be a non-deterministic automaton with state set S and input alphabet A. A state s is *coaccessible* if there is a string $x \in A^*$ such that $s \cdot x$ contains a terminal state. We say $\mathbf{A}$ is *coaccessible* if every state is coaccessible. The algorithm for finding the coaccessible states is simply a variant of the one for finding the accessible states. Notice that $\mathbf{A}$ is *coaccessible* if and only if $\mathbf{A}^{\text{rev}}$, the reverse of $\mathbf{A}$, is accessible. Thus to calculate the coaccessible states of $\mathbf{A}$, it is enough to calculate the accessible states of $\mathbf{A}^{\text{rev}}$. The *coaccessible part* of $\mathbf{A}$, denoted $\mathbf{A}^b$, is defined in the same way as $\mathbf{A}^a$ except that the states of $\mathbf{A}^b$ are the coaccessible states of $\mathbf{A}$. Clearly $L(\mathbf{A}^b) = L(\mathbf{A})$.

An automaton is *trim* if it is both accessible and coaccessible. A trim automaton is therefore one in which for each state s there is an initial state s' and a string x such that $s \in s' \cdot x$, and a string y and a terminal state t such that $t \in s \cdot y$; that is, in a trim automaton each state occurs on some path from an initial state to a terminal state. For every automaton $\mathbf{A}$ the automata $\mathbf{A}^{ab}$ and $\mathbf{A}^{ba}$ are the same and both are trim. See Exercises 3.4. We denote this common automaton by $\mathbf{A}^t$, called the *trim part of* $\mathbf{A}$. Clearly, $L(\mathbf{A}) = L(\mathbf{A}^t)$. We state this result formally as follows.

Proposition 3.4.1 *Each recognisable language is recognised by a trim automaton.* □

It is straightforward to construct the trim part of an automaton. However, there is one point where the reader has to be wary. The following example illustrates the difficulty.

Example 3.4.2 Consider the complete deterministic automaton $\mathbf{A}$ given by

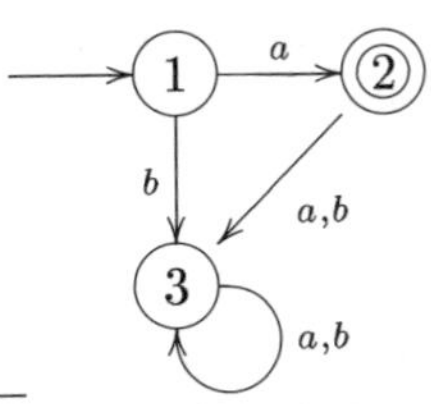

[1] This is a useful term common in mathematics. It is applied in situations where there are exactly two ways of progressing; for example, we might have a choice between left or right or, as here, between initial and terminal. If we choose one of the pair, then the other is the 'dual.' Here we have a choice between initial states and terminal states; accessibility was defined in terms of the initial states so the dual is defined in terms of the terminal states. The prefix 'co' is often prefixed to a word to denote the dual form if there is no other word readily available.

This automaton is already accessible but it is not coaccessible because there is no path from the state 3 to the terminal state. In this case, therefore, $\mathbf{A}^t$ is just

$$\longrightarrow (1) \xrightarrow{a} ((2))$$

This example shows that a complete automaton may become incomplete when the coaccessible part is taken. □

Accessibility and coaccessibility can be put to work to solve problems. We begin with three decision problems. The first is: given a non-deterministic automaton $\mathbf{A}$ is $L(\mathbf{A})$ empty?

Proposition 3.4.3 *Let* $\mathbf{A}$ *be a non-deterministic automaton. Then* $L(\mathbf{A}) = \emptyset$ *if and only if* $\mathbf{A}^t$ *is empty.*

Proof If $\mathbf{A}^t$ is empty then the result is immediate. Suppose now that the language $L(\mathbf{A})$ is empty and that $\mathbf{A}^t$ is not. Then any state in $\mathbf{A}^t$ is part of a path from an initial state to a terminal state in $\mathbf{A}$. Such a path would be labelled by a string in A^*, that would therefore belong to $L(\mathbf{A})$ by definition. Thus if $L(\mathbf{A})$ is empty then $\mathbf{A}^t$ is empty. □

If a language is not empty we want to know whether it is finite or infinite. Again this can easily be answered. By a *closed path* in an automaton we mean a path that begins and ends at the same state. The *length* of the path is the number of edges used.

Proposition 3.4.4 *Let* $\mathbf{A}$ *be a non-deterministic automaton. Then* $L(\mathbf{A})$ *is finite and non-empty if and only if* $\mathbf{A}^t$ *is non-empty and does not contain a closed path of length at least 1.*

Proof Suppose that $L(\mathbf{A})$ is finite and non-empty. Then by Proposition 3.4.3, we know that $\mathbf{A}^t$ is non-empty. Suppose $\mathbf{A}^t$ contained a closed path begining and ending at the vertex s. Let $x \in A^+$ be the non-empty string that labels the edges on this path. Since $\mathbf{A}^t$ is trim there is a string u such that $s' \cdot u = s$, where s' is an initial state. Likewise there is a string v such that $s \cdot v = t$ where t is a terminal state. Thus the string $uxv \in L(\mathbf{A}^t)$. However, $ux^*v \subseteq L(\mathbf{A}^t)$ by traversing the closed path an arbitrary number of times. It follows that $L(\mathbf{A}^t)$ is infinite, which is a contradiction. Thus $\mathbf{A}^t$ contains no closed paths of length 1 or more.

To prove the converse, suppose that A^t is non-empty and contains no closed paths. We prove that $L(\mathbf{A}^t)$ is finite. Suppose not. Let $\mathbf{A}^t$ have n states. Because $L(\mathbf{A}^t)$ is infinite we can find a string $x \in L(\mathbf{A}^t)$ such that $|x| \geq n$. By the same argument we used in proving the Pumping Lemma (Section 2.6), the path in $\mathbf{A}^t$ labelled by x must contain a repeated vertex. Thus $\mathbf{A}^t$ must contain a closed path, which is a contradiction. It follows that $L(\mathbf{A}^t)$ is finite.□

We can also use trim automata to compare the languages accepted by two different automata.

Proposition 3.4.5 *Given two complete deterministic automata* $\mathbf{A}$ *and* $\mathbf{B}$ *it is decidable whether* $L(\mathbf{A}) = L(\mathbf{B})$.

Proof Let $L = L(\mathbf{A})$ and $M = L(\mathbf{B})$. By simple set theory, $L = M$ if and only if $(L \setminus M) + (M \setminus L) = \emptyset$. We now use the constructions of Section 2.5. Construct the automaton $\mathbf{C} = \mathbf{A}' \times \mathbf{B}$. Then $L(\mathbf{C}) = L' \cap M = M \setminus L$. Similarly, we may construct an automaton $\mathbf{D}$ such that $L(\mathbf{D}) = L \setminus M$. Finally, we may construct an automaton $\mathbf{E}$ such that $L(\mathbf{E}) = L(\mathbf{C}) + L(\mathbf{D})$. By Proposition 3.4.3, it follows that $L = M$ if and only if $\mathbf{E}^t$ is empty. □

For each language $L \subseteq A^*$ we may define the following languages: Prefix(L), Suffix(L), and Factor(L) which are, respectively, the prefixes of elements of L, the suffixes of elements of L, and the factors of elements of L.

Proposition 3.4.6 *Let* L *be a recognisable language. Then each of* Prefix(L), Suffix(L), *and* Factor(L) *is recognisable.*

Proof We show first that Prefix(L) is recognisable. Let $\mathbf{A} = (S, A, I, \delta, T)$ be a coaccessible automaton that recognises L. Consider the automaton

$$\mathbf{B} = (S, A, I, \delta, S).$$

We show that $L(\mathbf{B}) = \text{Prefix}(L)$. Let $x \in L(\mathbf{B})$. Then x labels a path in $\mathbf{B}$, which starts at some initial state s and ends at a state t; that is, $t \in s \cdot x$. It follows that x labels a path from s to t in $\mathbf{A}$. By assumption, $\mathbf{A}$ is coaccessible and so there is a $y \in A^*$ such that $t \cdot y$ contains a terminal state. Hence $xy \in L(\mathbf{A})$ and so $x \in \text{Prefix}(L)$. Conversely, suppose that $x \in \text{Prefix}(L)$. Then there exists $y \in A^*$ such that $xy \in L(\mathbf{A})$. But then $x \in L(\mathbf{B})$ is immediate.

To show that Suffix(L) is recognisable, let $\mathbf{A} = (S, A, I, \delta, T)$ be an accessible automaton that recognises L. Consider the automaton

$$\mathbf{C} = (S, A, S, \delta, T).$$

We show that $L(\mathbf{C}) = \text{Suffix}(L)$. Let $x \in L(\mathbf{C})$. Then there is some state $s \in \mathbf{C}$ such that $s \cdot x$ contains a terminal state t. Now x labels a path in $\mathbf{A}$ from s to t. By assumption $\mathbf{A}$ is accessible, and so there is a string y from an initial state s' such that y labels a path from s' to s. Thus yx labels a path from an initial state of $\mathbf{A}$ to a terminal state of $\mathbf{A}$. Hence $yx \in L(\mathbf{A})$. Thus $x \in \text{Suffix}(L)$. Conversely, suppose that $x \in \text{Suffix}(L)$. By definition there exists a string y such that $yx \in L$. It is now immediate that $x \in L(\mathbf{C})$.

Finally, to show that Factor(L) is recognisable, let $\mathbf{A} = (S, A, I, \delta, T)$ be a trim automaton that recognises L. Consider the automaton

$$\mathbf{D} = (S, A, S, \delta, S).$$

We leave it as an exercise to check that $L(\mathbf{D}) = \text{Factor}(L)$. □

Exercises 3.4

1. Show that for each automaton $\mathbf{A}$ we have that $\mathbf{A}^{ab} = \mathbf{A}^{ba}$.

2. Complete the proof of Proposition 3.4.6, and prove that if L is recognisable, then Factor(L) is recognisable.

3.5 Grammars

Automata are devices for recognising strings belonging to a language. In this section, we introduce grammars that are devices for *generating* strings belonging to a language. We shall then use non-deterministic automata to show how to convert from automata to certain types of grammars and back again. We do not use grammars anywhere else in this book.

The motivation for the concept of a grammar comes from linguistics. Learning a language involves many different skills, however, at its most simplistic, we have to learn two things: the vocabulary of the language and the rules for combining the words in the vocabulary. These rules are usually couched in terms of words such as 'nouns,' 'adjectives,' 'verbs,' and so on. For example, we learn that in French most adjectives follow the noun they modify, whereas in German adjectives precede nouns. To expand on this we give a very simple grammar for a small portion of English.

Example 3.5.1 Consider the following fragment of English. Phrases in English can be classified according to the grammatical categories to which they belong. We shall use only the following:

- *Sentence* denoted by **S**.
- *Noun-phrase* denoted by **NP**.
- *Verb-phrase* denoted by **VP**.
- *Noun* denoted by **N**.
- *Definite article* denoted by **T**.
- *Verb* denoted by **V**.

In addition, there are rules that tell us how the grammatical categories are related to each other. We shall use an arrow $\rightarrow$ to indicate how a grammatical category on the left can be constructed from grammatical categories on the right; I have used the symbol $\cdot$ for concatenation for the sake of clarity:

(1) $\mathbf{S} \rightarrow \mathbf{NP} \cdot \mathbf{VP}$.

(2) $\mathbf{NP} \rightarrow \mathbf{T} \cdot \mathbf{N}$.

(3) $\mathbf{VP} \to \mathbf{V} \cdot \mathbf{NP}$.

We also include amongst these rules the specific English words that belong to those grammatical categories consisting only of words; the symbol '|' should be read 'or':

(4) $\mathbf{T} \to \text{the}$.

(5) $\mathbf{N} \to \text{girl} \mid \text{boy} \mid \text{ball} \mid \text{bat} \mid \text{frisbee}$.

(6) $\mathbf{V} \to \text{hit} \mid \text{took} \mid \text{threw} \mid \text{lost}$.

We can use this grammar to generate a language over the alphabet:

$$\{\text{the}, \text{girl}, \text{boy}, \text{ball}, \text{bat}, \text{frisbee}, \text{hit}, \text{took}, \text{threw}, \text{lost}\}.$$

The starting point is always the symbol **S**. Here is one example; we use the symbol $\Rightarrow$ when we are generating a sentence.

- $\mathbf{S} \Rightarrow \mathbf{NP} \cdot \mathbf{VP}$.
- $\mathbf{NP} \cdot \mathbf{VP} \Rightarrow \mathbf{T} \cdot \mathbf{N} \cdot \mathbf{VP}$.
- $\mathbf{T} \cdot \mathbf{N} \cdot \mathbf{VP} \Rightarrow \mathbf{T} \cdot \mathbf{N} \cdot \mathbf{V} \cdot \mathbf{NP}$.
- $\mathbf{T} \cdot \mathbf{N} \cdot \mathbf{V} \cdot \mathbf{NP} \Rightarrow \mathbf{T} \cdot \mathbf{N} \cdot \mathbf{V} \cdot \mathbf{T} \cdot \mathbf{N}$.
- $\mathbf{T} \cdot \mathbf{N} \cdot \mathbf{V} \cdot \mathbf{T} \cdot \mathbf{N} \Rightarrow$ the $\mathbf{N} \cdot \mathbf{V} \cdot \mathbf{T} \cdot \mathbf{N}$.
- the $\mathbf{N} \cdot \mathbf{V} \cdot \mathbf{T} \cdot \mathbf{N} \Rightarrow$ the boy $\mathbf{V} \cdot \mathbf{T} \cdot \mathbf{N}$.
- the boy $\mathbf{V} \cdot \mathbf{T} \cdot \mathbf{N} \Rightarrow$ the boy threw $\mathbf{T} \cdot \mathbf{N}$.
- the boy threw $\mathbf{T} \cdot \mathbf{N} \Rightarrow$ the boy threw the $\mathbf{N}$.
- the boy threw the $\mathbf{N} \Rightarrow$ the boy threw the ball.

We see that the string 'the boy threw the ball' belongs to the language generated by the grammar. □

The above example is not a very good representation of English; in fact, writing descriptions of natural languages is a complex business and the subject of intensive research. Our next example is much more representative of what we shall be dealing with. But first we need some terminology.

The symbols 'the,' 'man,' 'ball,' 'took,' 'hit' etc. are called terminals. It is important to note that we are thinking of the words 'the,' 'man,' 'ball,' etc. as being individual symbols not strings. The grammatical categories are called non-terminals. The rules are called productions, and **S** is called the start symbol. The language we constructed was over the alphabet consisting of the set of terminals.

Example 3.5.2 The following is a grammar for a fragment of a PASCAL-type language called the 'language of **while**-programs.' The set of non-terminals is:

$\langle$program$\rangle$, $\langle$statement sequence$\rangle$, $\langle$statement$\rangle$, $\langle$asgt$\rangle$, $\langle$test$\rangle$, $\langle$variable$\rangle$, $\langle$digit$\rangle$.

The set of terminals is

begin, **end**, *pred*, *succ*, **while**, **do**,

together with

$$:= \quad \neq \quad ; \quad (\quad)$$

and

$$0, 1, 2, 3, \ldots, 9$$

and

$$A, B, \ldots, Z,$$

and the start symbol is $\langle$program$\rangle$.
The productions are as follows:

(1) $\langle$program$\rangle$ $\rightarrow$ **begin end** | **begin** $\langle$statement sequence$\rangle$ **end**.

(2) $\langle$statement sequence$\rangle$ $\rightarrow$ $\langle$statement$\rangle$ | $\langle$statement sequence$\rangle$; $\langle$statement$\rangle$.

(3) $\langle$statement$\rangle$ $\rightarrow$ $\langle$asgt$\rangle$ | **while** $\langle$test$\rangle$ **do** $\langle$statement$\rangle$ | $\langle$program$\rangle$.

(4) $\langle$test$\rangle$ $\rightarrow$ $\langle$variable$\rangle$ $\neq$ $\langle$variable$\rangle$.

(5) $\langle$asgt$\rangle$ $\rightarrow$ $\langle$variable$\rangle$:= 0 | $\langle$variable$\rangle$:= *succ*($\langle$variable$\rangle$) | $\langle$variable$\rangle$:= *pred*($\langle$variable$\rangle$).

(6) $\langle$variable$\rangle$ $\rightarrow$ $\langle$alpha$\rangle$ | $\langle$variable$\rangle\langle$alpha$\rangle$ | $\langle$variable$\rangle\langle$digit$\rangle$.

(7) $\langle$digit$\rangle$ $\rightarrow 0|1|\ldots|9$.

(8) $\langle$alpha$\rangle$ $\rightarrow A|B|\ldots|Z$.

An example of a string in the language generated by this grammar is

begin while X $\neq$ Y **do begin** X := *succ*(X) **end end**.

which we leave to the reader to check. □

We now make precise what we mean by a 'grammar.' A *grammar* is a 4-tuple $\mathbf{G} = (N, A, P, S)$ consisting of the following

- N is a finite set of *non-terminal symbols*.
- A is a set of *terminal symbols*; we have that $N \cap A = \emptyset$.
- $S \in N$ is the *start symbol*.

- P is a finite set of *productions*; each production is of the form $x \to y$ where x, y are strings in $(N + A)^*$ such that x *contains at least one non-terminal symbol.*

Grammars are devices for *generating* the strings in a language. To see how they accomplish this, we need the following definitions and notation. Let $x \to y$ be a production in a grammar $\mathbf{G}$. Let $w, w' \in (N + A)^*$. Suppose that

$$w = uxv \text{ and } w' = uyv.$$

Then we say w' *is immediately derived from* w in the grammar $\mathbf{G}$. We denote this by

$$w \Rightarrow w'.$$

Let $w_1, \ldots, w_n$ be a sequence of strings in $(N + A)^*$ such that

$$w_1 \Rightarrow w_2 \Rightarrow \ldots \Rightarrow w_n.$$

Then we say that w_n *is derivable from* w_1 and denote this by

$$w_1 \stackrel{*}{\Rightarrow} w_n.$$

The sequence $w_1, \ldots, w_n$ is called a *derivation of* w_n *from* w_1 *in the grammar* $\mathbf{G}$. The *language generated by* $\mathbf{G}$ is:

$$L(\mathbf{G}) = \{x \in A^*\colon S \stackrel{*}{\Rightarrow} x\},$$

the set of all strings over the terminal alphabet, which can be derived from the start symbol. A language L is said to be *recursively enumerable (re)* if there is a grammar $\mathbf{G}$ such that $L = L(\mathbf{G})$. Recursively enumerable languages are very closely related to Turing machines: see the remarks at the end of this chapter.

We give three more examples of grammars before we make explicit the connection between finite automata and a special class of grammars.

Example 3.5.3 The grammar $\mathbf{G}_3 = (\{A, S\}, \{a, b, c\}, P, S)$ where P has the following productions:

$$S \xrightarrow{(1)} abASc \,|\, \varepsilon, \quad bAa \xrightarrow{(2)} abA, \quad bAc \xrightarrow{(3)} bc, \quad bAb \xrightarrow{(4)} bbA.$$

On this occasion, I have numbered the productions so that those used in a derivation may be recorded. In the production (1), we use the symbol $|$ to mean 'or', so that we really have two productions: $S \to abASc$ and $S \to \varepsilon$. Here is a derivation of $a^2b^2c^2$ from this grammar:

$$\begin{aligned} S &\Rightarrow abA\underline{S}c \Rightarrow abAabA\underline{S}c^2 \Rightarrow abAa\underline{bAc}c \Rightarrow a\underline{bAa}bc^2 \\ &\Rightarrow a^2\underline{bAb}c^2 \Rightarrow a^2b\underline{bAc}c \Rightarrow a^2b^2c^2. \end{aligned}$$

In this derivation, I have underlined, for the sake of clarity, those portions of a string to which a production is applied. □

Example 3.5.4 The grammar $\mathbf{G}_4 = (\{S\}, \{a, b\}, P, S)$ has productions:

$$S \rightarrow aSb \mid \varepsilon.$$

The language generated by this grammar is $\{a^n b^n : n \geq 0\}$. □

Example 3.5.5 The grammar $\mathbf{G}_5 = (\{S, A, B\}, \{0, 1\}, P, S)$ where P has the following productions:

$$S \rightarrow 1B \mid 1, \quad A \rightarrow 1B \mid 1, \quad B \rightarrow 0A.$$

The language generated by this grammar is $(10)^*1$. □

We have given five examples of grammars: Examples 3.5.1–3.5.5. The grammar in Example 3.5.3 stands out as being different: in all the other grammars, the left-hand side of each production consists of a single non-terminal, whereas in Example 3.5.3 there are productions where the left-hand side contains a mixture of terminals and non-terminals or contains more than one terminal. Grammars like Example 3.5.3 are usually much harder to work with. For this reason we single out the 'nice' cases. A grammar is said to be *context-free (CF)* if each production $x \rightarrow y$ is such that x is a single non-terminal. A language L is said to be *context-free* if there is a context-free grammar $\mathbf{G}$ such that $L(\mathbf{G}) = L$. Context-free languages are the basis for describing programming languages as Example 3.5.2 suggests. However, they are still too general for our purposes. Example 3.5.5 is an example of the kind of grammar that will interest us. A grammar is said to be *right linear* if each production has one of the following forms:

- $A \rightarrow bC$.
- $A \rightarrow b$.
- $A \rightarrow \varepsilon$.

where $A, C \in N$ and $b \in A$. From now on we shall deal solely with right linear grammars; this is because of the following result: a language can be described by a right linear grammar if and only if it is recognisable. To prove this result, we first need a lemma.

Lemma 3.5.6 *Let $\mathbf{G} = (N, V, P, S)$ be a right linear grammar. Then we can always assume that the only productions in P have the form $A \rightarrow bC$ or $A \rightarrow \varepsilon$.*

Proof Suppose we have a production of the form $A \rightarrow b$. Introduce a *new* non-terminal D such that $D \notin N$ and replace $A \rightarrow b$ by $A \rightarrow bD$ and $D \rightarrow \varepsilon$. It is clear that the new grammar $\mathbf{G}'$ we obtain satisfies $L(\mathbf{G}') = L(\mathbf{G})$. This procedure is repeated until all productions of the form $A \rightarrow b$ have been eradicated. In each case, the new symbol D can be re-used. □

We now come to the result that links right linear grammars and the languages that interest us.

Theorem 3.5.7 *A language is generated by a right linear grammar if and only if it is recognisable.*

Proof Let $\mathbf{G} = (N, V, P, S)$ be a right linear grammar, which by Lemma 3.5.6 we can assume to have productions only of the form $X \to aY$ or $X \to \varepsilon$. We shall construct a non-deterministic automaton $\mathbf{A} = (Q, A, i, \delta, T)$ from $\mathbf{G}$ and prove that $L(\mathbf{A}) = L(\mathbf{G})$. We define $\mathbf{A}$ as follows:

- The states Q are labelled by the non-terminals.
- The initial state i is labelled by S, the start symbol.
- The set of terminal states T is labelled by those non-terminals A for which $A \to \varepsilon$ is a production in P.

It only remains to define δ. This is done as follows: for each production $A \to cB \in P$ we construct a transition:

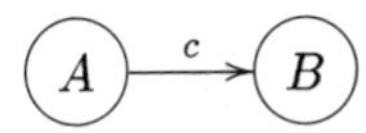

It is clear that $\mathbf{A}$ is a non-deterministic automaton (in general). It remains to prove that $L(\mathbf{A}) = L(\mathbf{G})$. Observe first that the initial state of $\mathbf{A}$ is also terminal if and only if the production $S \to \varepsilon$ belongs to P. Thus $\varepsilon \in L(\mathbf{A})$ iff $\varepsilon \in L(\mathbf{G})$. Let $x \in L(\mathbf{G})$ be a non-empty string where $x = a_1 \ldots a_n$. Then there is a derivation,

$$S \Rightarrow a_1 A_1 \Rightarrow a_1 a_2 A_2 \Rightarrow \ldots \Rightarrow a_1 \ldots a_n A_n \Rightarrow a_1 \ldots a_n,$$

where the productions used are

$$S \to a_1 A_1, \quad A_1 \to a_2 A_2, \quad \ldots \quad A_n \to \varepsilon.$$

This corresponds to the following sequence of transitions in $\mathbf{A}$:

$$\longrightarrow S \xrightarrow{a_1} A_1 \xrightarrow{a_2} A_2 \dashrightarrow A_n$$

Consequently $x \in L(\mathbf{A})$. It is clear that this argument works in reverse. We deduce that $L(\mathbf{A}) = L(\mathbf{G})$.

Now let $\mathbf{A} = (Q, V, I, \delta, T)$ be a deterministic automaton. We shall define a right linear grammar $\mathbf{G} = (N, V, P, S)$ such that $L(\mathbf{G}) = L(\mathbf{A})$. We define $\mathbf{G}$ as follows:

- The set of terminals N is labelled by the states Q.
- The start symbol S is the initial state I.

It only remains to define P. If $A \in T$ then $A \to \varepsilon$ is in P; if $\delta(A,c) = B$ then $A \to cB$ is in P. Clearly $\mathbf{G}$ is a right-linear grammar. Once again it is easy to check that $\varepsilon \in L(\mathbf{G})$ iff $\varepsilon \in L(\mathbf{A})$. Now let $x = a_1 \ldots a_n \in L(\mathbf{A})$. Then there is a path in $\mathbf{A}$ labelled by x, which begins at I and ends at a terminal state:

$$I \xrightarrow{a_1} S_1 \xrightarrow{a_2} S_2 \ldots \xrightarrow{a_n} S_n \in T.$$

This implies that we have the following productions in P:

$$I \to a_1 S_1, \quad S_1 \to a_2 S_2, \quad \ldots, \quad S_n \to \varepsilon.$$

Hence $x \in L(\mathbf{G})$. It is clear that this argument works in reverse. We deduce that $L(\mathbf{G}) = L(\mathbf{A})$. □

Exercises 3.5

1. In Example 3.5.1, derive the sentence: the boy lost the frisbee.

2. In Example 3.5.2, derive the string:

begin while X $\neq$ Y **do begin** X := *succ*(X) **end end**.

3. In Example 3.5.3, derive the string: $a^3b^3c^3$.

4. Find a right linear grammar that generates the language accepted by the following machine:

5. Find a non-deterministic automaton that recognises the language generated by the grammar,

$$\mathbf{G} = (\{S, T, U\}, \{a, b\}, P, S),$$

where P consists of the following productions:

$$S \to aS \,|\, bT, \quad T \to aU, \quad U \to a.$$

6. Consider the following context free grammar $\mathbf{G} = (N, V, P, S)$ where $N = \{S\}$ and P consists of the following productions:

$$S \to aSb \,|\, aSa \,|\, bSa \,|\, bSb \,|\, \varepsilon.$$

Is the language generated by this grammar recognisable or not? Justify your answer.

3.6 Summary of Chapter 3

- *Accessible automata*: A state s is accessible if there is an input string that labels a path starting at the initial state and ending at the state s. An automaton is accessible if every state is accessible. If $\mathbf{A}$ is an automaton, then the accessible part of $\mathbf{A}$, denoted by $\mathbf{A}^a$, is obtained by removing all inaccessible states and transitions to and from them. The language remains unaltered. There is an efficient algorithm for constructing $\mathbf{A}^a$ using the transition tree of $\mathbf{A}$.

- *Non-deterministic automata*: These are automata where the restrictions of completeness and determinism are renounced and where we are allowed to have a set of initial states. A string is accepted by such a machine if it labels at least one path from an initial state to a terminal state. If $\mathbf{A}$ is a non-deterministic automaton, then there is an algorithm, called the subset construction, which constructs a deterministic automaton, denoted $\mathbf{A}^d$, recognising the same language as $\mathbf{A}$. The disadvantage of this construction is that the states of $\mathbf{A}^d$ are labelled by the subsets of the set of states of $\mathbf{A}$. The accessible subset construction constructs the automaton $(\mathbf{A}^d)^a = \mathbf{A}^{da}$ directly from $\mathbf{A}$ and often leads to a much smaller automaton.

- *Applications of non-deterministic automata*: Let $\mathbf{A}$ be a non-deterministic automaton. Then $\mathbf{A}^{\mathrm{rev}}$, the reverse of $\mathbf{A}$, is obtained by reversing all the transitions in $\mathbf{A}$ and interchanging initial and terminal states. The language recognised by $\mathbf{A}^{\mathrm{rev}}$ is the reverse of the language recognised by $\mathbf{A}$. If L and M are recognisable, then we can prove that $L + M$, $L \cdot M$ and L^* are each recognisable using non-deterministic automata.

- *Trim automata*: If $\mathbf{A}$ is a non-deterministic automaton we can define its accessible part $\mathbf{A}^a$ and its co-accessible part $\mathbf{A}^b$. If we carry out both of these constructions, in either order, we obtain $\mathbf{A}^t$, the trim part of $\mathbf{A}$. None of these constructions changes the language.

- *Grammars*: Grammars are devices for generating the strings in a language. The class of right linear grammars generates precisely the recognisable languages.

3.7 Remarks on Chapter 3

I learnt about transition trees from Büchi's book [22]. Many of the results in Sections 3.2, 3.3, and 3.4, including the subset construction, can be found in [109]; Eilenberg [39] states (page 74) that virtually everything known about automata up to 1959 was presented in this paper. You can read about how Scott and Rabin came to work on non-deterministic automata in [119]. I learnt about trim automata and their applications from [39]. The grammatical approach to recognisable languages is from [29].

With the introduction of grammars, I can now sketch out the class of all languages. The largest class studied in automata theory are those that can be recognised by Turing machines; these turn out to be precisely the recursively enumerable languages. By the Church-Turing Thesis, the membership problem for languages that are not recursively enumerable cannot be solved by algorithmic means. The next class consists of those languages that can be generated by context-free grammars. The corresponding class of automata that recognise such languages are the so-called 'pushdown automata,' which can be regarded as consisting of a finite automaton with access to a memory in the form of a pushdown stack. Finally, we have the class of languages generated by right linear grammars which, as we have proved, are precisely the languages recognised by finite automata. This hierarchy of languages is called the *Chomsky hierarchy*. This book is solely concerned with the lowest rung of this hierarchy of languages.

If you want to learn more about languages, automata, and grammars, then [66] and [80] are good references. The source for Example 3.5.1 is Chomsky [28], and for Example 3.5.2 Kfoury et al. [73]. The book by Kfoury provides an alternative approach to the study of algorithms via this simple programming language. Finally, finite automata and context-free languages together provide the setting for designing compilers; see the 'red dragon book' [1].

Chapter 4

ε-automata

Non-deterministic and deterministic automata have something in common: both types of machines can only change state as a response to reading an input symbol. In the case of non-deterministic automata, a state and an input symbol lead to a set of possible states. The class of ε-automata, introduced in this chapter, can change state spontaneously without any input symbol being read. Although this sounds like a very powerful feature, we shall show that every non-deterministic automaton with ε-transitions can be converted into a non-deterministic automaton without ε-transitions that recognises the same language. Armed with ε-automata, we can construct automata to recognise all kinds of languages with great ease. Our main application of ε-automata will be in Section 5.3.

4.1 Automata with ε-transitions

In both deterministic and non-deterministic automata, transitions may only be labelled with elements of the input alphabet. No edge may be labelled with the empty string ε. We shall now waive this restriction. A *non-deterministic automaton with ε-transitions* or, more simply, an *ε-automaton*, is a 5-tuple,

$$\mathbf{A} = (S, A, I, \delta_\varepsilon, T),$$

where all the symbols have the same meanings as in the non-deterministic case except that

$$\delta_\varepsilon \colon S \times (A \cup \{\varepsilon\}) \to \mathsf{P}(S).$$

As before, we shall write $\delta_\varepsilon(s, a) = s \cdot a$. The only difference between such automata and non-deterministic automata is that we allow transitions:

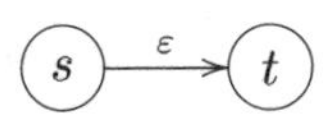

Such transitions are called *ε-transitions*.

In order to define what we mean by the language accepted by such a machine, we have to define an appropriate 'extended transition function.' This is slightly more involved than before, so I shall begin with an informal description. A path in an ε-automaton is a sequence of states each labelled by an element of the set $A \cup \{\varepsilon\}$. The string corresponding to this path is the *concatenation* of these labels in order. We say that a string x is accepted by an ε-automaton if there is a path from an initial state to a terminal state the concatenation of whose labels is x. I now have to put this idea on a sound footing.

Let A be an alphabet. If $a \in A$, then for all $m, n \in \mathbb{N}$ we have that $a = \varepsilon^m a \varepsilon^n$. However, $\varepsilon^m a \varepsilon^n$ is also a *string* consisting of m ε's followed by one a followed by a further n ε's. We call this string an *ε-extension* of the symbol a. The *value* of the ε-extension $\varepsilon^m a \varepsilon^n$ is a. More generally, we can define an ε-extension of a string $x \in A^*$ to be the product of ε-extensions of each symbol in x. The value of any such ε-extension is just x. For example, the string aba has ε-extensions of the form $\varepsilon^m a \varepsilon^n b \varepsilon^p a \varepsilon^q$, where $m, n, p, q \in \mathbb{N}$. Let $\mathbf{A}$ be a non-deterministic automaton with ε-transitions. We say that x is *accepted by* $\mathbf{A}$ if *some* ε-extension of x labels a path in $\mathbf{A}$ starting at some initial state and ending at some terminal state. As usual we write $L(\mathbf{A})$ to mean the set of all strings accepted by $\mathbf{A}$. It is now time for a concrete example.

Example 4.1.1 Consider the diagram below:

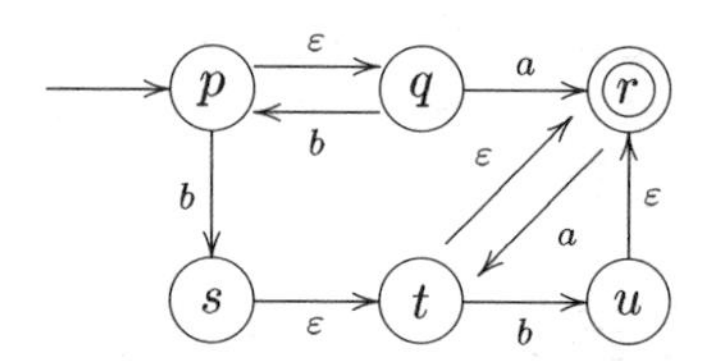

This is clearly a non-deterministic automaton with ε-transitions. We find some examples of strings accepted by this machine. First of all, the letter a is accepted. At first sight, this looks wrong, because there are no transitions from p to r labelled by a. However, this is not our definition of how a string is accepted. We have to check *all possible ε-extensions of a*. In this case, we immediately see that εa labels a path from p to r, and so a *is* accepted. Notice, by the way, that it is the *value* of the ε -extension that is accepted; so, if you said εa was accepted, I would have to say that you were wrong. The letter b is accepted, because $b\varepsilon\varepsilon$ labels a path from p to r. The string bb is accepted, because $b\varepsilon b\varepsilon$ labels a path from p to r. □

Now that we understand how ε-automata are supposed to behave, we can formally define the extended transition function δ_ε^*. To do this, we shall use the following definition. Let $\mathbf{A}$ be a non-deterministic automaton with ε-transitions, and let s be an arbitrary state of $\mathbf{A}$. The *ε-closure of s*, denoted by $E(s)$, consists of s itself together with all states in $\mathbf{A}$, which can be reached by following paths labelled *only* by ε's. If Q is a set of states, then we define the *ε-closure of* Q by

$$E(Q) = \sum_{q \in Q} E(q),$$

the union of the ε-closures of each of the states in Q. Observe that $E(\emptyset) = \emptyset$. Referring to Example 4.1.1, the reader should check that $E(p) = \{p, q\}$, $E(q) = \{q\}$, $E(r) = \{r\}$, $E(s) = \{s, t, r\}$, $E(t) = \{t, r\}$, and $E(u) = \{u, r\}$. The only point that needs to be emphasised is that the ε-closure of a state must contain that state, and so it can never be empty.

We are almost ready to define the extended transition function. We need one piece of notation.

Notation If Q is a set of states in an ε-automaton and $a \in A$ then we write $Q \cdot a$ to mean $\sum_{q \in Q} q \cdot a$; that is, a state s belongs to the set $Q \cdot a$ precisely when there is a state $q \in Q$ and a transition in $\mathbf{A}$ from q to s labelled by a.

The *extended transition function* of an ε-automaton δ_ε^* is the unique function from $S \times A^*$ to $\mathsf{P}(S)$ satisfying the following three conditions where $a \in A$, $x \in A^*$ and $s \in S$:

(ETF1) $\delta_\varepsilon^*(s, \varepsilon) = E(s)$.

(ETF2) $\delta_\varepsilon^*(s, a) = E(E(s) \cdot a)$.

(ETF3) $\delta_\varepsilon^*(s, ax) = \sum_{q \in E(E(s) \cdot a)} \delta_\varepsilon^*(q, x)$.

Once again, it can be shown that this defines a unique function. This definition agrees perfectly with our definition of the ε-extension of a string. To see why, observe that if $a \in A$, then $E(E(s) \cdot a)$ is the set of states that can be reached starting at s and following all paths labelled $\varepsilon^m a \varepsilon^n$. More generally, $\delta_\varepsilon^*(s, x)$, where $x \in A^*$, consists of all states that can be reached starting at s and following all paths labelled by ε-extensions of x. We conclude that the appropriate definition of the language accepted by an ε-automaton is

$$L(\mathbf{A}) = \{x \in A^* \colon \delta_\varepsilon^*(s, x) \cap T \neq \emptyset \text{ for some } s \in I\}.$$

Our goal now is to show that a language recognised by an ε-automaton can be recognised by an ordinary non-deterministic automaton. To do this, we shall use the following construction. Let $\mathbf{A} = (S, A, I, \delta_\varepsilon, T)$ be a non-deterministic automaton with ε-transitions. Define a non-deterministic automaton,

$$\mathbf{A}^s = (S \cup \{\diamondsuit\}, A, I \cup \{\diamondsuit\}, \Delta, T^s),$$

as follows:

- $\diamondsuit$ is a new state.

- $$T^s = \begin{cases} T \cup \{\Diamond\} & \text{if } \varepsilon \in L(\mathbf{A}) \\ T & \text{otherwise.} \end{cases}$$

- The function,
$$\Delta\colon (S \cup \{\Diamond\}) \times A \to \mathsf{P}(S \cup \{\Diamond\}),$$
is defined as follows: $\Delta(\Diamond, a) = \emptyset$ for all $a \in A$, and $\Delta(s, a) = E(E(s) \cdot a)$ for all $s \in S$ and $a \in A$.

It is clear that $\mathbf{A}^s$ is a well-defined, non-deterministic automaton. Observe that the role of the state $\Diamond$ is solely to accept ε if $\varepsilon \in L(\mathbf{A})$. If $\varepsilon \notin L(\mathbf{A})$, then you can omit $\Diamond$ from the construction of $\mathbf{A}^s$.

Theorem 4.1.2 *Let* $\mathbf{A} = (S, A, I, \delta_\varepsilon, T)$ *be a non-deterministic automaton with* ε*-transitions. Then* $L(\mathbf{A}^s) = L(\mathbf{A})$.

Proof The main plank of the proof is the following equation, which we shall prove below: for all $s \in S$ and $x \in A^+$ we have that

$$\Delta^*(s, x) \quad = \quad \delta_\varepsilon^*(s, x). \tag{4.1}$$

Observe that this equation holds for *non-empty* strings. We prove (4.1) by induction on the length of x.

For the base step, let $a \in A$. Then

$$\Delta^*(s, a) = \Delta(s, a) = \delta_\varepsilon^*(s, a),$$

simply following the definitions.

For the induction step, assume that the equality holds for all $x \in A^+$ where $|x| = n$. Let $y = ax$ where $a \in A$ and $|x| = n$. Then

$$\Delta^*(s, y) = \Delta^*(s, ax) = \sum_{q \in \Delta(s,a)} \Delta^*(q, x),$$

by the definition of Δ^*. By the base step and the induction step,

$$\sum_{q \in \Delta(s,a)} \Delta^*(q, x) = \sum_{q \in \delta_\varepsilon^*(s,a)} \delta_\varepsilon^*(q, x),$$

but by definition,

$$\sum_{q \in \delta_\varepsilon^*(s,a)} \delta_\varepsilon^*(q, x) = \delta_\varepsilon^*(s, ax) = \delta_\varepsilon^*(s, y),$$

and we have proved the equality.

Now we can prove that $L(\mathbf{A}^s) = L(\mathbf{A})$. Observe first that

$$\varepsilon \in L(\mathbf{A}) \Leftrightarrow \Diamond \in T^s \Leftrightarrow \varepsilon \in L(\mathbf{A}^s).$$

With this case out of the way, let $x \in A^+$. Then by definition $x \in L(\mathbf{A}^s)$ means that there is some $s \in I \cup \{\Diamond\}$ such that $\Delta^*(s,x) \cap T^s \neq \emptyset$. But since x is not empty, the state $\Diamond$ can play no role and so we can write that for some $s \in I$, we have $\Delta^*(s,x) \cap T \neq \emptyset$. By equation (4.1), $\Delta^*(s,x) = \delta_\varepsilon^*(s,x)$. Thus $x \in L(\mathbf{A}^s)$ if and only if $\delta_\varepsilon^*(s,x) \cap T \neq \emptyset$ for some $s \in I$. This of course says precisely that $x \in L(\mathbf{A})$ as required. □

Remark The meaning of the 's' in $\mathbf{A}^s$ is that of 'sans' since $\mathbf{A}^s$ is 'sans epsilons.'

The construction of the machine $\mathbf{A}^s$ is quite involved. It is best to set the calculations out in tabular form as suggested by the following example.

Example 4.1.3 We calculate $\mathbf{A}$ for the ε-automaton of Example 4.1.1.

state $\star$	$E(\star)$	$E(\star)\cdot a$	$E(\star)\cdot b$	$E(E(\star)\cdot a)$	$E(E(\star)\cdot b)$
p	$\{p,q\}$	$\{r\}$	$\{s,p\}$	$\{r\}$	$\{s,t,r,p,q\}$
q	$\{q\}$	$\{r\}$	$\{p\}$	$\{r\}$	$\{p,q\}$
r	$\{r\}$	$\{t\}$	$\emptyset$	$\{t,r\}$	$\emptyset$
s	$\{s,t,r\}$	$\{t\}$	$\{u\}$	$\{t,r\}$	$\{u,r\}$
t	$\{t,r\}$	$\{t\}$	$\{u\}$	$\{t,r\}$	$\{u,r\}$
u	$\{u,r\}$	$\{t\}$	$\emptyset$	$\{t,r\}$	$\emptyset$

The last two columns give us the information required to construct $\mathbf{A}^s$ below:

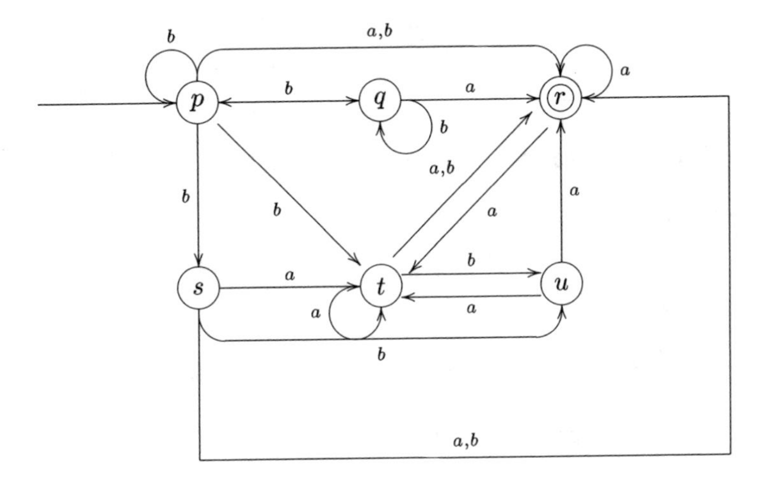

In this case, the state labelled $\Diamond$ is omitted because the original automaton does not accept the empty string. □

Exercises 4.1

1. For each of the ε-automata $\mathbf{A}$ below construct $\mathbf{A}^s$ and $\mathbf{A}^{sda} = ((\mathbf{A}^s)^d)^a$. In each case, describe $L(\mathbf{A})$.

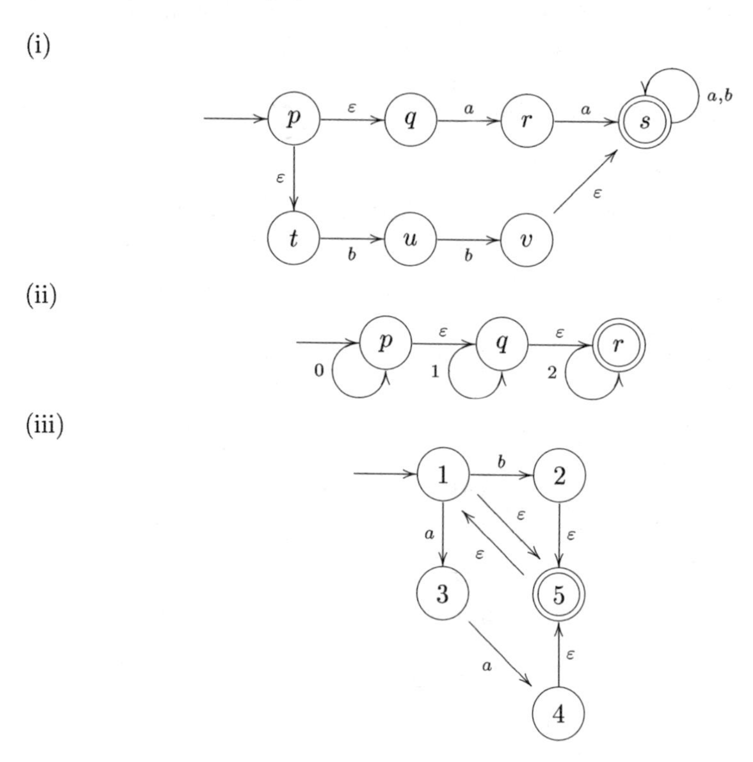

4.2 Applications of ε-automata

If L and M are both recognisable, then ε-automata provide a simple way of proving that $L + M$, LM and L^* are all recognisable. We shall use these versions of the proofs when we prove the algorithmic form of Kleene's theorem in Section 5.3.

Theorem 4.2.1 *Let A be an alphabet and L and M be languages over A.*

(i) *If L and M are recognisable then $L + M$ is recognisable.*

(ii) *If L and M are recognisable then LM is recognisable.*

(iii) *If L is recognisable then L^* is recognisable.*

Proof (i) By assumption, we are given two automata $\mathbf{A}$ and $\mathbf{B}$ such that $L(\mathbf{A}) = L$ and $L(\mathbf{B}) = M$. Construct the following ε-automaton: introduce a new state, which we label $\heartsuit$, to be the new initial state and draw ε-transitions to the initial state of **A** and the initial state of **B**; the initial states of **A** and **B** are now converted into ordinary states. Call the resulting ε-automaton **C**. It is clear that this machine recognises the language $L + M$. We now apply Theorem 4.1.2 to obtain a non-deterministic automaton recognising $L + M$. Thus $L + M$ is recognisable. If we picture **A** and **B** schematically as follows:

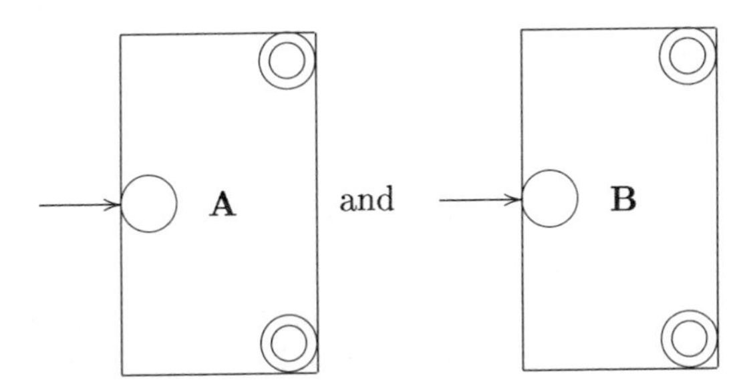

Then the machine **C** has the following form:

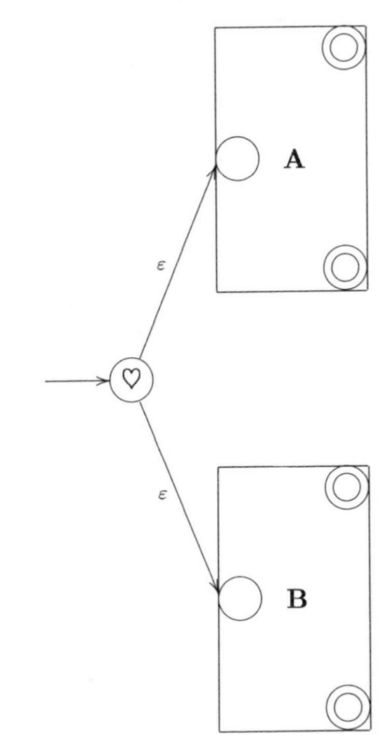

(ii) By assumption, we are given two automata **A** and **B** such that $L(\mathbf{A}) = L$ and $L(\mathbf{B}) = M$. Construct the following ε-automaton: from each terminal state of **A** draw an ε-transition to the initial state of **B**. Make each of the terminal states of **A** ordinary states and make the initial state of **B** an ordinary state. Call the resulting automaton **C**. It is easy to see that this ε-automaton recognises LM. We now apply Theorem 4.1.2 to obtain a non-deterministic automaton recognising LM. Thus LM is recognisable. If we picture **A** and **B** schematically as above then the machine **C** has the following form:

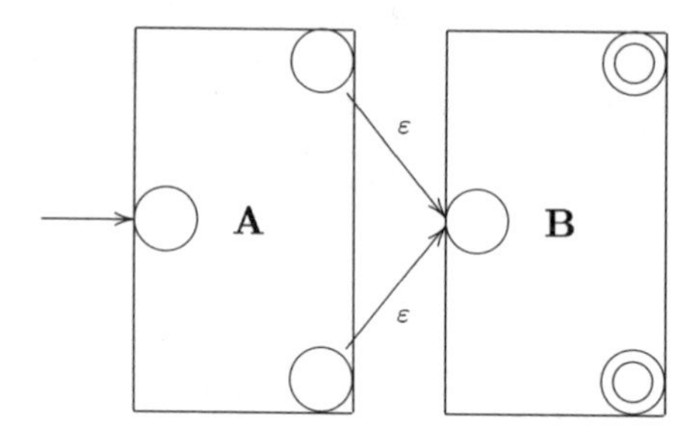

(iii) Let **A** be a deterministic automaton such that $L(\mathbf{A}) = L$. Construct an ε-automaton **D** as follows. It has two more states than **A**, which we shall label ♡ and ♠: the former will be initial and the latter terminal. Connect the state ♡ by an ε-transition to the initial state of **A** and then make the initial state of **A** an ordinary state. Connect all terminal states of **A** to the state labelled ♠ and then make all terminal states of **A** ordinary states. Connect the state ♡ by an ε-transition to the state ♠ and vice versa. If we picture **A** schematically as above, then **D** can be pictured schematically as follows:

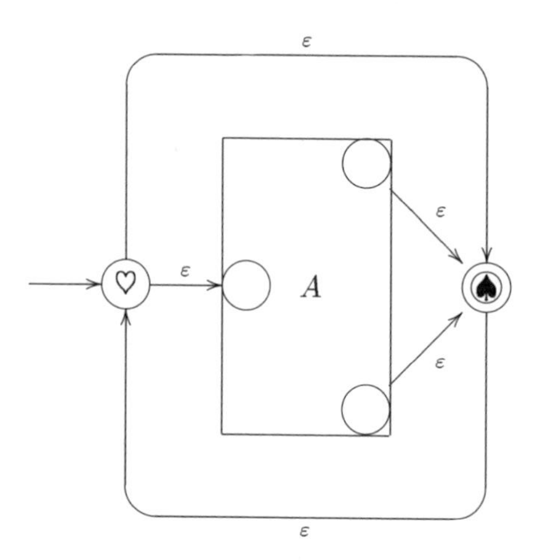

It is easy to check that $L(\mathbf{B}) = L^*$: first, by construction, ε is recognised. Second, the bottom ε-transition enables us to re-enter the machine **A** embedded in the diagram. The result now follows by Theorem 4.1.2 again. □

To motivate our next results, consider the following question: is the language $\{0^n1^n : n \geq 0\}$ recognisable? We know that the answer is no because we proved that the language $\{a^nb^n : n \geq 0\}$ is not recognisable and the only difference between these two languages is that we have replaced a by 0 and b by 1. We can similarly see that the language $\{(011)^n(10)^n : n \geq 0\}$ is not recognisable; this time we have replaced a by 011 and b by 10. We need to be able to prove these assertions and, in doing so, we will develop a powerful way of comparing different languages. Let A and B be two alphabets, not necessarily different. A function $\alpha: A^* \rightarrow B^*$ is said to be a *monoid homomorphism* if it satisfies two conditions:

(MM1) $\alpha(\varepsilon) = \varepsilon$.

(MM2) $\alpha(xy) = \alpha(x)\alpha(y)$ for all strings $x, y \in A^*$.

Although the definition is a little complex, it turns out to be very easy to define monoid homomorphisms. To see why, let $x = a_1 \dots a_n \in A^*$. Using (MM2) repeatedly we see that

$$\alpha(x) = \alpha(a_1) \dots \alpha(a_n).$$

In other words, once we know the values of $\alpha(a)$ for *each letter* $a \in A$, then we can compute the value of $\alpha(x)$ for *any string*. In fact, something stronger is true. Let $\alpha': A \rightarrow B^*$ be any function with no restrictions placed on α'. Then we can define a monoid homomorphism α from A^* to B^* as follows: we put $\alpha(\varepsilon) = \varepsilon$ because we have no choice; if $x = a_1 \dots a_n$ is an arbitrary non-empty string, then define

$$\alpha(x) = \alpha'(a_1) \dots \alpha'(a_n).$$

It is easy to check that α is a monoid homomorphism. It follows that to define a monoid homomorphism α we need only define α on the letters in A.

Example 4.2.2 Let $A = \{0, 1\}$ and $B = \{a, b\}$. Define $\alpha: A^* \rightarrow B^*$ by putting $\alpha(0) = a$ and $\alpha(1) = ba$. Then we can easily compute $\alpha(000111)$ as follows:

$$\alpha(000111) = \alpha(0)\alpha(0)\alpha(0)\alpha(1)\alpha(1)\alpha(1) = aaababababa.$$

□

Monoid homomorphisms enable us to compare languages over different alphabets. To see how, let $\alpha: A^* \rightarrow B^*$ be a monoid homomorphism, and let $L \subseteq A^*$. Define

$$\alpha(L) = \{\alpha(x): x \in L\},$$

the *image of* L *under* α. Clearly $\alpha(L)$ is a language over the alphabet B.

Theorem 4.2.3 *Let* $\alpha: A^* \rightarrow B^*$ *be a monoid homomorphism, and let* $L \subseteq A^*$ *be recognisable. Then* $\alpha(L) \subseteq B^*$ *is recognisable.*

Proof Let **A** be a deterministic automaton that recognises L. We shall construct an automaton **B** from **A**, which recognises $\alpha(L)$. In general, **B** will be an ε-automaton. The states of **B** will consist of the states of **A** together with possibly some extra states. For each $a \in A$ and for each transition $q \cdot a = q'$, carry out the following procedure. Suppose that $\alpha(a) = b_1 \dots b_n$ where $b_i \in B$;

we allow the possibility that $\alpha(a)$ is the empty-string. Then construct the sequence of transitions in $\mathbf{B}$:

$$q \xrightarrow{b_1} s_1 \xrightarrow{b_2} s_2 \dashrightarrow s_{n-1} \xrightarrow{b_n} q'$$

Here $s_1, \ldots, s_{n-1}$ are $n-1$ new states. Finally, the initial state of $\mathbf{B}$ is the same as the initial state of $\mathbf{A}$ and the terminal states of $\mathbf{B}$ are the same as the terminal states of $\mathbf{A}$. It is now easy to check that the language recognised by the non-deterministic ε-automaton $\mathbf{B}$ is precisely $\alpha(L)$. □

Let $\alpha\colon A^* \to B^*$ be a monoid morphism. For each $L \subseteq B^*$ define

$$\alpha^{-1}(L) = \{x \in A^*\colon \alpha(x) \in L\},$$

the *inverse image of* L *under* α. Clearly $\alpha^{-1}(L)$ is a language over the alphabet A.

Theorem 4.2.4 *Let* $\alpha\colon A^* \to B^*$ *be a monoid homomorphism. If* $L \subseteq B^*$ *is recognisable then* $\alpha^{-1}(L)$ *is recognisable.*

Proof Let $\mathbf{B} = (S, B, i, \delta, T)$ be a deterministic automaton recognising L. We construct a deterministic automaton $\mathbf{A}$ recognising $\alpha^{-1}(L)$ as follows. The set of states of $\mathbf{A}$ is the same as the set of states of $\mathbf{B}$; likewise the initial and terminal states. The difference between $\mathbf{A}$ and $\mathbf{B}$ lies in the transition function, which we denote by γ. This is defined as follows. For each $a \in A$, and for each state $s \in S$ define

$$\gamma(s, a) = \delta^*(s, \alpha(a)).$$

By construction $\mathbf{A}$ is deterministic. It is easy to check that

$$\gamma^*(s, x) = \delta^*(s, \alpha(x)).$$

It follows that

$$L(\mathbf{A}) = \alpha^{-1}(L(\mathbf{B})).$$

Hence $\alpha^{-1}(L)$ is recognisable as claimed. □

Here is one application of our results on monoid homomorphisms.

Example 4.2.5 The language $L = \{a^i b^j c^{i+j}\colon i, j \geq 0\}$ is not recognisable. To prove this, let $A = \{a, b, c\}$ and $B = \{b, c\}$. Define $\alpha\colon A^* \to B^*$ by $\alpha(a) = \varepsilon$, $\alpha(b) = b$ and $\alpha(c) = c$. The image of L under this monoid homomorphism is $M = \{b^k c^l\colon l \geq k \geq 0\}$. If L is recognisable then so is M. But we can show using the Pumping Lemma that M is not recognisable. It follows that L is not recognisable. □

Exercises 4.2

1. Construct ε-automata to recognise each of the followings languages:

(i) $(a^2)^*(b^3)^*$.

(ii) $(a(ab)^*b)^*$.

(iii) $(a^2b^* + b^2a^*)(ab + ba)$.

2. Complete the proof of Theorem 4.2.3.

3. Complete the proof of Theorem 4.2.4.

4. This question follows on from Question 3 of Exercises 2.6. Let $L \subseteq b^*$ be an arbitrary non-recognisable language. Put $M = a^+L+b^*$. Using Theorem 4.2.3, prove that M is not recognisable.

This example is taken from [136].

4.3 Summary of Chapter 4

- *ε-automata*: These are defined just as non-deterministic automata except that we also allow transitions to be labelled by ε. A string x over the input alphabet is accepted by such a machine **A** if there is at least one path in **A** starting at an initial state and finishing at a terminal state such that when the labels on this path are concatenated the string x is obtained.

- $\mathbf{A}^s$: There is an algorithm that converts an ε-automaton **A** into a non-deterministic automaton $\mathbf{A}^s$ recognising the same language. The 's' stands for 'sans' meaning 'without (epsilons).'

- *Applications*: Using ε-automata, simple proofs can be given of the recognisability of LM from the recognisability of L and M, and the recognisability of L^* from the recognisability of L.

- *Monoid homomorphisms*: A function $\alpha\colon A^* \to B^*$ is a monoid homomorphism if $\alpha(\varepsilon) = \varepsilon$ and $\alpha(xy) = \alpha(x)\alpha(y)$. If $L \subseteq A^*$ is recognisable, then $\alpha(L) \subseteq B^*$ is recognisable. If $M \subseteq B^*$ is recognisable, then $\alpha^{-1}(M) \subseteq A^*$ is recognisable.

4.4 Remarks on Chapter 4

The constructions of Theorem 4.2.1 are usually attributed to [86]. The closure of recognisable languages under monoid homomorphisms is a special case of [9], whereas the closure under inverse images is from [53].

It is interesting to compare the proofs of Theorems 4.2.3 and 4.2.4. In showing that the image of a recognisable language is recognisable, we had to use ε-automata because the image of a letter could be the empty string. Contrast this with the proof that the inverse image of a recognisable language is recognisable; here we could get away with using complete deterministic automata. In some sense, forming the inverse image of a recognisable language seems to be a more natural operation than taking its image. This idea will re-emerge in Chapter 12, when I come to describe 'varieties of languages.'

Chapter 5

Kleene's Theorem

Chapters 2 to 4 have presented us with an array of languages that we can show to be recognisable. At the same time, the Pumping Lemma has provided us with a tool for showing that specific languages are not recognisable. It is clearly time to find a characterisation of recognisable languages. This is exactly what Kleene's theorem does. The characterisation is in terms of regular expressions. Such expressions form a notation for describing languages in terms of finite languages, union, product, and Kleene star; it was informally introduced in Section 1.3. I give two different proofs of Kleene's theorem: in Section 5.2, I prove the bare fact that a language is recognisable if and only if it is regular; in Section 5.3, I describe two algorithms: one shows how to construct an ε-automaton from a regular expression, and the other shows how to construct a regular expression from an automaton. These two algorithms together give a constructive proof of Kleene's theorem. In the last section, I describe an algebraic method for constructing a regular expression from an automaton that involves solving language equations.

5.1 Regular languages

This is now a good opportunity to reflect on which languages we can now prove are recognisable. I want to pick out four main results:

- Finite languages are recognisable; this was proved in Proposition 2.2.4.
- The union of two recognisable languages is recognisable; this was proved in Proposition 2.5.6.
- The product of two recognisable languages is recognisable; this was proved in Proposition 3.3.4.
- The Kleene star of a recognisable language is recognisable; this was proved in Proposition 3.3.6.

We now analyse these results a little more deeply. A finite language that is neither empty nor consists of just the empty string is a finite union of strings, and each language consisting of a finite string is a finite product of languages each of which consist of a single letter. Call a language over an alphabet *basic* if it is either empty, consists of the empty string alone, or consists of a single symbol from the alphabet. Then what we have proved is the following: a language that can be constructed from the basic languages by using only the operations $+$, $\cdot$ and $*$ a finite number of times must be recognisable. The following two definitions give a precise way of describing such languages.

Let $A = \{a_1, \ldots, a_n\}$ be an alphabet. A *regular expression over* A (the term *rational expression* is also used) is a sequence of symbols formed by repeated application of the following rules:

(R1) $\emptyset$ is a regular expression.

(R2) ε is a regular expression.

(R3) $a_1, \ldots, a_n$ are each regular expressions.

(R4) If s and t are regular expressions then so is $(s + t)$.

(R5) If s and t are regular expressions then so is $(s \cdot t)$.

(R6) If s is a regular expression then so is (s^*).

(R7) Every regular expression arises by a finite number of applications of the rules (R1) to (R6).

We call $+$, $\cdot$, and $*$ the *regular operators*. As usual, we will generally write st rather than $s \cdot t$. It is easy to determine whether an expression is regular or not.

Example 5.1.1 We claim that $((0 \cdot (1^*)) + 0)$ is a regular expression over the alphabet $\{0, 1\}$. To prove that it is, we simply have to show that it can be constructed according to the rules above:

(1) 1 is regular by (R3).

(2) (1^*) is regular by (R6).

(3) 0 is regular by (R3).

(4) $(0 \cdot (1^*))$ is regular by (R5) applied to (2) and (3) above.

(5) $((0 \cdot (1^*)) + 0)$ is regular by (R4) applied to (4) and (3) above.

□

Each regular expression s describes a language, denoted by $L(s)$. This language is calculated by means of the following rules, which agree with the conventions we introduced in Section 1.3. Simply put, they tell us how to 'insert the curly brackets.'

(D1) $L(\emptyset) = \emptyset$.

(D2) $L(\varepsilon) = \{\varepsilon\}$.

(D3) $L(a_i) = \{a_i\}$.

(D4) $L(s + t) = L(s) + L(t)$.

(D5) $L(s \cdot t) = L(s) \cdot L(t)$.

(D6) $L(s^*) = L(s)^*$.

Now that we know how regular expressions are to be interpreted, we can introduce some conventions that will enable us to remove many of the brackets, thus making regular expressions much easier to read and interpret. The way we do this takes its cue from ordinary algebra. For example, consider the algebraic expression $a + bc^{-1}$. This can only mean $a + (b(c^{-1}))$, but $a + bc^{-1}$ is much easier to understand than $a + (b(c^{-1}))$. If we say that $*$, $\cdot$, and $+$ behave, respectively, like $^{-1}$, $\times$, and $+$ in ordinary algebra, then we can, just as in ordinary algebra, dispense with many of the brackets that the definition of a regular expression would otherwise require us to use. Using this convention, the regular expression $((0 \cdot (1^*)) + 0)$ would usually be written as $01^* + 0$. Our convention tells us that 01^* means $0(1^*)$ rather than $(01)^*$, and that $01^* + 0$ means $(01^*) + 0$ rather than $0(1^* + 0)$.

Example 5.1.2 We calculate $L(01^* + 0)$.

(1) $L(01^* + 0) = L(01^*) + L(0)$ by (D4).

(2) $L(01^*) + L(0) = L(01^*) + \{0\}$ by (D3).

(3) $L(01^*) + \{0\} = L(0) \cdot L(1^*) + \{0\}$ by (D5).

(4) $L(0) \cdot L(1^*) + \{0\} = \{0\} \cdot L(1^*) + \{0\}$ by (D3).

(5) $\{0\} \cdot L(1^*) + \{0\} = \{0\} \cdot L(1)^* + \{0\}$ by (D6).

(6) $\{0\} \cdot L(1)^* + \{0\} = \{0\} \cdot \{1\}^* + \{0\}$ by (D3).

□

Two regular expressions s and t are *equal*, written $s = t$, if and only if $L(s) = L(t)$. Two regular expressions can look quite different yet describe the same language and so be equal.

Example 5.1.3 Let $s = (0 + 1)^*$ and $t = (1 + 00^*1)^*0^*$. We shall show that these two regular expressions describe the same language. Consequently,

$$(0 + 1)^* = (1 + 00^*1)^*0^*.$$

We now prove this assertion. Because $(0+1)^*$ describes the language of all possible strings of 0's and 1's it is clear that $L(t) \subseteq L(s)$. We need to prove the reverse inclusion. Let $x \in (0+1)^*$, and let u be the longest prefix of x belonging to 1^*. Put $x = ux'$. Either $x' \in 0^*$, in which case $x \in L(t)$, or x' contains at least one 1. In the latter case, x' begins with a 0 and contains at least one 1. Let v be the longest prefix of x' from 0^+1. We can therefore write $x = uvx''$ where $u \in 1^*$, $v \in 0^+1$ and $|x''| < |x|$. We now replace x by x'' and repeat the above process. It is now clear that $x \in L(t)$. □

A language L is said to be *regular* (the term *rational* is also used) if there is a regular expression s such that $L = L(s)$.

Examples 5.1.4 Here are a few examples of regular expressions and the languages they describe over the alphabet $A = \{a, b\}$.

(1) Let $L = \{x \in (a+b)^* : |x| \text{ is even}\}$. A string of even length is either just ε on its own or can be written as the concatenation of strings each of length 2. Thus this language is described by the regular expresssion $((a+b)^2)^*$.

(2) Let $L = \{x \in (a+b)^* : |x| \equiv 1 \pmod 4\}$. A string belongs to this language if its length is one more than a multiple of 4. A string of length a multiple of 4 can be described by the regular expression $((a+b)^4)^*$. Thus a regular expression for L is $((a+b)^4)^*(a+b)$.

(3) Let $L = \{x \in (a+b)^* : |x| < 3\}$. A string belongs to this language if its length is $0, 1$, or 2. A suitable regular expression is therefore $\varepsilon + (a+b) + (a+b)^2$. The language L', the complement of L, consists of all strings whose length is at least 3. This language is described by the regular expression $(a+b)^3(a+b)^*$.

□

We have seen that two regular expressions s and t may look different but describe the same language $L(s) = L(t)$ and so be equal as regular expressions. The collection of all languages $\mathsf{P}(A^*)$ has a number of properties that are useful in showing that two regular expressions are equal. The simplest ones are described in the proposition below. The proofs are left as exercises.

Proposition 5.1.5 *Let A be an alphabet, and let $L, M, N \in \mathsf{P}(A^*)$. Then the following properties hold:*

(i) $L + (M + N) = (L + M) + N$.

(ii) $\emptyset + L = L = \emptyset + L$.

(iii) $L + L = L$.

(iv) $L \cdot (M \cdot N) = (L \cdot M) \cdot N$.

(v) $\varepsilon \cdot L = L = L \cdot \varepsilon$.

(vi) $\emptyset \cdot L = \emptyset = L \cdot \emptyset$.

(vii) $L \cdot (M + N) = L \cdot M + L \cdot N$, *and* $(M + N) \cdot L = M \cdot L + N \cdot L$. □

Result (i) above is called the ***associativity law*** for unions of languages, whereas result (iv) is the associativity law for products of languages. Result (vii) contains the two ***distributity laws (left*** and ***right*** respectively***)*** for product over union.

Because equality of regular expressions $s = t$ is defined in terms of the equality of the corresponding languages $L(s) = L(t)$ it follows that the seven properties above also hold for regular expressions.[1] A few examples are given below.

Examples 5.1.6 Let r, s and t be regular expressions. Then

(1) $r + (s + t) = (r + s) + t$.

(2) $(rs)t = r(st)$.

(3) $r(s + t) = rs + rt$.

□

The relationship between the Kleene star and the other two regular operators is much more complex. Here are two examples.

Examples 5.1.7 Let $A = \{a, b\}$.

(1) $(a + b)^* = (a^*b)^*a^*$. To prove this we apply the usual method for showing that two sets X and Y are equal: we show that $X \subseteq Y$ and $Y \subseteq X$. It is clear that the language on the right is a subset of the language on the left. We therefore need only explicitly prove that the language on the left is a subset of the language on the right. A typical term of $(a + b)^*$ consists of a finite product of a's and b's. Either this product consists entirely of a's, in which case it is clearly a subset of the right-hand side, or it also contains at least one b: in which case, we can split the product

[1] The set of regular languages forms a 'semiring' in the following sense; we use the term 'monoid,' which is defined in Chapter 8. A ***semiring*** $(S, +, \cdot, 0, 1)$ consists of a commutative monoid $(S, +, 0)$, a monoid $(S, \cdot, 1)$ such that 0 is the zero for multiplication, and multiplication distributes over addition on the left and on the right. Semirings in which addition is also idempotent are termed ***idempotent semirings*** or ***dioids***. See [107] for examples of applications of semirings.

into sequences of a's followed by a b, and possibly a sequence of a's at the end. This is also a subset of the right-hand side. For example,

$$aaabbabaaabaaa$$

can be written as

$$(aaab)(a^0b)(ab)(aaab)aaa,$$

which is clearly a subset of $(a^*b)^*a^*$.

(2) $(ab)^* = \varepsilon + a(ba)^*b$. The left-hand side is

$$\varepsilon + (ab) + (ab)^2 + (ab)^3 + \dots.$$

However, for $n \geq 1$, the string $(ab)^n$ is equal to $a(ba)^{n-1}b$. Thus the left-hand side is equal to the right-hand side.

□

Exercises 5.1

1. Find regular expressions for each of the languages over $A = \{a, b\}$.

(i) All strings in which a always appears in multiples of 3.

(ii) All strings that contain exactly 3 a's.

(iii) All strings that contain exactly 2 a's or exactly 3 a's.

(iv) All strings that do not contain aaa.

(v) All strings in which the total number of a's is divisible by 3.

(vi) All strings that end in a double letter.

(vii) All strings that have exactly one double letter.

2. Let r and s be regular expressions. Prove that each of the following equalities holds between the given pair of regular expressions.

(i) $r^* = (rr)^* + r(rr)^*$.

(ii) $(r + s)^* = (r^*s^*)^*$.

(iii) $(rs)^*r = r(sr)^*$.

3. Prove Proposition 5.1.5.

5.2 Kleene's theorem: proof

We can now prove the first major result in automata theory.

Theorem 5.2.1 (Kleene) *A language is recognisable if and only if it is regular.*

Proof Throughout this proof A will be a fixed alphabet.

We prove first that every regular language is recognisable. To do this, we shall use induction on the number of regular operators in a regular expression. Regular expressions containing no regular operators can only describe languages of the form ε, $\emptyset$ or $\{a\}$ where $a \in A$. Each of these languages is recognisable. This is the base step of our induction. Our induction hypothesis is that if r is a regular expression containing at most $n-1$ regular operators then $L(r)$ is recognisable. Now let r be a regular expression containing n regular operators. We shall use the induction hypothesis to show that $L(r)$ is recognisable. There are three cases to consider: $r = s + t$, $r = s \cdot t$, and $r = s^*$ where s and t are regular expressions containing at most $n-1$ regular operators. By the induction hypothesis, $L(s)$ and $L(t)$ are both recognisable. We now apply Propositions 2.5.6, 3.3.4, and 3.3.6 to deduce that $L(r)$ is recognisable, as required. This proves one direction of Kleene's theorem.

We now prove that every recognisable language is regular. To do this, it is convenient to use non-deterministic automata. We shall use the following idea. Given a non-deterministic automaton $\mathbf{A}$, the total number of edges in the directed graph representing $\mathbf{A}$ will be called the *transition number* of $\mathbf{A}$. Our proof will be by induction on this number. If $\mathbf{A}$ has transition number zero, then $L(\mathbf{A})$ is either $\emptyset$ or ε, the latter occurring if one of the initial states is terminal. This is the base step of our induction. Our induction hypothesis is that if $\mathbf{A}$ is a non-deterministic automaton with transition number of at most $n-1$, then $L(\mathbf{A})$ is regular. Let $\mathbf{A} = (S, A, I, \delta, T)$ be a non-deterministic automaton with transition number n. We prove that $L(\mathbf{A})$ is regular. By assumption, there is at least one edge in the directed graph representing $\mathbf{A}$. Choose one, and denote it by $p \xrightarrow{a} q$. We now construct four non-deterministic automata $\mathbf{A}_1, \mathbf{A}_2, \mathbf{A}_3$, and $\mathbf{A}_4$. These automata have the same transition functions: in each case, the transition function is identical to the one in $\mathbf{A}$ except that we erase the transition from p to q chosen above, but retain all the states of $\mathbf{A}$. The automata therefore only differ in the choice of initial and terminal states:

- $\mathbf{A}_1$ has initial states I and terminal states T.
- $\mathbf{A}_2$ has initial states I and terminal state $\{p\}$.
- $\mathbf{A}_3$ has initial state $\{q\}$ and terminal state $\{p\}$.
- $\mathbf{A}_4$ has initial state $\{q\}$ and terminal states T.

By construction, the transition numbers of each of these automata is $n-1$. It follows by the induction hypothesis that each of the languages,

$$L(\mathbf{A}_1), L(\mathbf{A}_2), L(\mathbf{A}_3), \text{ and } L(\mathbf{A}_4),$$

is regular. We shall show that $L(\mathbf{A})$ can be written in terms of these four languages using the regular operators. This will prove that $L = L(\mathbf{A})$ is regular. In fact, I claim that

$$L = L_1 + (L_2 a)(L_3 a)^* L_4.$$

It is easy to check that

$$L_1 \subseteq L \text{ and } (L_2 a)(L_3 a)^* L_4 \subseteq L.$$

We prove the reverse inclusion. Let $x \in L$. Then x labels a path in $\mathbf{A}$, which starts at one of the initial states and ends at one of the terminal states. This path either includes the transition $p \xrightarrow{a} q$ or avoids it. If it avoids it then $x \in L(A_1)$. So we may suppose that it includes this transition. Locate those occurrences of the letter a in the string x that correspond to the transition $p \xrightarrow{a} q$. We may therefore factorise x as follows:

$$x = (ua)(v_1 a) \dots (v_n a) w,$$

where u labels a path from an initial state to the state p; each of the strings v_i labels a path from q to p; and w labels a path from q to a terminal state. Thus $x \in (L_2 a)(L_3 a)^* L_4$, and we have proved the reverse inclusion. □

Kleene's theorem describes languages over an arbitrary alphabet. In the case where the alphabet contains exactly one letter, it is possible to say more about their structure. Let $A = \{a\}$ be a one-letter alphabet. Our first result describes the recognisable subsets of a^* in terms of regular expressions.

Theorem 5.2.2 *A language $L \subseteq a^*$ is recognisable if and only if*

$$L = X + Y(a^p)^*,$$

where X and Y are finite sets and $p \geq 0$.

Proof There is only one direction that needs proving. Let L be recognisable. Because the alphabet contains only one letter, an accessible automaton recognising L must have a particular form, which we now describe. Let the initial state be q_1. Then either $q_1 \cdot a = q_1$ in which case q_1 is the only state, since the automaton is accessible, or $q_1 \cdot a$ is some other state, q_2 say. For each state q, either $q \cdot a$ is a previously constructed state or a new state. Since the automaton is finite there must come a point where $q \cdot a$ is a previously occurring state.

It follows that an accessible automaton recognising L consists of a *stem* of s states $q_1, \ldots, q_s$, and a *cycle* of p states $r_1, \ldots r_p$ connected together as follows:

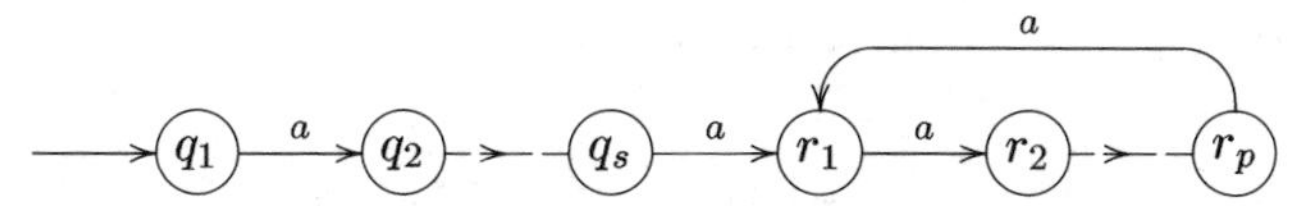

The terminal states therefore form two sets: the terminal states T' that occur in the stem and the terminal states T'' that occur in the cycle. Let X be the set of strings recognised by the stem states: each string in X corresponds to exactly one terminal state T' in the stem. Let T'' consist of n terminal states, which we number 1 to n. For each terminal state i let y_i be the shortest string required to reach it from q_1. Then $y_i(a^p)^*$ is recognised by the automaton for all $1 \leq i \leq n$. Put $Y = \{y_i \colon 1 \leq i \leq n\}$. Then the language recognised by the automaton is $X + Y(a^p)^*$. □

Working over a one-letter alphabet involves the arbitrary choice of what that one letter should be. There is a much more natural way of thinking about languages over such alphabets. There is a bijection τ from A^* to $\mathbb{N}$ given by $x \mapsto |x|$. Using this bijection, we define a subset $L \subseteq \mathbb{N}$ to be recognisable if $\tau^{-1}(L)$ is a recognisable subset of A^*. We shall now describe the recognisable subsets of $\mathbb{N}$ referring only to properties of natural numbers. Recall that an *arithmetic progression* in $\mathbb{N}$ is a sequence of numbers of the form $m + np$ where m and $p \geq 1$ are fixed and $n \in \mathbb{N}$. The number p is called the *period* of the progression.

Theorem 5.2.3 *A subset of the natural numbers is recognisable if and only if it is the union of a finite set, and a finite number of arithmetic progressions all having the same period.*

Proof Let L be a recognisable subset of $\mathbb{N}$. Then

$$\tau^{-1}(L) = X + Y(a^p)^*$$

for some finite sets X and Y and for some natural number p by Theorem 5.2.2. If p is zero, then L is just a finite set, so we can assume that p is not zero in what follows. Now $\tau(X)$ is simply a finite subset $U \subseteq \mathbb{N}$. Put $\tau(Y) = \{n_1, \ldots, n_k\}$. Then $\tau(Y(a^p)^*)$ is equal to the union of the sets $\{n_i + np \colon n \geq 0\}$ where $1 \leq i \leq k$. Thus L is the union of a finite set, and a finite number of arithmetic progressions all having the same period, as required.

Let L be a subset of the natural numbers that is the union of a finite set, and a finite number of arithmetic progressions all having the same period. Arithmetic progressions correspond under τ to regular, and so recognisable, languages of the form $a^m(a^p)^*$ where $p \geq 1$. The union of any finite set of such languages is recognisable, as is their union with a finite set. Thus L is recognisable. □

A subset of $\mathbb{N}$ is said to be *ultimately periodic* if it is the union of a finite set and a finite number of arithmetic progressions all having the same period. The above theorem can therefore be stated in the following terms: the recognisable subsets of the natural numbers are precisely the ultimately periodic ones.

5.3 Kleene's theorem: algorithms

In this section, we shall describe two algorithms that together provide an algorithmic proof of Kleene's theorem: our first algorithm will show explicitly how to construct an ε-automaton from a regular expression, and our second will show explicitly how to construct a regular expression from an automaton.

In the proof below we shall use a class of ε-automata. A *normalised ε-automaton* is just an ε-automaton having exactly one initial state and one terminal state, and the property that there are no transitions into the initial state or out of the terminal state.

Theorem 5.3.1 (Regular expression to ε-automaton) *Let r be a regular expression over the alphabet A. Let m be the sum of the following two numbers: the number of symbols from A occurring in r, counting repeats, and the number of regular operators occurring in r, counting repeats. Then there is an ε-automaton $\mathbf{A}$ having at most $2m$ states such that $L(\mathbf{A}) = L$.*

Proof We shall prove that each regular language is recognised by some normalised ε-automaton satisfying the conditions of the theorem. Base step: prove that if $L = L(r)$ where r is a regular expression without regular operators, then L can be recognised by a normalised ε-automaton with at most 2 states. However, in this case L is either $\{a\}$ where $a \in A$, $\emptyset$, or $\{\varepsilon\}$. The normalised ε-automata, which recognise each of these languages, are

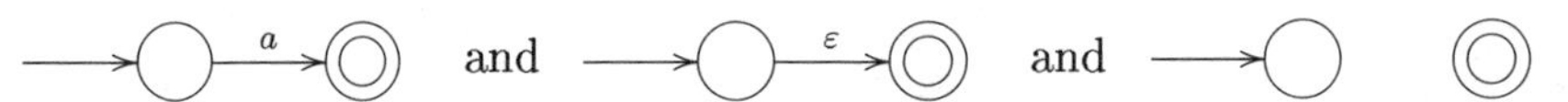

Induction hypothesis: assume that if r is a regular expression, using at most $n-1$ regular operators and containing p occurrences of letters from the underlying alphabet, then $L(r)$ can be recognised by a normalised ε-automaton using at most $2(n-1)+2p$ states. Now let r be a regular expression having n regular operators and q occurrences of letters from the underlying alphabet. We shall prove that $L(r)$ can be recognised by a normalised ε-automaton containing at most $2n+2q$ states. From the definition of a regular expression, r must have one of the following three forms: (1) $r = s + t$, (2) $r = s \cdot t$ or (3) $r = s^*$. Clearly, s and t each use at most $n-1$ regular operators; let n_s and n_t be the number of regular operators occurring in s and t, respectively, and let q_s and q_t be the number of occurrences of letters from the underlying alphabet in s and t, respectively. Then $n_s + n_t = n - 1$ and $q_s + q_t = q$. So by the induction hypothesis $L(s)$ and $L(t)$ are recognised by normalised ε-automata $\mathbf{A}$ and $\mathbf{B}$, respectively, which have at most $2(n_s + q_s)$ and $2(n_t + q_t)$ states apiece. We can picture these as follows:

A and **B**

We now show how **A** and **B** can be used to construct ε-automata to recognise each of the languages described by the regular expressions (1), (2), and (3), respectively; our constructions are just mild modifications of the ones used in Theorem 4.2.1.

(1) A normalised ε-automaton recognising $L(s + r)$ is

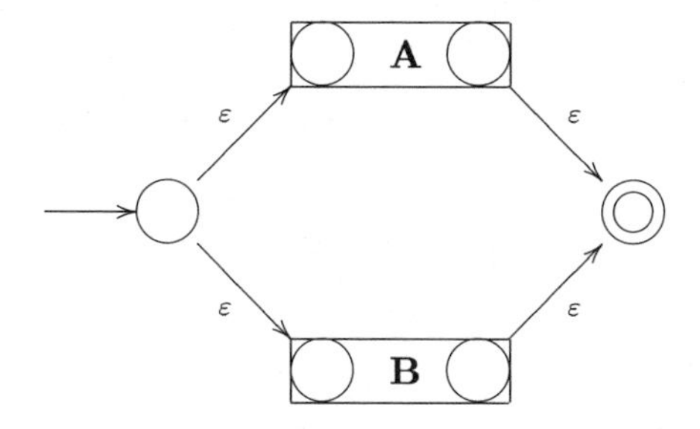

(2) A normalised ε-automaton recognising $L(s \cdot t)$ is

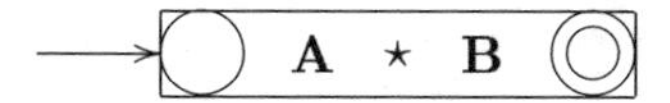

This automaton is obtained by merging the terminal state of **A** with the initial state of **B** and making the resulting state, marked with a $\star$, an ordinary state.

(3) A normalised ε-automaton recognising $L(s^*)$ is

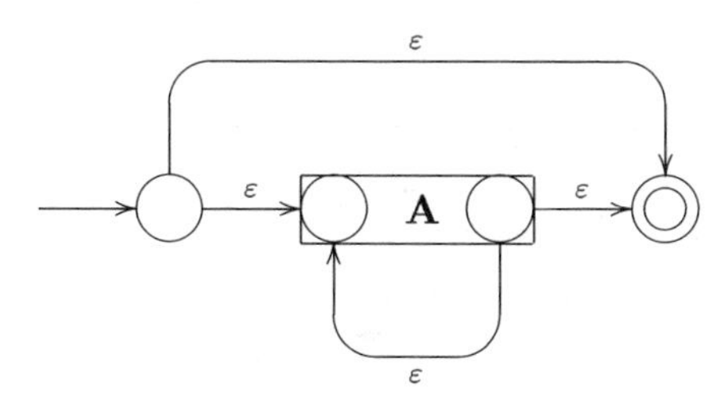

In all three cases, the number of states in the resulting machine is at most

$$2(n_s + q_s) + 2(n_t + q_t) + 2 = 2(n + q),$$

which is the answer required. □

Example 5.3.2 Here is an example of Theorem 5.3.1 in action. Consider the regular expression $01^* + 0$. To construct an ε-automaton recognising the language described by this regular expression, we begin with the two automata

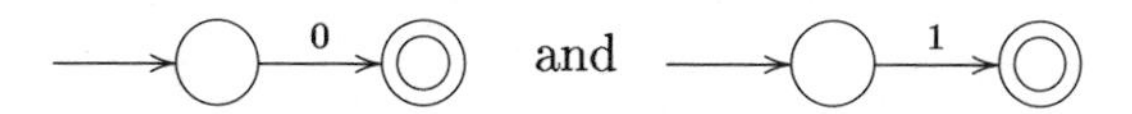

We convert the second machine into one recognising 1^*:

We then combine this machine with our first to obtain a machine recognising 01^*:

We now combine this machine with the one for 0 to obtain the following machine respresenting $01^* + 0$:

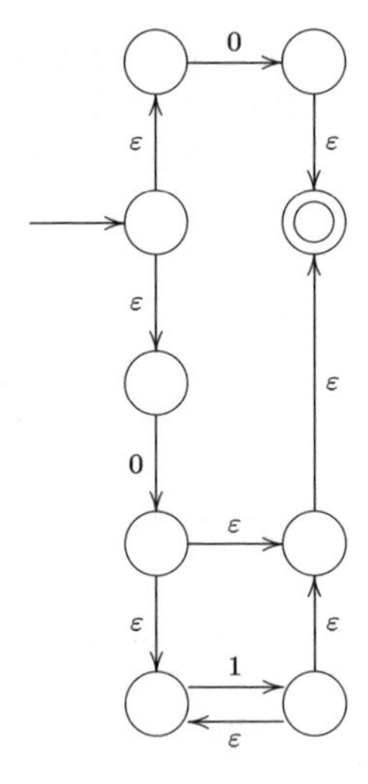

□

We shall now show how to construct a regular expression from an automaton. To do so, it is convenient to introduce a yet more general class of automata than even the ε-automata.

A *generalised automaton* over the input alphabet A is the same as an ε-automaton except that we allow the transitions to be labelled by arbitrary regular expressions over A. The language $L(\mathbf{A})$ recognised by a generalised automaton $\mathbf{A}$ is defined as follows. Let $x \in A^*$. Then $x \in L(\mathbf{A})$ if there is a path in $\mathbf{A}$, which begins at one of the initial states, ends at one of the terminal states, and whose labels are, in order, the regular expressions $r_1, \ldots, r_n$, such that x can be factorised $x = x_1 \ldots x_n$ in such a way that each $x_i \in L(r_i)$. The definition of $L(\mathbf{A})$ generalises the definition of the language recognised by an ε-automaton. To use generalised automata to find the regular expression describing the language recognised by an automaton, we shall need the following. A *normalised (generalised) automaton* is a generalised automaton with exactly one initial state, which I shall always label α, and exactly one terminal state, which I shall always label ω; in addition, there are no transitions into α nor any transitions out of ω. A normalised generalised automaton is therefore a substantial generalisation of a normalised ε-automaton used in the proof of Theorem 5.3.1. Every generalised automaton can easily be normalised in such a way that the language is not changed: adjoin a new initial state with transitions labelled ε pointing at the old initial states, and adjoin a new terminal state with transitions from each of the old terminal states labelled ε pointing at the new terminal state.

Terminology For the remainder of this section, 'normalised automaton' will always mean a 'normalised generalised automaton'.

The simplest kinds of normalised automata are those with exactly one initial state, one terminal state, and at most one transition:

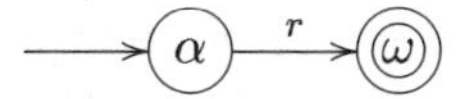

We call such a normalised automaton *trivial*. If a trivial automaton has a transition, then the label of that transition will be a regular expression, and this regular expression obviously describes the language accepted by the automaton; if there is no transition then the language is $\emptyset$.

We shall describe an algorithm that converts a normalised automaton into a trivial normalised automaton recognising the same language. The algorithm depends on three operations, which may be carried out on a normalised automaton that we now describe.

(T) transition elimination Given any two states p and q, where $p = q$ is not excluded, all the transitions from p to q can be replaced by a single transition by applying the following rule:

$$p \xrightarrow{r} q,\quad p \xrightarrow{s} q \quad \Rightarrow \quad p \xrightarrow{r+s} q$$

In the case where $p = q$, this rule takes the following form:

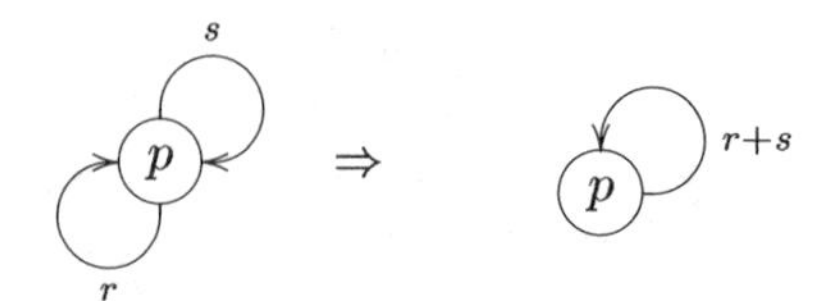

(L) loop elimination Let q be a state that is neither α nor ω, and suppose this state has a single loop labelled r. If there are no transitions entering this state or no transitions leaving this state, then q can be erased together with any transitions from it or any transitions to it. We may therefore restrict our attention to the case where q has at least one transition entering it and at least one transition leaving it. In this case, the loop at q can be erased, and for each transition leaving q labelled by s we change the label to r^*s. This operation is summarised in the diagram below:

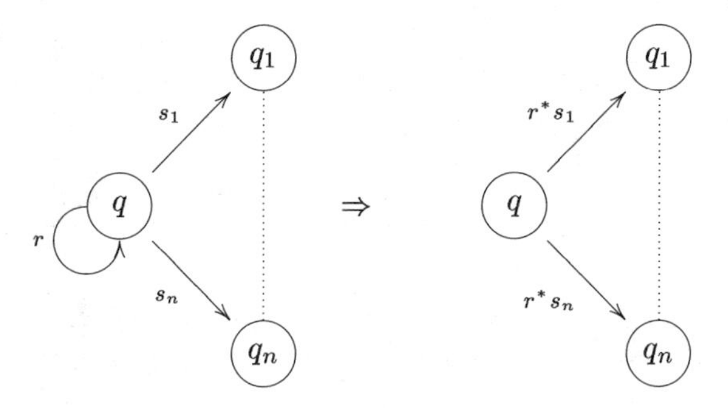

(S) state elimination Let q be a state that is neither α nor ω and that has no loop. If there are no transitions entering this state or no transitions leaving this state, then q can be erased together with any transitions from it or any transitions to it. We may therefore restrict our attention to the case where q has at least one transition entering it and at least one transition leaving it. In this case, we do the following: for each transition $p' \xrightarrow{r} q$ and for each transition $q \xrightarrow{s} p''$, both of which I shall call 'old,' we construct a 'new' transition $p' \xrightarrow{rs} p''$. At the end of this process the state q and all the old transitions are erased. This operation is summarised in the diagram below:

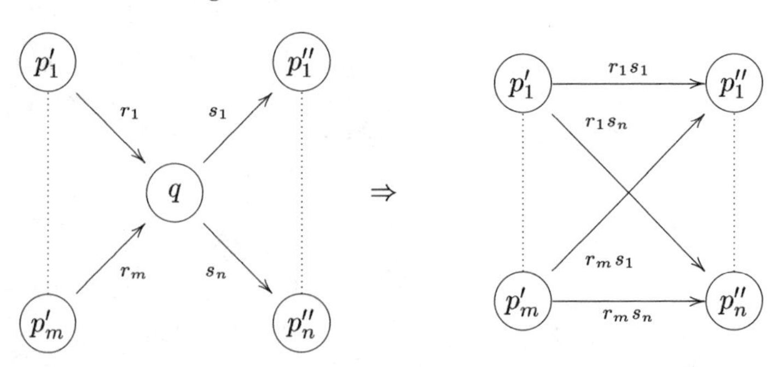

Lemma 5.3.3 *Let* **A** *be a normalised automaton, and let* **B** *be the normalised automaton that results when one of the rules* (T), (L) *or* (S) *is applied. Then* $L(\mathbf{B}) = L(\mathbf{A})$.

Proof We simply check each case in turn. This is left as an exercise. □

These operations are the basis of the following algorithm.

Algorithm 5.3.4 (Automaton to regular expression) The input to this algorithm is a normalised automaton **A**. The output is a regular expression r such that $L(r) = L(\mathbf{A})$.

Repeat the following procedure until there are only two states and at most one transition between them, at which point the algorithm terminates.

Procedure: repeatedly apply rule (T) if necessary until the resulting automaton has the property that between each pair of states there is at most one transition; now repeatedly apply rule (L) if necessary to eliminate all loops; finally, apply rule (S) to eliminate a state.

When the algorithm has terminated, a regular expression describing the language recognised by the original machine is given by the label of the unique transition, if there is a transition, otherwise the language is the empty set. □

Example 5.3.5 Consider the automaton:

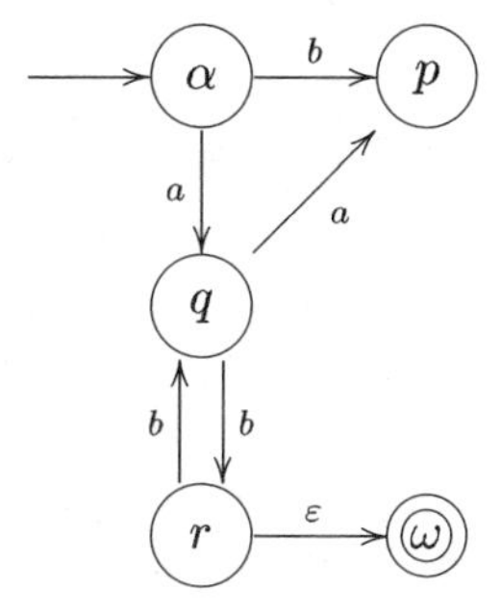

The rules (T) and (L) are superfluous here, so we shall go straight to rule (S) applied to the state q. The pattern of incoming and outgoing transitions for this state is

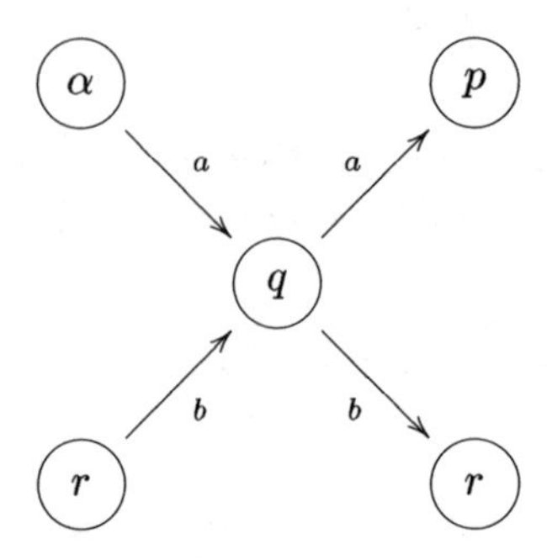

If we now apply rule (S), we obtain the following pattern of transitions:

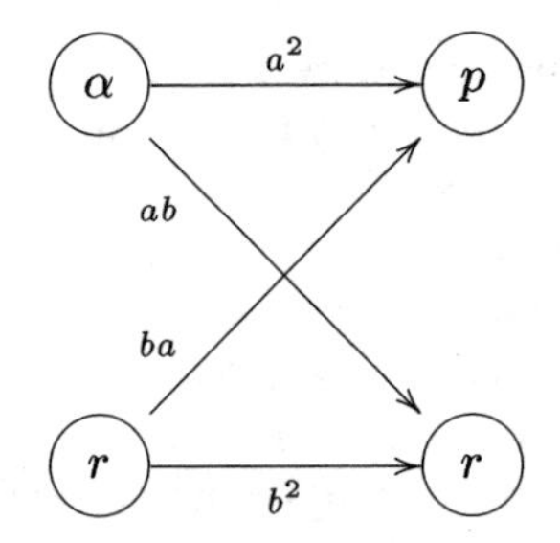

The resulting generalised automaton is therefore

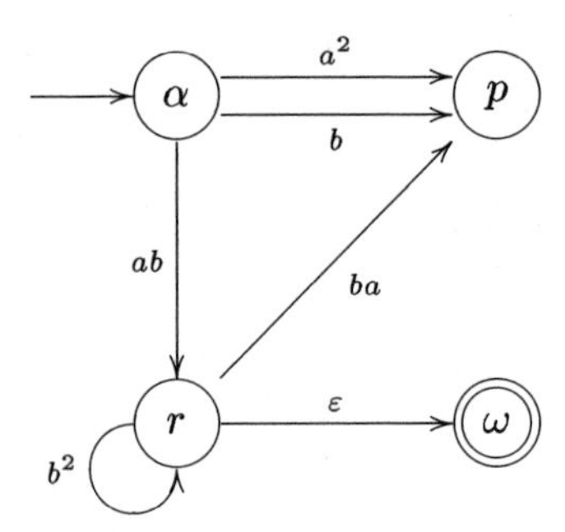

If we apply the rules (T) and (L) to this generalised automaton we get

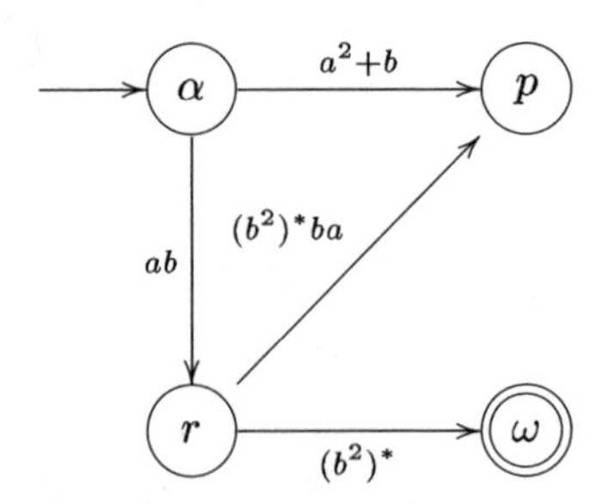

We can now eliminate the vertices p and r in turn. As a result, we end up with the following trivial generalised automaton:

$$\rightarrow (\alpha) \xrightarrow{ab(b^2)^*} ((\omega))$$

Thus the language recognised by our original machine is $ab(b^2)^*$. □

We now prove that Algorithm 5.3.4 is correct.

Theorem 5.3.6 *Algorithm 5.3.4 computes a regular expression for the language recognised by a normalised automaton.*

Proof Lemma 5.4.3 tells us that the three operations we apply do not change the language, so we need only prove that the algorithm will always lead to a trivial normalised automaton. Each application of the procedure reduces the number of states by one. In addition, none of the three rules can ever lead to a loop appearing on α or ω. The result is now clear. □

Exercises 5.3

1. For each of the regular expressions below, construct an ε-automaton recognising the corresponding language using Theorem 5.3.1. The alphabet in question is $A = \{a, b\}$.

(i) $ab + ba$.

(ii) $a^2 + b^2 + ab$.

(iii) $a + b^*$.

2. Convert each of the following automata **A** into a normalised automaton and then use Algorithm 5.3.4 to find a regular expression describing $L(\mathbf{A})$.

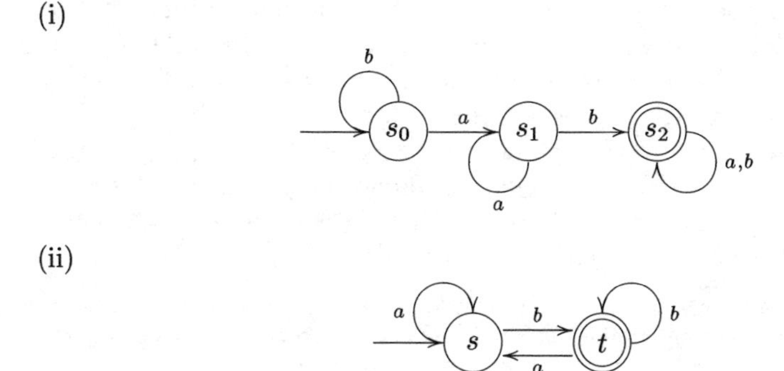

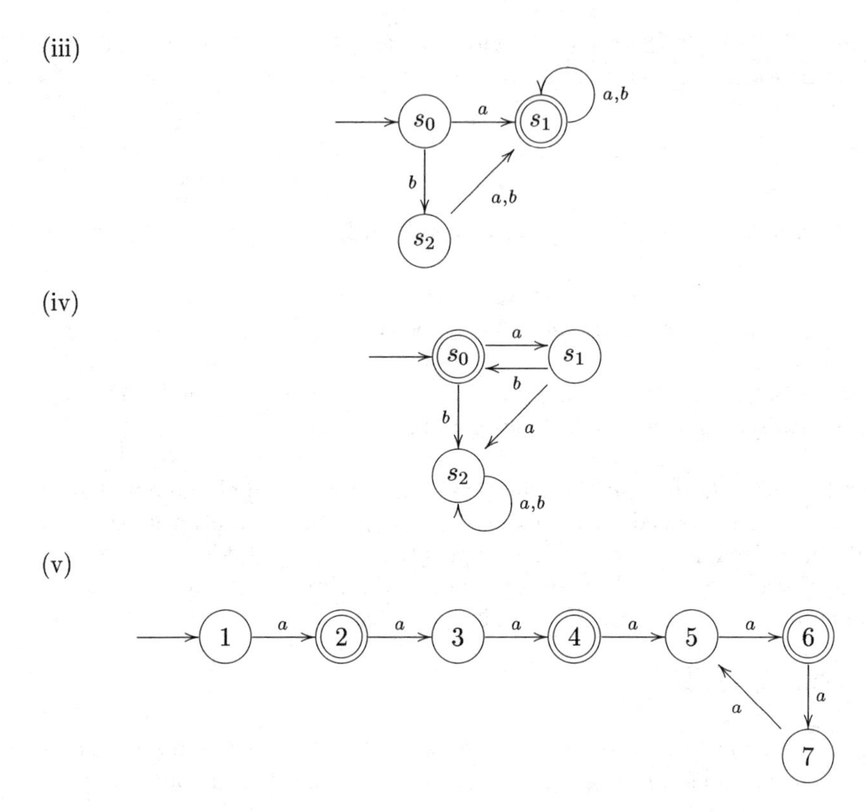

3. Prove Lemma 5.3.3.

5.4 Language equations

In this section, we describe an algebraic way of proving that every recognisable language is regular.

Let A be a fixed alphabet. In Section 5.1, we described some of the algebraic properties of the two operations $+$ and $\cdot$ on $\mathsf{P}(A^*)$. Apart from the fact that $L + L = L$, these properties are reminiscent of the behaviour of real numbers with respect to addition and multiplication, where 0 plays the role of $\emptyset$ and 1 plays the role of $\{\varepsilon\}$. The main difference is that we have no notion of subtraction for languages that corresponds to subtraction of numbers.[2]

In ordinary algebra, we replace numbers by letters. Typically, we have to solve algebraic equations: this means that an unknown number x is described by means of some equations and we have to find out what x is. For example, if the number x satisfies $3x = 6$, then you can easily see that in fact $x = 2$. The

[2] The set $A \setminus B$ is sometimes denoted by $A - B$, which makes it look as if it might be the language-theoretic version of subtraction. It is true, for example, that $A - A = \emptyset$. However, $(A - B) + B$ is *not* equal to A.

most general type of (linear) equation for numbers is of the form $a = bx + c$ where b and c are known and x is to be determined. This is very easily solved: if $b \neq 0$ then $x = (a - c)/b$; if $b = 0$ then x can assume any value, provided $a = c$; if $b = 0$ and $a \neq c$ then there is no solution.

We shall now consider equations where the unknown is a language, and we shall try to find out what that language is. The most general (linear) equation for languages is

$$X = CX + R$$

where $C, R \in \mathsf{P}(A^*)$ are known languages and X is our unknown language that we would like to find. We call such an equation a *right linear* equation. The word 'right' refers to the fact that the unknown X appears on the right-hand side of the coefficient C. As always, order matters so writing XC instead of CX is wrong.

The only odd feature of the above equation is that X appears on both sides. In ordinary algebra, this is unnecessary because if $x = ax + b$, then we can subtract x from both sides to obtain $0 = (a - 1)x + b$, which is just $-b = (a - 1)x$, and this has the form $b' = a'x$. But as we said above, for languages there is no operation that behaves like subtraction for numbers.

Examples 5.4.1 To get a feel for right linear language equations, we describe two simple examples.

(1) Suppose X is a language over the alphabet $\{0, 1\}$ such that

$$X = 1X + \varepsilon.$$

What is X? Remember that the equation really means

$$X = \{1\} \cdot X + \{\varepsilon\}.$$

First, any solution L must contain ε, because

$$L = \{1\} \cdot L + \{\varepsilon\}$$

and so $\{\varepsilon\} \subseteq L$. Second, because $\{1\} \cdot L \subseteq L$ then whenever $x \in L$ we must have $1x \in L$. Since $\varepsilon \in L$ it follows that $1 \in L$. But $1 \in L$ implies $1^2 \in L$, and so on. Because ε, 1, 1^2, 1^3 ... all belong to L it follows that $1^* \subseteq L$. We are therefore led to guess that 1^* is *a* solution. We can verify this directly:

$$1 \cdot 1^* + \varepsilon = 1^*.$$

In fact, in this case, there is no other solution.

(2) Now suppose Y is a language over the same alphabet such that $Y = \varepsilon Y + 0$. Then $\{0\}$ is a solution, but so too is $\{0, 1\}$. In fact, any subset of $(0+1)^*$ that contains 0 as an element is a solution, and every solution has this property.

□

The two examples above illustrate the two ways that a right linear equation can behave: there is either exactly one solution or infinitely many solutions all of which contain a unique smallest solution. The following theorem contains all we need to know about solving right linear equations.

Theorem 5.4.2 (Arden) *Let $X = CX + R$, where C, R are languages over the alphabet A. Then*

(i) *C^*R is a solution.*

(ii) *If Y is any solution then $C^*R \subseteq Y$.*

(iii) *If $\varepsilon \notin C$ then C^*R is the unique solution.*

Proof (i) To prove this, we substitute C^*R into the right-hand side and then check that we get the left-hand side. Thus

$$CX + R = C(C^*R) + R = CC^*R + R = (CC^* + \varepsilon)R = C^*R.$$

(ii) Let Y be any solution to $X = CX + R$. Thus $Y = CY + R$. Hence both $R \subseteq Y$ and $CY \subseteq Y$. Now from these inclusions, we have that $CR \subseteq CY \subseteq Y$. Thus $CR \subseteq Y$. From this, we obtain $C^2R \subseteq CY \subseteq Y$. Thus $C^2R \subseteq Y$. By an induction argument, we have that $C^nR \subseteq Y$ for all $n \in \mathbb{N}$. Hence $C^*R \subseteq Y$, as required.

(iii) Suppose that $\varepsilon \notin C$, and let W be any solution to $X = CX + R$. We shall prove that $W = C^*R$. By (ii), we certainly have that $C^*R \subseteq W$. Thus we have to show that $W \setminus C^*R$ is empty. Assume for the sake of argument that $W \setminus C^*R \neq \emptyset$. Let $z \in W \setminus C^*R$ be of smallest possible length. Thus by design $z \in W$ but $z \notin C^*R$. Because $R \subseteq C^*R$, it is immediate that $z \notin R$. By assumption, $W = CW + R$ and so $z \in CW + R$. But $z \notin R$ and so $z \in CW$. It follows that there exist $c \in C$ and $w \in W$ such that $z = cw$. Now we invoke our assumption that $\varepsilon \notin C$. Thus $c \neq \varepsilon$. It follows that $|w| < |z|$. We claim that $w \notin C^*R$. Suppose on the contrary that $w \in C^*R$. Then

$$z = cw \in C(C^*R) \subseteq C^*R.$$

But this contradicts the assumption that $z \notin C^*R$. Hence $w \notin C^*R$. But $w \in W$. Thus $w \in W \setminus C^*R$ and $|w| < |z|$, contradicting the choice of z. We conclude that $W \setminus C^*R$ is empty and so $W = C^*R$. □

The above results explain the phenomena we observed in Examples 5.4.1. In the first example, $C = \{1\}$ does not contain ε and so the equation has a unique solution. In the second example, $C = \{\varepsilon\}$ obviously does contain ε and consequently the equation does not have a unique solution.

Arden's theorem tells us how to solve one equation in one unknown. We shall show later how it can be used to solve n equations in n unknowns. First we need to define what we mean by 'right linear equations in many unknowns.'

Let A be an alphabet. A *set of n right linear equations in n unknowns* X_i, where $i = 1, \ldots, n$, is a set of equations of the form,

$$\begin{aligned} X_1 &= C_{11}X_1 + \ldots + C_{1n}X_n + R_1 \\ &\vdots \\ X_n &= C_{n1}X_1 + \ldots + C_{nn}X_n + R_n, \end{aligned}$$

where the C_{ij} and R_i are languages over A. These equations can be put into matrix form, which is sometimes more convenient: let $\mathbf{X}$ be the $n \times 1$ matrix,

$$\begin{pmatrix} X_1 \\ \vdots \\ X_n \end{pmatrix},$$

and let $\mathbf{R}$ be the $n \times 1$ matrix,

$$\begin{pmatrix} R_1 \\ \vdots \\ R_n \end{pmatrix},$$

and finally let $\mathbf{C}$ be the $n \times n$ matrix,

$$\begin{pmatrix} C_{11} & \ldots & C_{1n} \\ \vdots & & \vdots \\ C_{n1} & \ldots & C_{nn} \end{pmatrix}.$$

Then our set of n equations in n unknowns takes the form

$$\mathbf{X} = \mathbf{CX} + \mathbf{R}.$$

Systems of equations such as this can be constructed from any automaton. Let

$$\mathbf{A} = (S = \{s_1, \ldots, s_n\}, A = \{a_1, \ldots a_r\}, s_1, \delta, T)$$

be an automaton. For each $i = 1, \ldots, n$ define

$$X_i = \{x \in A^*: s_i \cdot x \in T\};$$

that is, X_i is the language recognised by the automaton (S, A, s_i, δ, T). Observe that $X_1 = L(\mathbf{A})$. The languages $X_1, \ldots, X_n$ are the principal actors in the following algorithm.

Algorithm 5.4.3 (Equations associated with a finite automaton) Let

$$\mathbf{A} = (S = \{s_1, \ldots, s_n\}, A = \{a_1, \ldots a_r\}, s_1, \delta, T)$$

be a finite automaton. Let $\mathbf{C}$ be the $n \times n$-matrix defined as follows: $\mathbf{C}_{i,j}$ is the union of all the $a \in A$ such that $s_i \cdot a = s_j$. Let $\mathbf{R}$ be the $n \times 1$-matrix defined as follows:

$$\mathbf{R}_i = \begin{cases} \emptyset & \text{if } s_i \text{ is not terminal} \\ \varepsilon & \text{if } s_i \text{ is terminal.} \end{cases}$$

The *equations associated with* $\mathbf{A}$ are $\mathbf{X} = \mathbf{CX} + \mathbf{R}$. □

Notice that the matrix C is related to the adjacency matrix of the underlying directed graph: in the adjacency matrix the entry in the ith row and jth column would give the number of arrows from i to j, whereas the entry C_{ij} is the union of the labels of the transitions from i to j. Before we justify our algorithm we give an example.

Example 5.4.4 We return to our first example of an automaton: Example 1.5.1.

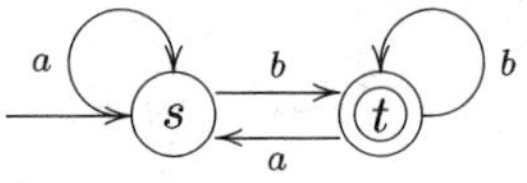

a b b
s t
a

This machine has two states, and so there will be two unknowns:

$$X = \{x \in (a+b)^* \colon s \cdot x = t\} \text{ and } Y = \{x \in (a+b)^* \colon t \cdot x = t\}.$$

The language accepted by this machine is X. The matrix of coefficients $\mathbf{C}$ derived from this automaton is

$$\begin{pmatrix} a & b \\ a & b \end{pmatrix}.$$

The matrix $\mathbf{R}$ encoding terminal and non-terminal states is

$$\begin{pmatrix} \emptyset \\ \varepsilon \end{pmatrix}.$$

Thus the equations associated with the automaton in matrix form are

$$\begin{pmatrix} X \\ Y \end{pmatrix} = \begin{pmatrix} a & b \\ a & b \end{pmatrix} \begin{pmatrix} X \\ Y \end{pmatrix} + \begin{pmatrix} \emptyset \\ \varepsilon \end{pmatrix}.$$

That is,

$$\begin{aligned} X &= aX + bY \\ Y &= aX + bY + \varepsilon. \end{aligned}$$

□

We now justify Algorithm 5.4.3.

Theorem 5.4.5 *Let*

$$\mathbf{A} = (S = \{s_1, \ldots, s_n\}, A = \{a_1, \ldots a_r\}, s_1, \delta, T)$$

be an automaton. For each $i = 1, \ldots, n$ *define*

$$X_i = \{x \in A^*: s_i \cdot x \in T\}.$$

(i) *Let* s_i *be a non-terminal state. Suppose that* $s_i \cdot a_j = s_{i_j}$*, where* $j = 1, \ldots, r$. *Then*

$$X_i = a_1 X_{i_1} + \ldots + a_r X_{i_r}.$$

(ii) *Let* s_i *be a terminal state. Suppose that* $s_i \cdot a_j = s_{i_j}$*, where* $j = 1, \ldots, r$. *Then*

$$X_i = a_1 X_{i_1} + \ldots + a_r X_{i_r} + \varepsilon.$$

Proof (i) Let $x \in X_i$. Then by definition, $s_i \cdot x \in T$. Since s_i is not a terminal state, the string x has at least one symbol. Thus we can write $x = ay$ where $a \in A$ and $y \in A^*$. We must have $a = a_j$ for some $j = 1 \ldots r$. Thus

$$s_i \cdot x = s_i \cdot (a_j y) = (s_i \cdot a_j) \cdot y = s_{i_j} \cdot y \in T.$$

It follows that $y \in X_{i_j}$ and so $x = a_j y \in a_j X_{i_j}$. We have therefore shown that

$$X_i \subseteq a_1 X_{i_1} + \ldots + a_r X_{i_r}.$$

To prove the reverse inclusion, let $x \in a_1 X_{i_1} + \ldots + a_r X_{i_r}$. Say $x \in a_j X_{i_j}$. Then $x = a_j y$ where $y \in X_{i_j}$. By definition $s_{i_j} \cdot y \in T$. Thus

$$s_i \cdot x = s_i \cdot (a_j y) = s_{i_j} \cdot y \in T.$$

Hence $x \in X_i$. This proves the result.

(ii) The proof of this case is the same as (i), except that if s_i is terminal, the empty string will take s_i to a terminal state, namely s_i itself. □

Thus from a deterministic automaton with n states, we obtain n right linear equations in n unknowns. One of these unknowns is the language recognised by the automaton. The algorithm for solving these equations makes repeated use of Arden's Theorem. The condition we impose on the coefficients of the matrix is one that automatically holds for the equations arising from deterministic automata.

Algorithm 5.4.6 (Solving systems of right linear equations) Let

$$\mathbf{X} = \mathbf{C}\mathbf{X} + \mathbf{R}$$

be a set of n right linear equations in n unknowns $X_1, \ldots, X_n$. Assume in addition that no entry of $\mathbf{C}$ contains ε. The algorithm computes the value of X_1, and then the values of $X_2, \ldots, X_n$ in turn.

The algorithm works iteratively to successively remove each unknown X_n, $X_{n-1}, \ldots, X_2$. Thus at the end of the first stage, we have $n-1$ equations in the unknowns $X_1 \ldots, X_{n-1}$; at the end of the second stage, we have $n-2$ equations in the unknowns $X_1 \ldots, X_{n-2}$; and so forth, until at the end of the final stage we have one equation in the unknown X_1. Once we have one equation in X_1 we can apply Arden's Theorem to obtain an explicit description of X_1. The values of $X_2, \ldots, X_n$ can then be calculated in turn by back substitution.

The key procedure is as follows:

- The last equation in the system has the form

$$X_n = C_{n1}X_1 + \ldots + C_{nn}X_n + R_n.$$

- Rearrange this equation to obtain

$$X_n = C_{nn}X_n + (C_{n1}X_1 + \ldots + C_{n,n-1}X_{n-1} + R_n).$$

- Apply Arden's Theorem to express X_n in terms of $X_1, \ldots, X_{n-1}$:

$$X_n = C_{nn}^*(C_{n1}X_1 + \ldots + C_{n,n-1}X_{n-1} + R_n).$$

- Substitute this value for X_n into the preceding $n-1$ equations, to obtain $n-1$ equations in $n-1$ unknowns, the coefficient matrix of which has the property that no entry contains ε.

□

Before we justify this algorithm we work through two examples.

Example 5.4.7 We return to the two equations we obtained from Example 5.4.4:

$$X = aX + bY, \quad (5.1)$$
$$Y = aX + bY + \varepsilon. \quad (5.2)$$

To solve these equations using Algorithm 5.4.6, we take the last equation first:

$$Y = aX + bY + \varepsilon,$$

and rearrange it:

$$Y = bY + (aX + \varepsilon).$$

We now apply Arden's Theorem to express Y in terms of X:

$$Y = b^*(aX + \varepsilon).$$

We now substitute this expression for Y into equation (5.1) to obtain

$$X = aX + b(b^*(aX + \varepsilon)),$$

which we rearrange to obtain

$$X = (a + bb^*a)X + bb^*.$$

By Arden's Theorem we obtain the following explicit description of X:

$$X = (a + bb^*a)^*bb^*.$$

This is the language recognised by our automaton. We originally found that the language recognised by this automaton was $(a + b)^*b$. It follows that

$$(a + b)^*b = (a + bb^*a)^*bb^*.$$

□

In our next example, we consider three equations in three unknowns.

Example 5.4.8 Consider the following three equations in X, Y, and Z below:

$$X = aX + bY + aZ, \quad (5.3)$$
$$Y = bX + aY + aZ, \quad (5.4)$$
$$Z = aX + aY + bZ + \varepsilon. \quad (5.5)$$

Our aim is to find an explicit expression for Z. We take the last equation first:

$$Z = aX + aY + bZ + \varepsilon$$

and rearrange it:

$$Z = bZ + (aX + aY + \varepsilon).$$

Apply Arden's Theorem to obtain a description of Z in terms of X and Y:

$$Z = b^*(aX + aY + \varepsilon).$$

Substitute this expression for Z into equations (5.3) and (5.4):

$$X = aX + bY + a(b^*(aX + aY + \varepsilon)), \quad (5.6)$$
$$Y = bX + aY + a(b^*(aX + aY + \varepsilon)). \quad (5.7)$$

Rearrange these equations to obtain

$$X = (a + ab^*a)X + (b + ab^*a)Y + ab^*, \quad (5.8)$$
$$Y = (b + ab^*a)X + (a + ab^*a)Y + ab^*. \quad (5.9)$$

We started with three equations and three unknowns X, Y, Z; we now have two equations in two unknowns X, Y together with an expression for Z in terms of X and Y. We can apply our procedure again to these two equations to obtain one equation in one unknown X and two expressions: one of Y in terms of X, and one of Z in terms of X and Y. □

We now justify Algorithm 5.4.6.

Theorem 5.4.9 *Let* $\mathbf{X} = \mathbf{CX} + \mathbf{R}$ *be a set of* n *right linear equations in* n *unknowns. Assume in addition that no entry of* $\mathbf{C}$ *contains* ε. *Then the equations have a unique solution. If, in addition, all the entries of* $\mathbf{C}$ *and* $\mathbf{R}$ *are regular, then the solution is regular.*

Proof Before we present the formal proof of the first part of the theorem, we describe the idea that lies behind it. This is just a re-iteration of Algorithm 5.4.6. Let $\mathbf{X} = \mathbf{CX} + \mathbf{R}$ be a set of n right linear equations in n unknowns such that each entry of $\mathbf{C}$ does not contain ε. The last equation in the system has the form

$$X_n = C_{n1}X_1 + \ldots + C_{nn}X_n + R_n.$$

By rearranging this equation, we obtain

$$X_n = C_{nn}X_n + (C_{n1}X_1 + \ldots + C_{n,n-1}X_{n-1} + R_n).$$

If solutions to $X_1, \ldots, X_{n-1}$ existed, *then* we could use Arden's theorem to find X_n:

$$X_n = C_{nn}^*(C_{n1}X_1 + \ldots + C_{n,n-1}X_{n-1} + R_n).$$

If we substitute this value for X_n into the preceding $n-1$ equations, we obtain $n-1$ equations in $n-1$ unknowns, whose coefficient matrix has the property that no entry contains ε. Iterating this procedure, we would eventually obtain one equation in X_1. By Arden's theorem we could solve the equation explicitly. The values of $X_2, \ldots, X_n$ could then be explicitly calculated in turn. We now have to make this idea precise.

Let $P(n)$ denote the following statement:

> "Every set of n right linear equations in n unknowns has a unique solution if the coefficient matrix has the property that no entry contains the empty string."

The statement $P(1)$ is true by Theorem 5.4.2(iii). As our induction hypothesis, assume that $P(n-1)$ is true; we shall prove that $P(n)$ is true. Let $\mathbf{X} = \mathbf{CX}+\mathbf{R}$ be a set of n right linear equations in n unknowns such that each entry of $\mathbf{C}$ does not contain ε. We associate with this system a set of $n-1$ equations in $n-1$ unknowns as follows: let

$$\mathbf{X}' = \begin{pmatrix} X_1 \\ \vdots \\ X_{n-1} \end{pmatrix}$$

and

$$\mathbf{R}' = \begin{pmatrix} R_1 + C_{1n}C_{nn}^*R_n \\ \vdots \\ R_{n-1} + C_{n-1,n}C_{nn}^*R_n \end{pmatrix},$$

and finally let

$$\mathbf{C}' = \begin{pmatrix} C_{11} + C_{1n}C_{nn}^*C_{n1} & \dots & C_{1,n-1} + C_{1n}C_{nn}^*C_{n,n-1} \\ \vdots & & \vdots \\ C_{n-1,1} + C_{n-1n}C_{nn}^*C_{n1} & \dots & C_{n-1,n-1} + C_{n-1,n}C_{nn}^*C_{n,n-1} \end{pmatrix}.$$

Observe that no coefficient of $\mathbf{C}'$ contains ε. By the induction hypothesis, the system $\mathbf{X}' = \mathbf{C}'\mathbf{X}' + \mathbf{R}'$ has a unique solution:

$$X_1 = L_1, \dots, X_{n-1} = L_{n-1}.$$

Define

$$L_n = C_{nn}^*(C_{n1}L_1 + \dots + C_{n,n-1}L_{n-1} + R_n).$$

We claim that

$$X_1 = L_1, \dots, X_n = L_n$$

is a solution to $\mathbf{X} = \mathbf{C}\mathbf{X} + \mathbf{R}$. First we calculate the ith row of the right-hand side,

$$C_{i1}L_1 + \dots + C_{in}L_n + R_i,$$

for $1 \leq i \leq n-1$. This is achieved by replacing L_n by

$$C_{nn}^*(C_{n1}L_1 + \dots + C_{n,n-1}L_{n-1} + R_n)$$

and then collecting together like terms. By using the ith row of the equation $\mathbf{X} = \mathbf{C}\mathbf{X} + \mathbf{R}$, it is easy to check that we obtain L_i, as required. It remains to calculate

$$C_{n1}L_1 + \dots + C_{nn}L_n + R_n.$$

As before, replace L_n by

$$C_{nn}^*(C_{n1}L_1 + \dots + C_{n,n-1}L_{n-1} + R_n)$$

and then collect together like terms. The coefficient of L_i is $C_{ni} + C_{nn}C_{nn}^*C_{ni}$, which is equal to

$$(\varepsilon + C_{nn}C_{nn}^*)C_{ni} = C_{nn}^*C_{ni}.$$

Consequently we get L_n, as required.

We have proved that the system $\mathbf{X} = \mathbf{C}\mathbf{X} + \mathbf{R}$ has at least one solution. It remains to prove that it has exactly one solution. Suppose that there were two solutions: $L_1, \dots, L_n$ and $M_1, \dots, M_n$. Both $L_1, \dots, L_{n-1}$ and $M_1, \dots, M_{n-1}$ are solutions to $\mathbf{X}' = \mathbf{C}'\mathbf{X}' + \mathbf{R}'$ and so by the induction hypothesis

$$L_1 = M_1, \dots, L_{n-1} = M_{n-1}.$$

We can now use the idea introduced at the beginning of the proof. Consider the final equation:

$$X_n = C_{n1}X_1 + \ldots C_{nn}X_n + R_n.$$

We can bracket the terms in the following way:

$$X_n = C_{nn}X_n + (C_{n1}X_1 + \ldots + C_{n,n-1}X_{n-1} + R_n).$$

By assumption, $\varepsilon \notin C_{nn}$ and $X_1 = L_1, \ldots, X_{n-1} = L_{n-1}$. Thus by Arden's theorem the equation has the unique solution,

$$X_n = C_{nn}^*(C_{n1}L_1 + \ldots + C_{n,n-1}L_{n-1} + R_n),$$

and so $L_n = M_n$, as required.

The second part of the theorem follows immediately because we only use regular operators to calculate the solution. □

Exercises 5.4

1. For each automaton **A** below, write down the corresponding set of language equations, solve them, and so find a regular expression for the language recognised by **A**.

(i)

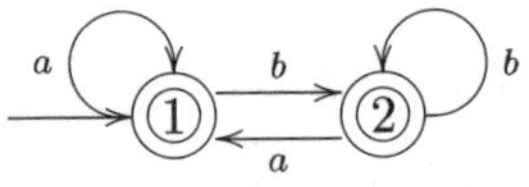

(ii)

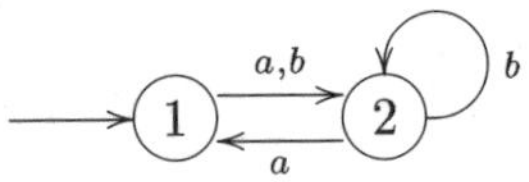

(iii)

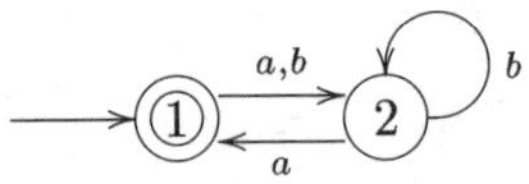

(iv)

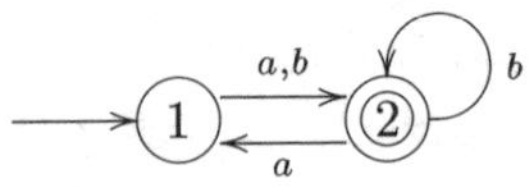

(v)

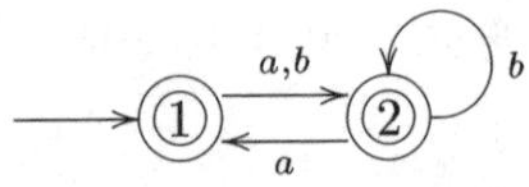

(vi)

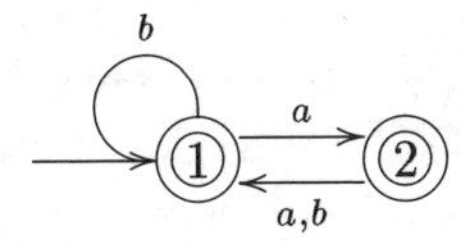

2. For each of the following sets of language equations write down the automaton that produced the equations (assume that X corresponds to the initial state). Also find a regular expression for X in each case.

(i) $X = (a + b)Y + \varepsilon$, $Y = (a + b)Z$, $Z = (a + b)X$.

(ii) $X = aX + bY$, $Y = aY + bZ$, $Z = aZ + bX + \varepsilon$.

(iii) $X = bY + aZ$, $Y = bY + aX + \varepsilon$, $Z = aZ + bX$.

5.5 Summary of Chapter 5

- *Regular expressions*: Let A be an alphabet. A regular expression is constructed from the symbols ε, $\emptyset$ and a, where $a \in A$, together with the symbols $+, \cdot$, and * and left and right brackets according to the following rules: ε, $\emptyset$, and a are regular expressions, and if s and t are regular expressions so are $(s + t)$, $(s \cdot t)$ and (s^*).

- *Regular languages*: Every regular expression r describes a language $L(r)$. A language is regular if it can be described by a regular expression.

- *Kleene's theorem*: A language is recognisable if and only if it is regular. The proof of this theorem can be made algorithmic: there is an algorithm that constructs an ε-automaton $\mathbf{A}$ from each regular expression r such that $L(\mathbf{A}) = L(r)$; there is also an algorithm that constructs a regular expression r from an automaton $\mathbf{A}$ by graphical means such that $L(r) = L(\mathbf{A})$. There is another algorithm that associates with each automaton $\mathbf{A}$ with n states n right linear equations in n unknowns. These equations may be solved to find a regular expression r such that $L(\mathbf{A}) = L(r)$.

5.6 Remarks on Chapter 5

Both [66] and [75] give fuller accounts of regular expressions than I have done at about the same level as this book. Kleene's Theorem was first stated and proved in [74]. My proof that the language recognised by a non-deterministic automaton is regular is taken from [51]. The two algorithms of Section 5.3 are due to McNaughton and Yamada [86] and Brzozowski and McCluskey [20], respectively. I learnt about the second algorithm from [16]. The theory of language equations is due to Arden [7].

Regular expressions form a precise notation for describing languages with the added advantage over automata that they can be written as strings on the line. Furthermore, they can be implemented: for example, by means of the `egrep` command in UNIX (see Section 4.3 of [44], or Section 3.3 of [66] for example). A detailed account of how regular expressions can be used in solving practical programming problems may be found in [47]. See also the bibliographic notes on pages 157 and 158 of [1]. Finite automata and regular expressions are also useful in information extraction. See [72] pages 577–583.

The problem of converting a regular expression into an automaton is an important one. In this book, we describe two further such algorithms: one in Chapter 6 via linearisation and local automata due to Berry and Sethi [11]; and the other is the 'Method of Quotients' described in Chapter 7 due to Brzozowski [19]. More about algorithms to convert automata into regular expressions can be found in [17], and further references in [13].

Chapter 6

Local languages

In Chapter 5, we showed how to construct an ε-automaton $\mathbf{A}$ from a regular expression r. To complete this process, and find a deterministic automaton recognising $L(r)$, we need to convert $\mathbf{A}$ into a non-deterministic automaton and then apply the accessible subset construction. The main aim of this chapter is to show how to construct a non-deterministic automaton directly from the regular expression r.

6.1 Myhill graphs

In this section, we are going to focus on a simple-looking class of recognisable languages. We begin with an example.

Example 6.1.1 Consider the following directed graph:

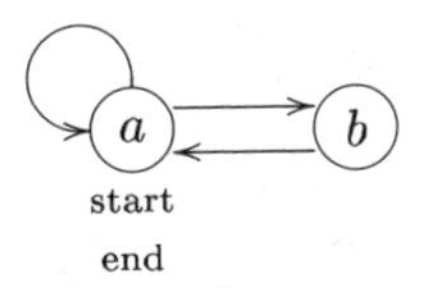

This is not an automaton, because there are no labels on the edges. However, this graph can still be used to recognise strings over the alphabet $A = \{a, b\}$. The reason is that each *non-empty* string over A labels at most one path in this graph beginning at the vertex labelled 'start' and ending at the vertex labelled 'end.' Specifically, the string a labels the unique path of length 0 starting and ending at the vertex a, the string $x = a_1 \dots a_n$, where $n \geq 2$, labels a path beginning at a if $a_1 = a$, and if for each pair $a_i a_{i+1}$, where $1 \leq i \leq n-1$, there is an edge from a_i to a_{i+1}. By definition, the language recognised by this graph consists of those non-empty strings that label a path beginning at a and ending at a. We can explicitly describe this language. Observe that strings containing consecutive b's cannot be recognised, because there is no loop at the vertex b. It is now easy to see that the language recognised by this graph is $L = (aA^* \cap A^*a) \setminus A^*b^2A^*$. □

The graph in our example is an instance of the following. A *Myhill graph* **G** *(over the alphabet A)* is a directed graph satisfying the following three conditions:

(MG1) For each ordered pair of vertices v_1 and v_2 there is at most one edge from v_1 to v_2.

(MG2) Some vertices are designated *start vertices*, and some vertices are designated *end vertices*. A vertex may be both a start and an end vertex.

(MG3) Each vertex is labelled with a different symbol from a finite alphabet A. This means that we can refer to a vertex unambiguously by its label.

Let **G** be a Myhill graph over the alphabet A. An individual letter of A is said to be *allowable* if it labels a vertex that is both a start and an end vertex. A non-empty string $a_1, \ldots, a_n$ over A of length at least two is said to be *allowable* if a_1 labels a start vertex, a_n labels an end vertex, and for each $1 \leq i \leq n-1$ there is an edge in **G** joining the vertex a_i to the vertex a_{i+1}. The *language*, $L(\mathbf{G})$, *recognised* by a Myhill graph **G** consists of all allowable strings in A^+.

Myhill graphs provide a simple way of describing languages, and we would obviously like to characterise what these languages are. The first step in answering this question is to show that the language recognised by a Myhill graph is recognisable. To do this, we shall prove that Myhill graphs can be converted into special kinds of automata. We first indicate how this can be done with Example 6.1.1.

Example 6.1.2 Consider the following deterministic but incomplete automaton:

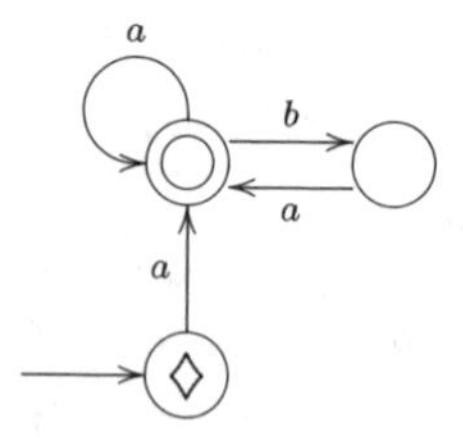

It is easy to check that it recognises the same language as Example 6.1.1. □

Let $\mathbf{A} = (S, A, i, \delta, T)$ be a deterministic automaton not necessarily complete. We say that **A** is a *local automaton* if for each $a \in A$ the set $\{s \cdot a \colon s \in S\}$ contains at most one element; we say that it is a *standard local automaton* if, in addition, there is no transition arriving at the initial state. Thus an automaton is local if for each $a \in A$ either there are no transitions labelled by a

or if there are they all arrive at the same state. We show that Myhill graphs can be converted into standard local automata and vice versa in such a way that languages are not changed.

Theorem 6.1.3 *A language can be recognised by a Myhill graph if and only if it can be recognised by a standard local automaton whose initial state is not terminal.*

Proof Let **G** be a Myhill graph. We construct an automaton **A** as follows. Adjoin an extra vertex $\diamondsuit$ to **G** with arrows from $\diamondsuit$ to each of the start vertices of **G**; label $\diamondsuit$ as the initial state; label the end vertices as terminal states; label each arrow of the resulting directed graph with the symbol labelling the vertex to which it points. Thus

is replaced by

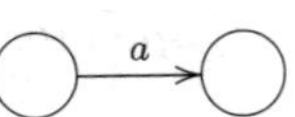

It is clear that every transition is labelled. Call the resulting machine **A**. By construction, the initial state is not terminal. I claim that **A** is deterministic. First the arrows emerging from the initial state are all differently labelled because they point to states that were originally differently labelled by (MG3). If we now consider any other state s, then two transitions emerging from s could only have the same label if in **G** both arrows pointed to the same vertex. However, this cannot occur by (MG1). Thus **A** is a deterministic automaton, possibly incomplete. By construction it is both standard and local. It is now easy to check that $L(\mathbf{A}) = L(\mathbf{G})$.

To prove the converse, let $\mathbf{A} = (S, A, i, \delta, T)$ be a standard local automaton whose initial state is not terminal. We construct a Myhill graph **G** as follows. First, label all states of **A**, except the initial state, with the input symbol of all transitions that enter that state. This is well-defined from the definition of 'local.' Erase all labels on directed edges. Next, mark all states s as start vertices for which there are transitions from i to s, and mark all terminal states as end vertices. Finally, erase the vertex i and all arrows that leave it. Call the resulting graph **G**. It is straightforward to check that **G** is a Myhill graph and that $L(\mathbf{G}) = L(\mathbf{A})$. □

We shall now describe the languages recognised by standard local automata. A language $L \subseteq A^*$ is said to be a *local language* if $L \setminus \varepsilon$ can be described in the following way: there are subsets $P, S \subseteq A$ and $N \subseteq A^2$ such that

$$L \setminus \varepsilon = (PA^* \cap A^*S) \setminus A^*NA^*.$$

This definition simply says that a local language is one in which determining whether a non-empty string belongs to it boils down to some very elementary checks:

- Check that the string begins with a letter from P.
- Check that no consecutive pair of letters in the string belongs to the set N of 'forbidden factors.'
- Check that the string ends with a letter from S.

Thus to determine whether a string belongs to a local language it is enough to scan the string from left to right through a window two-letters long; this is the meaning of the word 'local.' The set N is equal to $A^2 \setminus F$ where F is the set of all factors of length 2 in the language L.

Theorem 6.1.4 *Let*

$$L = (PA^* \cap A^*S) \setminus A^*NA^*$$

be a local language. Define an automaton $\mathbf{A}$ *as follows:*

- *The set of states* $Q = A \cup \{\varepsilon\}$.
- *The initial state is* ε.
- *The terminal states are* S.
- *The transitions are defined as follows:*

$$\varepsilon \cdot a = \begin{cases} a & \text{if } a \in P \\ \text{undefined} & \text{else} \end{cases}$$

and

$$a \cdot b = \begin{cases} b & \text{if } ab \notin N \\ \text{undefined} & \text{else.} \end{cases}$$

Then $\mathbf{A}$ *is a standard local automaton recognising* L. *If* L *contains the empty string, then* $\mathbf{A}$ *is defined as above except that the terminal states are defined to be* $S \cup \{\varepsilon\}$. *Conversely, every language recognised by a (standard) local automaton is local.*

Proof The automaton is local because for each state s and each input letter a, either $s \cdot a$ is not defined or it equals a. By construction the automaton is standard. It remains to prove that $L(\mathbf{A}) = L$. Let $x = a_1 \ldots a_n \in L(\mathbf{A})$. Then there is a path in $\mathbf{A}$ labelled as follows:

$$\varepsilon \xrightarrow{a_1} a_1 \xrightarrow{a_2} a_2 \ldots a_{n-1} \xrightarrow{a_n} a_n,$$

where a_n is a terminal state. It follows that $a_n \in S$. Likewise, $\varepsilon \cdot a_1$ is defined and so $a_1 \in P$. Finally, for each j such that $1 \leq j \leq n-1$ the fact that $a_j \cdot a_{j+1}$ is defined implies that $a_j \cdot a_{j+1} \notin N$. It follows that $x \in L$. Conversely, let

$x = a_1 \ldots a_n \in L$. Then $a_1 \in P$, $a_n \in S$ and for each j such that $1 \leq j \leq n-1$ we have that $a_j \cdot a_{j+1} \notin N$. It follows that

$$\varepsilon \xrightarrow{a_1} a_1 \xrightarrow{a_2} a_2 \ldots a_{n-1} \xrightarrow{a_n} a_n$$

is a path in $\mathbf{A}$ from the initial state to a terminal state. Thus $x \in L(\mathbf{A})$.

We conclude by proving that every language recognised by a local automaton is local. Let $\mathbf{A} = (Q, A, q_0, \delta, T)$ be a local automaton recognising the language $L = L(\mathbf{A})$. Define the following sets based on the definition of $\mathbf{A}$:

- $P = \{a \in A\colon q_0 \cdot a \text{ is defined}\}$.
- $S = \{a \in A\colon q \cdot a \in T \text{ for some } q \in Q\}$.
- $N = \{x \in A^2\colon x \text{ labels no path in } \mathbf{A}\}$.
- $K = (PA^* \cap A^*S) \setminus A^*NA^*$.

Let $x = a_1 \ldots a_n$ be a non-empty string in $L = L(\mathbf{A})$. It is easy to check that $x \in K$. Conversely, let $x = a_1 \ldots a_n \in K$. We show that x is accepted by $\mathbf{A}$, so that $x \in L$. By assumption, $a_1 \in P$ and $a_n \in S$ and for each $1 \leq j \leq n-1$ we have that $a_j a_{j+1} \notin N$. Since $a_1 \in P$ we have that $q_0 \cdot a_1$ is defined. Put $q_1 = q_0 \cdot a_1$. Since $a_1 a_2 \notin N$ we have that $a_1 a_2$ must label some path in $\mathbf{A}$. Thus there are states p, q and r such that

$$p \xrightarrow{a_1} q \xrightarrow{a_2} r.$$

Now $p \cdot a_1$ is defined as is $q_0 \cdot a_1$. But $\mathbf{A}$ is local and so $p \cdot a_1 = q_0 \cdot a_1 = q_1$. Thus $q_1 \cdot a_2$ is defined. Put $q_2 = q_1 \cdot a_2$. This same idea can be repeated with $a_2 a_3$. Continuing in this way, we see that $a_1 \ldots a_n$ labels a path in $\mathbf{A}$ starting at q_0 and finishing at a terminal state. Thus $x \in L(\mathbf{A}) = L$. It follows that $L \setminus \{\varepsilon\} = K$, and so L is a local language. □

Thus the languages recognised by Myhill graphs are the local languages without ε.

Example 6.1.5 Let $A = \{a, b, c\}$, $P = \{a, b\}$, $S = \{a, c\}$ and $N = \{ab, bc, ca\}$. We use Theorem 6.1.4, to construct a deterministic automaton to recognise

$$L = (PA^* \cap A^*S) \setminus A^*NA^*.$$

There are four states labelled ε, a, b, c; the initial state is ε and the terminal states are a, c, the elements of S. Thus we have so far constructed

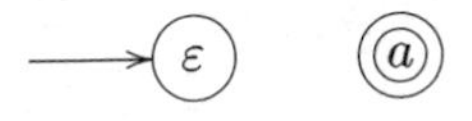

We now have to calculate the transitions. The only transitions emerging from ε are those labelled a and b, the elements of P, which go to the states a and b, respectively. We now calculate the transitions emerging from the state a. Observe that $aa, ac \notin N$ whereas $ab \in N$. Thus there are two transitions emerging from state a one labelled a and one labelled c. To calculate the transitions emerging from state b, observe that $ba, bb \notin N$ whereas $bc \in N$. Thus there are two transitions emerging from state b one labelled a and one labelled b. Finally, we calculate the transitions emerging from state c. We have that $cb, cc \notin N$ but $ca \in N$. Thus there are two transitions emerging from state c labelled b and c. The resulting standard local automaton is therefore

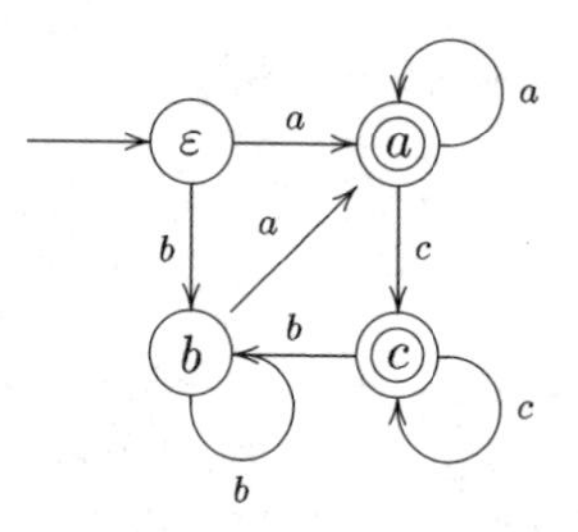

□

Local languages are special kinds of recognisable languages. However, they can be used to obtain another characterisation of arbitrary recognisable languages. To do this, we shall use monoid homomorphisms introduced in Section 4.2. In fact, we need monoid homomorphisms having the following property. A monoid homomorphism $\alpha\colon A^* \to B^*$ is *strictly alphabetic* if $\alpha(a)$ is a letter in B for each letter $a \in A$.

Theorem 6.1.6 *Let $L \subseteq B^*$ be a recognisable language such that $\varepsilon \notin L$. Then there is an alphabet A, a local language $K \subseteq A^*$, and a strictly alphabetic monoid homomorphism $\alpha\colon A^* \to B^*$ such that $\alpha(K) = L$. If L contains ε then $\alpha(K + \varepsilon) = L$.*

Proof Let $\mathbf{B} = (S, B, s_0, \delta, T)$ be an automaton that recognises L. The alphabet A is defined to be

$$A = \{(q, a, q \cdot a)\colon q \in S \text{ and } a \in B\},$$

the set of all transitions of $\mathbf{B}$. The monoid homomorphism $\alpha\colon A^* \to B^*$ is defined by

$$\alpha((q, a, q \cdot a)) = a,$$

which is evidently strictly alphabetic. We shall now define the language K. Two triples,

$$(q_1, a_1, q_1 \cdot a_1) \text{ and } (q_2, a_2, q_2 \cdot a_2),$$

describe a consecutive pair of transitions in **B** if and only $q_1 \cdot a_1 = q_2$. We define N as follows:

$$N = \{(q_1, a_1, q_1 \cdot a_1)(q_2, a_2, q_2 \cdot a_2)\colon \text{ where } q_1 \cdot a_1 \neq q_2\},$$

the *non-consecutive* transitions in **B**. Define

$$P = \{(s_0, a, s_0 \cdot a)\colon a \in A\},$$

the set of transitions beginning at the initial state, and

$$S = \{(q, a, q \cdot a)\colon a \in A \text{ and } q \cdot a \in T\},$$

the set of transitions that end at a terminal state. Then the language,

$$K = (PA^* \cap A^*S) \setminus A^*NA^*,$$

consists of all sequences of transitions in **B** that begin at s_0 and end at a terminal state. Thus if $\varepsilon \notin L$, we have immediately that $\alpha(K) = L$. If $\varepsilon \in L$ then we need to add ε to our language K. □

The theorem above can be paraphrased in the following terms: the language needed to describe an automaton is simpler than the language the automaton describes; this makes sense: the syntax of a programming language is easier to understand than what the programs written in that language accomplish. If this were not true, why write the programs?

Example 6.1.7 Here is a concrete example of Theorem 6.1.6. Consider the automaton **B** given by the transition table below:

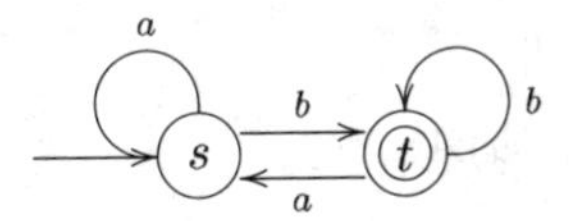

The new alphabet A is given by

$$A = \{(s, a, s), (s, b, t), (t, b, t), (t, a, s)\}.$$

The set P is given by

$$P = \{(s, a, s), (s, b, t)\}$$

and the set S is given by

$$S = \{(s, b, t), (t, b, t)\}.$$

Finally, N consists of the following strings of length 2 over the alphabet A:

$$(s, a, s)(t, b, t),$$

$$(s, a, s)(t, a, s),$$
$$(s, b, t)(s, a, s),$$
$$(s, b, t)(s, b, t),$$
$$(t, b, t)(s, a, s),$$
$$(t, b, t)(s, b, t),$$
$$(t, a, s)(t, b, t),$$
$$(t, a, s)(t, a, s).$$

Define

$$K = (PA^* \cap A^*S) \setminus A^*NA^*.$$

Consider now the string *abab*, which is accepted by **B**. This corresponds to the sequence of transitions,

$$(s, a, s)(s, b, t)(t, a, t)(t, b, t),$$

which clearly belongs to the language K. □

Exercises 6.1

1. Complete the proof of Theorem 6.1.3 by checking that the two constructions described do not alter the languages involved.

2. Construct standard local automata to recognise the following languages over the alphabet $A = \{a, b\}$.

(i) $(aA^* \cap A^*b) \setminus A^*a^2A^*$.

(ii) $(aA^* \cap A^*a) \setminus A^*(a^2 + b^2)A^*$.

(iii) $(aA^* \cap A^*b) \setminus A^*(ab + ba)A^*$.

3. Complete the proof of Theorem 6.1.4 by checking that the theorem holds when the local language contains the empty string.

6.2 Linearisation

Regular expressions describe recognisable languages and, as we have seen, each recognisable language is an image of a local language; in addition, it is easy to construct an automaton recognising a local language. This raises the possibility of finding an algorithm for converting regular expressions into finite automata that makes use of local automata. The goal of this section is to show how this can be achieved. The linchpin of the construction is the notion of a 'linear regular expression.' A regular expression r over an alphabet A is said to be *linear* if each letter $a \in A$ occurs at most once in r. For example, $(ab+c)^*dg$ is a linear regular expression over the alphabet $\{a,b,c,d,e,f,g\}$.

Linear regular expressions describe only a limited class of languages; we shall say more about which languages these are below. However, with any regular expression we can easily associate a linear regular expression over a larger alphabet by means of a simple-minded trick. Let r be a regular expression over the alphabet A. Define a new regular expression r' as follows: r' is constructed from r by replacing the ith letter a occurring in r from the left by a_i. We call r' the *linearisation* of r. For example, let $r = [ab(ba)^* + (ac)^*b]^*$. Then the linearisation of r is

$$r' = [a_1b_2(b_3a_4)^* + (a_5c_6)^*b_7]^*,$$

which is a regular expression over the alphabet $\{a_1, b_2, b_3, a_4, a_5, c_6, b_7\}$.

The first plank of our argument is that from a deterministic automaton $\mathbf{A}'$ recognising r', one can easily construct a (non-deterministic) automaton $\mathbf{A}$ recognising r in the following way: simply replace each letter a_i labelling a transition of $\mathbf{A}'$ by the corresponding letter a. Thus the problem of constructing an automaton from a regular expression r reduces to the problem of constructing a deterministic automaton from the linearisation of r. The second plank of our argument is provided by the following theorem.

Theorem 6.2.1 *Let r be a linear regular expression. Then the language $L(r)$ is local.*

Proof Linear regular expressions are special kinds of regular expressions and so we can prove things about them inductively. First we need three subsidiary results.

(1) *Let $A_1, A_2 \subseteq A$ be two disjoint subsets of our alphabet A, and let $L_1 \subseteq A_1^*$ and $L_2 \subseteq A_2^*$ be two local languages. Then $L_1 + L_2$ is a local language.*

Let $\mathbf{A}_1 = (Q_1, A_1, i_1, \delta_1, T_1)$ be a standard local automaton recognising L_1 and let $\mathbf{A}_2 = (Q_2, A_2, i_2, \delta_2, T_2)$ be a standard local automaton recognising L_2. We construct a standard local automaton $\mathbf{A} = (Q, A, i, \delta, T)$ to recognise $L_1 + L_2$ as follows. Let Q consist of the elements of $Q_1 \setminus \{i_1\}$ and $Q_2 \setminus \{i_2\}$ together with a new initial state i. The transitions of $\mathbf{A}$ are as follows: all transitions of $\mathbf{A}_1$ except those transitions that begin at i_1, together with all the transitions of $\mathbf{A}_2$ except those transitions that begin at i_2, together with the following new transitions: $i \xrightarrow{a} q$ whenever $i_1 \xrightarrow{a} q$ is a transition in $\mathbf{A}_1$ or $i_2 \xrightarrow{a} q$ is a transition in $\mathbf{A}_2$. Finally, the set of terminal states is defined to be T as long as neither i_1 nor i_2 is terminal, otherwise it is defined to be

$$(T_1 \setminus \{i_1\} \cup T_2 \setminus \{i_2\}) \cup \{i\}.$$

It is easy to see that $\mathbf{A}$ is a standard local automaton accepting $L_1 + L_2$.

(2) *Let $A_1, A_2 \subseteq A$ be two disjoint subsets of our alphabet A, and let $L_1 \subseteq A_1^*$ and $L_2 \subseteq A_2^*$ be two local languages. Then L_1L_2 is a local language.*

Let $\mathbf{A}_1 = (Q_1, A_1, i_1, \delta_1, T_1)$ be a standard local automaton recognising L_1 and let $\mathbf{A}_2 = (Q_2, A_2, i_2, \delta_2, T_2)$ be a standard local automaton recognising L_2. We construct a standard local automaton $\mathbf{A} = (Q, A, i, \delta, T)$ recognising L_1L_2 as follows. The set of states is the set $(Q_1 + Q_2) \setminus \{i_2\}$ and the initial state is i_1. The transitions consist of all the transitions of $\mathbf{A}_1$, all the transitions of $\mathbf{A}_2$ that do not begin at i_2, together with the following new transitions: $q_1 \xrightarrow{a} q_2$ if $q_1 \in T_1$ and if $i_2 \xrightarrow{a} q_2$ is a transition of $\mathbf{A}_2$. The set of terminal states is T_2 if $i_2 \notin T_2$ and is $T_1 \cup (T_2 \setminus \{i_2\})$ if $i_2 \in T_2$. It is easy to check that $\mathbf{A}$ is a standard local automaton accepting L_1L_2. This construction is simply a variant of the construction used to prove Proposition 3.3.4.

(3) *Let L be a local language. Then L^* is a local language.*

Let $\mathbf{A} = (Q, A, i, \delta, T)$ be a standard local automaton recognising L. Define $\mathbf{A}' = (Q, A, i, \delta', T \cup i)$ as follows: the transitions of $\mathbf{A}'$ consist of all the transitions of $\mathbf{A}$ together with the transitions $q \xrightarrow{a} q'$ where $q \in T$ and where $i \xrightarrow{a} q'$ is a transition in $\mathbf{A}$. Then $\mathbf{A}'$ is a standard local automaton recognising L^*. This construction is simply a variant of the one used to prove Proposition 3.3.6.

We can now prove our claim that if r is linear then $L(r)$ is local. We do so by induction on the number of operators in r. The languages $\emptyset$, ε, and a where $a \in A$ are each local. Suppose that all linear regular expressions with at most n operators represent local languages. Let r be a regular linear expression with $n+1$ operators. We prove that $L(r)$ is local. The expression r has one of three possible forms: $r = s + t$, $r = st$, or $r = s^*$. Clearly s and t are both linear regular expressions with at most n operators apiece. Let B be the set of letters occurring in s and C be the set of letters occurring in t. Then since r is linear B and C are disjoint. It follows from results (1) and (2) above that $L(r) = L(s) + L(t)$ and $L(r) = L(s)L(t)$ are both local. The fact that $L(r) = L(s)^*$ is local follows by result (3). □

The above theorem suggests the following algorithm.

Algorithm 6.2.2 This algorithm takes as input a regular expression r, and produces as output a non-deterministic automaton $\mathbf{A}$ recognising $L(r)$. First, linearise r to obtain a linear regular expression r'. Then construct a standard local automaton $\mathbf{A}'$ recognising $L(r')$. Finally, convert $\mathbf{A}'$ into a non-deterministic automaton $\mathbf{A}$ by relabelling the edges of $\mathbf{A}'$. □

This algorithm will be complete if we can find an algorithm for constructing the automaton described in Theorem 6.1.4 from a linear regular expression r. This requires that we construct the sets $P(L(r))$, $S(L(r))$, and $F(L(r))$; in addition, we need to know whether $\varepsilon \in L(r)$. To do this we shall use the following definitions. Let A be an alphabet and let r be a regular expression over A. We make the following definitions:

- $\delta(r) = \begin{cases} \varepsilon & \text{if } \varepsilon \in L(r) \\ \emptyset & \text{else.} \end{cases}$
- $P(r) = \{a \in A\colon aA^* \cap L(r) \neq \emptyset\}$.
- $S(r) = \{a \in A\colon A^*a \cap L(r) \neq \emptyset\}$.
- $F(r) = \{x \in A^2\colon A^* \cap L(r) \neq \emptyset\}$.

These four sets can easily be calculated from a regular expression by using the inductive nature of the construction of regular expressions. The following lemma accomplishes this. We leave the proofs as exercises.

Lemma 6.2.3 *Let A be an alphabet and let L and M be languages over A.*

(i)

- $\delta(a) = \emptyset$ *for each* $a \in A$;
- $\delta(\emptyset) = \emptyset$; $\delta(\varepsilon) = \varepsilon$;
- $\delta(L + M) = \delta(L) + \delta(M)$;
- $\delta(LM) = \delta(L) \cap \delta(M)$;
- $\delta(L^*) = \varepsilon$.

(ii)

- $P(a) = a$ *for each* $a \in A$;
- $P(\emptyset) = \emptyset$; $P(\varepsilon) = \emptyset$;
- $P(L + M) = P(L) + P(M)$;
- $P(LM) = P(L) + \delta(L)P(M)$;
- $P(L^*) = P(L)$.

(iii)

- $S(a) = a$ *for each* $a \in A$;
- $S(\emptyset) = \emptyset$; $S(\varepsilon) = \emptyset$;
- $S(L + M) = S(L) + S(M)$;
- $S(LM) = S(M) + S(L)\delta(M)$;
- $S(L^*) = S(L)$.

(iv)

- $F(a) = \emptyset$ *for each* $a \in A$;

- $F(\emptyset) = \emptyset$; $F(\varepsilon) = \emptyset$;
- $F(L + M) = F(L) + F(M)$;
- $F(LM) = F(L) + F(M) + S(L)P(M)$;
- $F(L^*) = F(L) + S(L)P(L)$.

□

Example 6.2.4 We use Algorithm 6.2.2 together with Lemma 6.2.3 to construct a non-deterministic automaton recognising the language described by $r = (ab + b)^*ba$. First put $r' = (a_1b_2 + b_3)^*b_4a_5$. We can easily calculate the following:

- $\delta(r') = \emptyset$.
- $P(r') = a_1 + b_3 + b_4$.
- $S(r') = a_5$.
- $F(r') = a_1b_2 + b_2b_3 + b_3a_1 + b_4a_5 + b_2b_4 + b_3b_4 + b_2a_1 + b_3b_3$.

We now have the data necessary to construct a standard local automaton recognising $L(r')$:

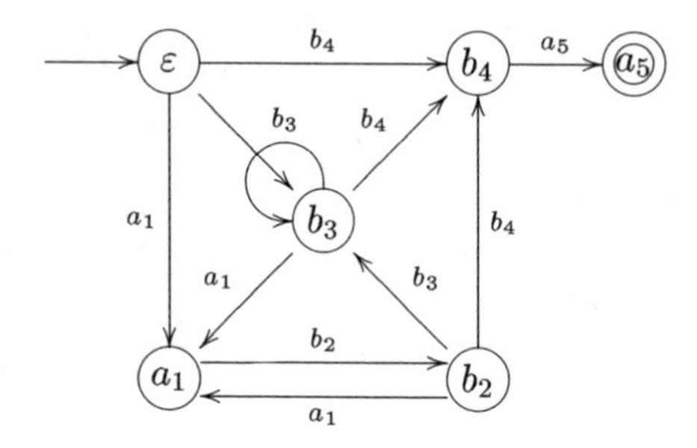

The automaton for $L(r)$ is now obtained from this automaton by simply erasing all the subscripts on the transition labels. □

Exercises 6.2

1. Prove Lemma 6.2.3.

2. Use Algorithm 6.2.2 to find non-deterministic automata that recognise the following languages.

(i) $(a^*b^*)^*$.

(ii) $(a^* + ba)^*$.

(iii) $a^*(bca^*)^*$.

6.3 Summary of Chapter 6

- *Local languages*: A local language is one in which all strings, apart from possibly the empty string, begin with certain letters, do not contain consecutive pairs of letters from a list of forbidden factors, and end with certain letters. Such languages are accepted by Myhill graphs (apart from any identity) or by local automata. Every recognisable language is the image of a local language under a strictly alphabetic monoid homomorphism.

- *Linear regular expressions*: A regular expression is linear if each letter occurs exactly once. The languages described by such regular expressions are local. There is an algorithm that converts a linear regular expression r into a deterministic local automaton $\mathbf{A}$ recognising $L(r)$. This algorithm can be used to construct a non-deterministic automaton recognising $L(r)$ where r is any regular expression.

6.4 Remarks on Chapter 6

I have relied heavily on the paper by Berstel and Pin [13] to write this chapter. Algorithm 6.2.2 is due to Berry and Sethi [11]. Local languages themselves seem to have been independently discovered by Myhill [97] and Chomsky and Schützenberger [30]; local automata are also termed 'Glushkov automata' [55]. My source for Myhill graphs was Brauer [16]. The paper by McNaughton and Yamada [86] introduced the idea of labelling the letters occurring in a regular expression.

Chapter 7

Minimal automata

We have so far only been concerned with the question of whether or not a language can be recognised by a finite automaton. If it can be, then we have not been interested in how efficiently the job can be done. In this chapter, we shall show that for each recognisable language there is a smallest complete deterministic automaton that recognises it. By 'smallest' we simply mean one having the smallest number of states. As we shall prove later in this section, two deterministic automata that recognise the same language each having the smallest possible number of states must be essentially the same; in mathematical terms, they are isomorphic. This means that with each recognisable language we can associate an automaton that is unique up to isomorphism: this is known as the minimal automaton of the language. This automaton plays a crucial role in Chapters 9–12.

There are two different ways of constructing the minimal automaton of a recognisable language L. The first starts with a deterministic automaton **A** recognising L and converts it into the minimal automaton by means of two steps: removal of inaccessible states, introduced in Section 3.1, and 'reduction,' a new process, which involves combining states. The second starts with a regular expression for L and constructs the minimal automaton using a process called the 'Method of Quotients.'

7.1 Partitions and equivalence relations

A collection of individuals can be divided into disjoint groups in many different ways. This simple idea is the main mathematical tool needed in this chapter and forms one of the most important ideas in algebra.

Let X be a set. A *partition* of X is a set P of subsets of X satisfying the following three conditions:

(P1) Each element of P is a non-empty subset of X.

(P2) Distinct elements of P are disjoint.

(P3) Every element X belongs to at least one (and therefore by (P2) exactly one) element of P.

The elements of P are called the *blocks* of the partition.

Examples 7.1.1 Some examples of partitions.

(1) Let

$$X = \{0, 1, \ldots, 9\}$$

and

$$P = \{\{0, 1, 2\}, \{3, 4\}, \{5, 6, 7, 8\}, \{9\}\}.$$

Then P is a partition of X containing four blocks.

(2) The set $\mathbb{N}$ of natural numbers can be partitioned into two blocks: the set of even numbers, and the set of odd numbers.

(3) The set $\mathbb{N}$ can be partitioned into three blocks: those numbers divisible by 3, those numbers that leave remainder 1 when divided by 3, and those numbers that leave remainder 2 when divided by 3.

(4) The set $\mathbb{R}^2$ can be partitioned into infinitely many blocks: consider the set of all lines l_a of the form $y = x + a$ where a is any real number. Each point of $\mathbb{R}^2$ lies on exactly one line of the form l_a.

□

A partition is defined in terms of the set X and the set of blocks P. However, there is an alternative way of presenting this information that is often useful. With each partition P on a set X, we can define a binary relation $\sim_P$ on X as follows:

$$x \sim_P y \Leftrightarrow x \text{ and } y \text{ belong to the same block of } P.$$

The proof of the following is left as an exercise.

Lemma 7.1.2 *The relation $\sim_P$ is reflexive, symmetric, and transitive.* □

Any relation on a set that is reflexive, symmetric, and transitive is called an *equivalence relation*. Thus from each partition we can construct an equivalence relation. In fact, the converse is also true.

Lemma 7.1.3 *Let $\sim$ be an equivalence relation on the set X. For each $x \in X$ put*

$$[x] = \{y \in X : x \sim y\}$$

and

$$X/\sim = \{[x] : x \in X\}.$$

Then $X/\sim$ is a partition of X.

Proof For each $x \in X$, we have that $x \sim x$, because $\sim$ is reflexive. Thus (P1) and (P3) hold. Suppose that $[x] \cap [y] \neq \emptyset$. Let $z \in [x] \cap [y]$. Then $x \sim z$ and $y \sim z$. By symmetry $z \sim y$, and so by transitivity $x \sim y$. It follows that $[x] = [y]$. Hence (P2) holds. $\square$

The set

$$[x] = \{y \in X \colon x \sim y\}$$

is called the *$\sim$-equivalence class* containing x.

Lemma 7.1.2 tells us how to construct equivalence relations from partitions, and Lemma 7.1.3 tells us how to construct partitions from equivalence relations. The following theorem tells us what happens when we perform these two constructions one after the other.

Theorem 7.1.4 *Let X be a non-empty set.*

(i) *Let P be a partition on X. Then the partition associated with the equivalence relation $\sim_P$ is P.*

(ii) *Let $\sim$ be an equivalence relation on X. Then the equivalence relation associated with the partition $X/\sim$ is $\sim$.*

Proof (i) Let P be a partition on X. By Lemma 7.1.2, we can define the equivalence relation $\sim_P$. Let $[x]$ be a $\sim_P$-equivalence class. Then $y \in [x]$ iff $x \sim_P y$ iff x and y are in the same block of P. Thus each $\sim_P$-equivalence class is a block of P. Now let $B \in P$ be a block of P and let $u \in B$. Then $v \in B$ iff $u \sim_P v$ iff $v \in [u]$. Thus $B = [u]$. It follows that each block of P is a $\sim_P$-equivalence class and vice versa. We have shown that P and $X/\sim_P$ are the same.

(ii) Let $\sim$ be an equivalence relation on X. By Lemma 7.1.3, we can define a partition $X/\sim$ on X. Let $\equiv$ be the equivalence relation defined on X by the partition $X/\sim$ according to Lemma 7.1.2. We have that $x \equiv y$ iff $y \in [x]$ iff $x \sim y$. Thus $\sim$ and $\equiv$ are the same relation. $\square$

Notation Let ρ be an equivalence relation on a set X. Then the ρ-equivalence class containing x is often denoted $\rho(x)$.

Theorem 7.1.4 tells us that partitions on X and equivalence relations on X are two ways of looking at the same thing. In applications, it is the partition itself that is interesting, but checking that we have a partition is usually done indirectly by checking that a relation is an equivalence relation.

The following example introduces some notation that we shall use throughout this chapter.

Example 7.1.5 Let $X = \{1, 2, 3, 4\}$ and let $P = \{\{2\}, \{1, 3\}, \{4\}\}$. Then P is a partition on X. The equivalence relation $\sim$ associated with P can be described by a set of ordered pairs, and these can be conveniently described by a table. The table has rows and columns labelled by the elements of X. Thus each square can be located by means of its co-ordinates: (a, b) means the square in row a and column b. The square (a, b) is marked with $\surd$ if $a \sim b$ and marked with $\times$ otherwise. Strictly speaking we need only mark the squares corresponding to pairs which are $\sim$-related, but I shall use both symbols.

	1	2	3	4
1	$\surd$	$\times$	$\surd$	$\times$
2	$\times$	$\surd$	$\times$	$\times$
3	$\surd$	$\times$	$\surd$	$\times$
4	$\times$	$\times$	$\times$	$\surd$

In fact, this table contains redundant information because if $a \sim b$ then $b \sim a$. It follows that the squares beneath the leading diagonal need not be marked. Thus we obtain

	1	2	3	4
1	$\surd$	$\times$	$\surd$	$\times$
2	$*$	$\surd$	$\times$	$\times$
3	$*$	$*$	$\surd$	$\times$
4	$*$	$*$	$*$	$\surd$

We call this the *table form* of the equivalence relation. □

Exercises 7.1

1. List all equivalence relations on the set $X = \{1, 2, 3, 4\}$ in:

(i) Partition form.

(ii) As sets of ordered pairs.

(iii) In table form.

2. Prove Lemma 7.1.2.

3. Let ρ and σ be two equivalence relations on a set X.

(i) Show that if $\rho \subseteq \sigma$ then each σ-class is a disjoint union of ρ-classes.

(ii) Show that $\rho \cap \sigma$ is an equivalence relation. Describe the equivalence classes of $\rho \cap \sigma$ in terms of the ρ-equivalence classes and the σ-equivalence classes.

7.2 The indistinguishability relation

In Section 3.1, we described one way of removing unnecessary states from an automaton: the construction of the accessible part of **A**, denoted $\mathbf{A}^a$, from **A**. In this section, we shall describe a different way of reducing the number of states in an automaton without changing the language it recognises. On a point of notation: if T is the set of terminal states of a finite automaton, then T' is the set of non-terminal states. Let $\mathbf{A} = (S, A, s_0, \delta, T)$ be an automaton. Two states $s, t \in S$ are said to be *distinguishable* if there exists $x \in A^*$ such that

$$(s \cdot x, t \cdot x) \in (T \times T') \cup (T' \times T).$$

In other words, for some string x, the states $s \cdot x$ and $t \cdot x$ are not both terminal or both non-terminal. The states s and t are said to be *indistinguishable* if they are not distinguishable. This means that for each $x \in A^*$ we have that

$$s \cdot x \in T \Leftrightarrow t \cdot x \in T.$$

Define the relation $\simeq_{\mathbf{A}}$ on the set of states S by

$$s \simeq_{\mathbf{A}} t \Leftrightarrow s \text{ and } t \text{ are indistinguishable.}$$

We call $\simeq_{\mathbf{A}}$ the *indistinguishability relation.* We shall often write $\simeq$ rather than $\simeq_{\mathbf{A}}$ when the machine $\mathbf{A}$ is clear. The relation $\simeq$ will be our main tool in constructing the minimal automaton of a recognisable language. The following result is left as an exercise.

Lemma 7.2.1 *Let* $\mathbf{A}$ *be an automaton. Then the relation* $\simeq_{\mathbf{A}}$ *is an equivalence relation on the set of states of* $\mathbf{A}$. □

The next lemma will be useful in the proof of Theorem 7.2.3.

Lemma 7.2.2 *If* $s \simeq t$*, then* s *is terminal if and only if* t *is terminal.*

Proof Suppose that s is terminal and $s \simeq t$. Then s terminal means that $s \cdot \varepsilon \in T$. But then $t \cdot \varepsilon \in T$, and so $t \in T$. The converse is proved similarly. □

Let $s \in S$ be a state in an automaton $\mathbf{A}$. Then the $\simeq$-equivalence class containing s will be denoted by $[s]$ or sometimes by $[s]_{\mathbf{A}}$. The set of $\simeq$-equivalence classes will be denoted by $S/\simeq$.

It can happen, of course, that each pair of states in an automaton is distinguishable. This is an important case that we single out for a definition. An automaton $\mathbf{A}$ is said to be *reduced* if the relation $\simeq_{\mathbf{A}}$ is equality.

Theorem 7.2.3 (Reduction of an automaton) *Let* $\mathbf{A} = (S, A, s_0, \delta, T)$ *be a finite automaton. Then there is an automaton* $\mathbf{A}/\simeq$*, which is reduced and recognises* $L(\mathbf{A})$*. In addition, if* $\mathbf{A}$ *is accessible then* $\mathbf{A}/\simeq$ *is accessible.*

Proof Define the machine $\mathbf{A}/\simeq$ as follows:

- The set of states is $S/\simeq$.

- The input alphabet is A.
- The initial state is $[s_0]$.
- The set of terminal states is $\{[s]: s \in T\}$.
- The transition function is defined by $[s] \cdot a = [s \cdot a]$ for each $a \in A$.

To show that the transition function is well-defined, we need the following result: if $[s] = [s']$ and $a \in A$ then $[s \cdot a] = [s' \cdot a]$. To verify this is true let $x \in A^*$. Then $(s \cdot a) \cdot x \in T$ precisely when $s \cdot (ax) \in T$. But $s \simeq s'$ and so

$$s \cdot (ax) \in T \Leftrightarrow s' \cdot (ax) \in T.$$

Hence $(s \cdot a) \cdot x \in T$ precisely when $(s' \cdot a) \cdot x \in T$. It follows that $s \cdot a \simeq s' \cdot a$. Thus $[s \cdot a] = [s' \cdot a]$. We have therefore proved that $\mathbf{A}/\simeq$ is a well-defined automaton. A simple induction argument shows that $[s] \cdot x = [s \cdot x]$ for each $x \in A^*$.

We can now prove that $\mathbf{A}/\simeq$ is reduced. Let $[s]$ and $[t]$ be a pair of indistinguishable states in $\mathbf{A}/\simeq$. By definition, $[s] \cdot x$ is terminal if and only if $[t] \cdot x$ is terminal for each $x \in A^*$. Thus $[s \cdot x]$ is terminal if and only if $[t \cdot x]$ is terminal. However, by Lemma 7.2.2, $[q]$ is terminal in $\mathbf{A}/\simeq$ precisely when q is terminal in $\mathbf{A}$. It follows that

$$s \cdot x \in T \Leftrightarrow t \cdot x \in T.$$

But this simply means that s and t are indistinguishable in $\mathbf{A}$. Hence $[s] = [t]$, and so $\mathbf{A}/\simeq$ is reduced.

Next we prove that $L(\mathbf{A}/\simeq) = L(\mathbf{A})$. By definition, $x \in L(\mathbf{A}/\simeq)$ precisely when $[s_0] \cdot x$ is terminal. This means that $[s_0 \cdot x]$ is terminal and so $s_0 \cdot x \in T$ by Lemma 7.2.2. Thus

$$x \in L(\mathbf{A}/\simeq) \Leftrightarrow x \in L(\mathbf{A}).$$

Hence $L(\mathbf{A}/\simeq) = L(\mathbf{A})$.

Finally, we prove that if $\mathbf{A}$ is accessible then $\mathbf{A}/\simeq$ is accessible. Let $[s]$ be a state in $\mathbf{A}/\simeq$. Because $\mathbf{A}$ is accessible there exists $x \in A^*$ such that $s_0 \cdot x = s$. Thus $[s] = [s_0 \cdot x] = [s_0] \cdot x$. It follows that $\mathbf{A}/\simeq$ is accessible. □

We denote the automaton $\mathbf{A}/\simeq$ by $\mathbf{A}^r$ and call it $\mathbf{A}$-*reduced*. For each automaton $\mathbf{A}$, the machine $\mathbf{A}^{ar} = (\mathbf{A}^a)^r$ is both accessible and reduced.

Before we describe an algorithm for constructing $\mathbf{A}^r$, we give an example.

Example 7.2.4 Consider the automaton $\mathbf{A}$ below:

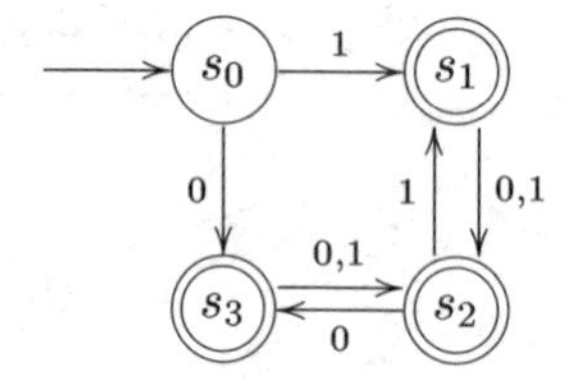

We shall calculate $\simeq$ first, and then $\mathbf{A}/\simeq$ using Theorem 7.2.3. To compute $\simeq$ we shall need to locate the elements of

$$\{s_0, s_1, s_2, s_3\} \times \{s_0, s_1, s_2, s_3\},$$

which belong to $\simeq$. To do this, we shall use the table we described in Example 7.1.5:

	s_0	s_1	s_2	s_3
s_0	$\surd$			
s_1	$*$	$\surd$		
s_2	$*$	$*$	$\surd$	
s_3	$*$	$*$	$*$	$\surd$

Because each pair of states in $(T \times T') \cup (T' \times T)$ is distinguishable we mark the squares (s_0, s_1), (s_0, s_2) and (s_0, s_3) with a $\times$:

	s_0	s_1	s_2	s_3
s_0	$\surd$	$\times$	$\times$	$\times$
s_1	$*$	$\surd$		
s_2	$*$	$*$	$\surd$	
s_3	$*$	$*$	$*$	$\surd$

To fill in the remaining squares, observe that in this case once the machine reaches the set of terminal states it never leaves it. Thus we obtain the following:

	s_0	s_1	s_2	s_3
s_0	$\surd$	$\times$	$\times$	$\times$
s_1	$*$	$\surd$	$\surd$	$\surd$
s_2	$*$	$*$	$\surd$	$\surd$
s_3	$*$	$*$	$*$	$\surd$

From the table we see that the $\simeq$-equivalence classes are $\{s_0\}$ and $\{s_1, s_2, s_3\}$. We now use the construction described in the proof of Theorem 7.2.3 to construct $\mathbf{A}/\simeq$. This is just

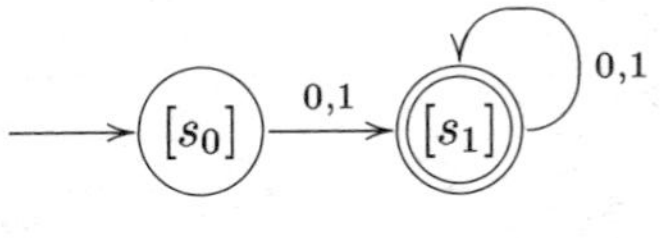

$[s_0]$ 0,1 $[s_1]$ 0,1

□

We shall now describe an algorithm for constructing $\mathbf{A}^r$.

Algorithm 7.2.5 (Reduction of an automaton) Let $\mathbf{A}$ be an automaton with set of states $S = \{s_1, \ldots, s_n\}$, initial state s_1, terminal states T, and input alphabet A. The algorithm calculates the equivalence relation $\simeq$. To do so we shall use two tables: table 1 will display the indistinguishability relation at the end of the algorithm, and table 2 will be used for side calculations.

(1) Initialisation: draw up a table (table 1) with rows and columns labelled by the elements of S. Mark main diagonal squares with $\surd$, and squares below the main diagonal with $*$. Mark with $\times$ all squares above the main diagonal in $(T \times T') \cup (T' \times T)$. Squares above the diagonal, which contain neither $\times$ nor $\surd$, are said to be 'empty.'

(2) Main procedure: construct an auxiliary table (table 2) as follows: working from left to right and top to bottom of table 1, label each row of table 2 with the pair (s, t) whenever the (s, t)-entry in table 1 is empty; the columns are labelled by the elements of A.

Now work from top to bottom of table 2: for each pair (s, t) labelling a row calculate the states $(s \cdot a, t \cdot a)$ for each $a \in A$ and enter them in table 2:

- If any of these pairs of states or $(t \cdot a, s \cdot a)$ labels a square marked with a $\times$ in table 1 then mark (s, t) with a $\times$ in table 1.
- If all the pairs $(s \cdot a, t \cdot a)$ are diagonal, mark (s, t) with a $\surd$ in table 1.
- Otherwise do not mark (s, t) and move to the next row.

(3) Finishing off: work from left to right and top to bottom of table 1. For each empty square (s, t) use table 2 to find all the squares $(s \cdot a, t \cdot a)$:

- If any of these squares in table 1 contains $\times$, then mark (s, t) with a $\times$ in table 1 and move to the next empty square.
- If all of these squares in table 1 contain $\surd$, then mark (s, t) with $\surd$ in table 1 and move to the next empty square.
- In all other cases move to the next empty square.

When an iteration of this procedure is completed we say that a 'pass' of table 1 has been completed. This procedure is repeated until a pass occurs in which no new squares are marked with $\times$, or until there are no empty squares. At this point, all empty squares are marked with $\surd$ and the algorithm terminates.

□

Before we prove that the algorithm works, we give an example.

Example 7.2.6 Consider the automaton **A** below:

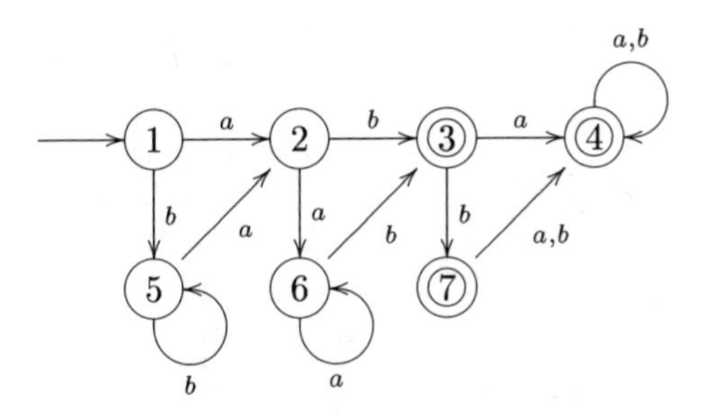

We shall use the algorithm to compute $\simeq$. The first step is to draw up the initialised table 1:

	1	2	3	4	5	6	7
1	$\surd$		$\times$	$\times$			$\times$
2	$*$	$\surd$	$\times$	$\times$			$\times$
3	$*$	$*$	$\surd$		$\times$	$\times$	
4	$*$	$*$	$*$	$\surd$	$\times$	$\times$	
5	$*$	$*$	$*$	$*$	$\surd$		$\times$
6	$*$	$*$	$*$	$*$	$*$	$\surd$	$\times$
7	$*$	$*$	$*$	$*$	$*$	$*$	$\surd$

We now construct table 2 and at the same time modify table 1:

	a	b
$(1,2)$	$(2,6)$	$(5,3)$
$(1,5)$	$(2,2)$	$(5,5)$
$(1,6)$	$(2,6)$	$(5,3)$
$(2,5)$	$(6,2)$	$(3,5)$
$(2,6)$	$(6,6)$	$(3,3)$
$(3,4)$	$(4,4)$	$(7,4)$
$(3,7)$	$(4,4)$	$(7,4)$
$(4,7)$	$(4,4)$	$(4,4)$
$(5,6)$	$(2,6)$	$(5,3)$

As a result the squares,

$$(1,2),(1,6),(2,5),(5,6),$$

are all marked with $\times$ in table 1, whereas the squares,

$$(1,5),(2,6),(4,7),$$

are marked with $\surd$. The squares,

$$(3,4),(3,7),$$

are left unchanged. Table 1 now has the following form:

	1	2	3	4	5	6	7
1	$\surd$	$\times$	$\times$	$\times$	$\surd$	$\times$	$\times$
2	$*$	$\surd$	$\times$	$\times$	$\times$	$\surd$	$\times$
3	$*$	$*$	$\surd$		$\times$	$\times$	
4	$*$	$*$	$*$	$\surd$	$\times$	$\times$	$\surd$
5	$*$	$*$	$*$	$*$	$\surd$	$\times$	$\times$
6	$*$	$*$	$*$	$*$	$*$	$\surd$	$\times$
7	$*$	$*$	$*$	$*$	$*$	$*$	$\surd$

To finish off, we check each empty square (s,t) in table 1 in turn to see if the corresponding entries in table 2 should cause us to mark this square. When we do this we find that no squares are changed. Thus the algorithm terminates. We now place $\surd$'s in all blank squares. We arrive at the following table:

	1	2	3	4	5	6	7
1	$\surd$	$\times$	$\times$	$\times$	$\surd$	$\times$	$\times$
2	$*$	$\surd$	$\times$	$\times$	$\times$	$\surd$	$\times$
3	$*$	$*$	$\surd$	$\surd$	$\times$	$\times$	$\surd$
4	$*$	$*$	$*$	$\surd$	$\times$	$\times$	$\surd$
5	$*$	$*$	$*$	$*$	$\surd$	$\times$	$\times$
6	$*$	$*$	$*$	$*$	$*$	$\surd$	$\times$
7	$*$	$*$	$*$	$*$	$*$	$*$	$\surd$

We can read off the $\simeq$-equivalence classes from this table. They are $\{1,5\}$, $\{2,6\}$ and $\{3,4,7\}$. The automaton $\mathbf{A}/\simeq$ is therefore

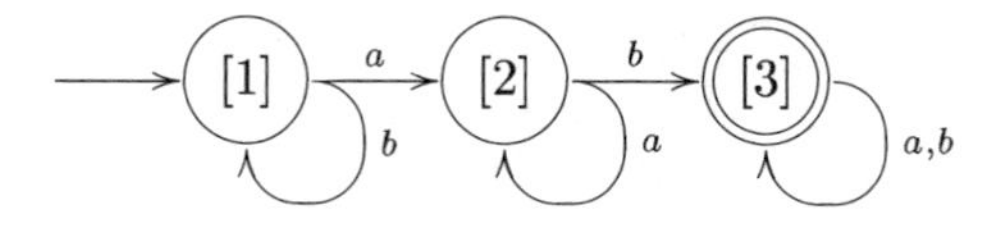

$\square$

We now justify that this algorithm works.

Theorem 7.2.7 *Algorithm 7.2.5 is correct*

Proof Let $\mathbf{A} = (S, A, s_0, \delta, T)$ be an automaton. By definition, the pair of states (s,t) is distinguishable if and only if there is a string $x \in A^*$ such that

$$(s \cdot x, t \cdot x) \in (T \times T') \cup (T' \times T);$$

I shall say that x *distinguishes* s and t. Those states distinguished by the empty string are precisely the elements of

$$(T \times T') \cup (T' \times T).$$

Suppose that (s,t) is distinguished by a string y of length $n > 0$. Put $y = ax$ where $a \in A$ and $x \in A^*$. Then $(s \cdot a, t \cdot a)$ is distinguished by the string x of length $n-1$. It follows that the pair (s,t) is distinguishable if and only if there is a sequence of pairs of states,

$$(s_0, t_0), (s_1, t_1), \ldots, (s_n, t_n),$$

such that $(s,t) = (s_0, t_0)$, and $(s_n, t_n) \in (T \times T') \cup (T' \times T)$ and

$$(s_i, t_i) = (s_{i-1} \cdot a_i, t_{i-1} \cdot a_i)$$

for $1 \leq i \leq n$ for some $a_i \in A$. The algorithm marks the pairs of states in $(T \times T') \cup (T' \times T)$ with a cross, and marks (s,t) with a cross whenever the square $(s \cdot a, t \cdot a)$ (or the square $(t \cdot a, s \cdot a)$) is marked with a cross for some $a \in A$. It is now clear that if the algorithm marks a square (s,t) with a cross, then s and t are distinguishable.

It therefore remains to prove that if a pair of states is distinguishable, then the corresponding square (or the appropriate one above the diagonal) is marked with a cross by the algorithm. We shall prove this by induction on the length of the strings that distinguish the pair. If the pair can be distinguished by the empty string then the corresponding square will be marked with a cross during initialisation. Suppose now that the square corresponding to any pair of states that can be distinguished by a string of length n is marked with a cross by the algorithm. Let (s,t) be a pair of states that can be distinguished by a string y of length $n+1$. Let $y = ax$ where $a \in A$ and x has length n. Then the pair $(s \cdot a, t \cdot a)$ can be distinguished by the string x, which has length n. By the induction hypothesis, the square $(s \cdot a, t \cdot a)$ will be marked with a cross by the algorithm. But then the square (s,t) will be marked with a cross either during the main procedure or whilst finishing off. □

Exercises 7.2

1. Let **A** be a finite automaton. Prove Lemma 7.2.1 that $\simeq_{\mathbf{A}}$ is an equivalence relation on the set of states of **A**.

2. Complete the proof of Theorem 7.2.3, by showing that $[s] \cdot x = [s \cdot x]$ for each $x \in A^*$.

3. For each of the automata **A** below find $\mathbf{A}^r$. In each case, we present the automaton by means of its transition table turned on its side. This helps in the calculations.

(i)

1	2	3	4	5	6	7	
2	2	5	6	5	6	5	a
4	3	3	4	7	7	7	b

The initial state is 1 and the terminal states are $3, 5, 6, 7$.

(ii)

0	1	2	3	4	5	
1	3	4	5	5	2	a
2	4	3	0	0	2	b

The initial state is 0 and the terminal states are 0 and 5.

(iii)

1	2	3	4	5	6	7	8	
2	7	1	3	8	3	7	7	a
6	3	3	7	6	7	5	3	b

The initial state is 1 and the terminal state is 3.

4. Let $\mathbf{A} = (S, A, i, \delta, \{t\})$ be an automaton with exactly one terminal state, and the property that for each $s \in S$ there is a string $x \in A^*$ such that $s \cdot x = t$. Suppose that for each $a \in A$ the function τ_a, defined by τ_a maps s to $s \cdot a$ for each $s \in S$, is a bijection. Prove that $\mathbf{A}$ is reduced.

7.3 Isomorphisms of automata

We begin with an example. Consider the following two automata, which we denote by $\mathbf{A}$ and $\mathbf{B}$, respectively:

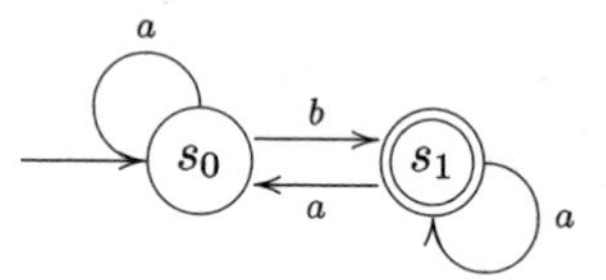

and

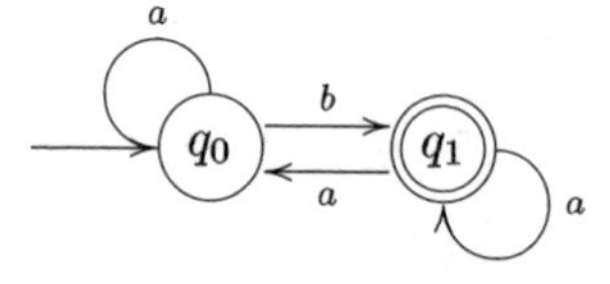

These automata are different because the labels on the states are different. But in every other respect, $\mathbf{A}$ and $\mathbf{B}$ are 'essentially the same.' In this case, it was easy to see that the two automata were essentially the same, but if they each had more states then it would have been much harder. In order to realise the main goal of this chapter, we need to have a precise mathematical definition of when two automata are essentially the same, one that we can check in a systematic way however large the automata involved. The definition below provides the answer to this question.

Let $\mathbf{A} = (S, A, s_0, \delta, F)$ and $\mathbf{B} = (Q, A, q_0, \gamma, G)$ be two automata with the same input alphabet A. An *isomorphism* θ *from* $\mathbf{A}$ *to* $\mathbf{B}$, denoted by $\theta \colon \mathbf{A} \to \mathbf{B}$, is a function $\theta \colon S \to Q$ satisfying the following four conditions:

(IM1) The function θ is bijective.

(IM2) $\theta(s_0) = q_0$.

(IM3) $s \in F \Leftrightarrow \theta(s) \in G$.

(IM4) $\theta(\delta(s, a)) = \gamma(\theta(s), a)$ for each $s \in S$ and $a \in A$.

If we use our usual notation for the transition function in an automaton, then (IM4) would be written as

$$\theta(s \cdot a) = \theta(s) \cdot a.$$

If there is an isomorphism from $\mathbf{A}$ to $\mathbf{B}$ we say that $\mathbf{A}$ is *isomorphic* to $\mathbf{B}$, denoted by $\mathbf{A} \equiv \mathbf{B}$. Isomorphic automata may differ in their state labelling and may look different when drawn as directed graphs, but by suitable relabelling, and by moving states and bending transitions, they can be made to look identical.

Lemma 7.3.1 *Let* $\mathbf{A} = (S, A, s_0, \delta, F)$ *and* $\mathbf{B} = (Q, A, q_0, \gamma, G)$ *be automata, and let* $\theta \colon \mathbf{A} \to \mathbf{B}$ *be an isomorphism. Then*

$$\theta(\delta^*(s, x)) = \gamma^*(\theta(s), x)$$

for each $s \in S$ *and* $x \in A^*$. *In particular,* $L(\mathbf{A}) = L(\mathbf{B})$.

Proof Using our usual notation for the extended state transition function, the lemma states that

$$\theta(s \cdot x) = \theta(s) \cdot x.$$

We prove the first assertion by induction on the length of x. Base step: we check the result holds when $x = \varepsilon$:

$$\theta(s \cdot \varepsilon) = \theta(s) \text{ whereas } \theta(s) \cdot \varepsilon = \theta(s),$$

as required. Induction hypothesis: assume the result holds for all strings of length at most n. Let u be a string of length $n + 1$. Then $u = ax$ where $a \in A$ and x has length n. Now

$$\theta(s \cdot u) = \theta(s \cdot (ax)) = \theta((s \cdot a) \cdot x).$$

Put $s' = s \cdot a$. Then

$$\theta((s \cdot a) \cdot x) = \theta(s' \cdot x) = \theta(s') \cdot x$$

by the induction hypothesis. However,

$$\theta(s') = \theta(s \cdot a) = \theta(s) \cdot a$$

by (IM4). Hence

$$\theta(s \cdot u) = (\theta(s) \cdot a) \cdot x = \theta(s) \cdot (ax) = \theta(s) \cdot u,$$

as required.

We now prove that $L(\mathbf{A}) = L(\mathbf{B})$. By definition

$$x \in L(\mathbf{A}) \Leftrightarrow s_0 \cdot x \in F.$$

By (IM3),

$$s_0 \cdot x \in F \Leftrightarrow \theta(s_0 \cdot x) \in G.$$

By our result above,

$$\theta(s_0 \cdot x) = \theta(s_0) \cdot x.$$

By (IM2), we have that $\theta(s_0) = q_0$, and so

$$s_0 \cdot x \in F \Leftrightarrow q_0 \cdot x \in G.$$

Hence $x \in L(\mathbf{A})$ if and only if $x \in L(\mathbf{B})$ and so $L(\mathbf{A}) = L(\mathbf{B})$ as required. □

Exercises 7.3

1. Let **A**, **B** and **C** be automata. Prove the following:

(i) $\mathbf{A} \equiv \mathbf{A}$; each automaton is isomorphic to itself.

(ii) If $\mathbf{A} \equiv \mathbf{B}$ then $\mathbf{B} \equiv \mathbf{A}$; if **A** is isomorphic to **B** then **B** is isomorphic to **A**.

(iii) If $\mathbf{A} \equiv \mathbf{B}$ and $\mathbf{B} \equiv \mathbf{C}$ then $\mathbf{A} \equiv \mathbf{C}$; if **A** is isomorphic to **B**, and **B** is isomorphic to **C** then **A** is isomorphic to **C**.

2. Let $\theta\colon \mathbf{A} \to \mathbf{B}$ be an isomorphism from $\mathbf{A} = (S, A, s_0, \delta, F)$ to $\mathbf{B} = (Q, A, q_0, \gamma, G)$. Prove that:

(i) The number of states of **A** is the same as the number of states of **B**.

(ii) The number of terminal states of **A** is the same as the number of terminal states of **B**.

(iii) **A** is accessible if and only if **B** is accessible.

(iv) **A** is reduced if and only if **B** is reduced.

3. Let **A** be an accessible automaton. Show that if $\theta, \phi\colon \mathbf{A} \to \mathbf{B}$ are both isomorphisms then $\theta = \phi$.

7.4 The minimal automaton

We now come to a fundamental definition. Let L be a recognisable language. A complete deterministic automaton $\mathbf{A}$ is said to be *minimal (for L)* if $L(\mathbf{A}) = L$ and if $\mathbf{B}$ is any complete deterministic automaton such that $L(\mathbf{B}) = L$, then the number of states of $\mathbf{A}$ is less than or equal to the number of states of $\mathbf{B}$. Minimal automata for a language L certainly exist. The problem is to find a way of constructing them. Our first result narrows down the search.

Lemma 7.4.1 *Let L be a recognisable language. If $\mathbf{A}$ is minimal for L, then $\mathbf{A}$ is both accessible and reduced.*

Proof If $\mathbf{A}$ is not accessible, then $\mathbf{A}^a$ has fewer states than $\mathbf{A}$ and $L(\mathbf{A}^a) = L$. But this contradicts the definition of $\mathbf{A}$. It follows that $\mathbf{A}$ is accessible. A similar argument shows that $\mathbf{A}$ is reduced. □

If $\mathbf{A}$ is minimal for L, then $\mathbf{A}$ must be both reduced and accessible. The next result tells us that any reduced accessible automaton recognising L is in fact minimal.

Theorem 7.4.2 *Let L be a recognisable language.*

(i) *Any two reduced accessible automata recognising L are isomorphic.*

(ii) *Any reduced accessible automaton recognising L is a minimal automaton for L.*

(iii) *Any two minimal automata for L are isomorphic.*

Proof (i) Let $\mathbf{A} = (S, A, s_0, \delta, F)$ and $\mathbf{B} = (Q, A, q_0, \gamma, G)$ be two reduced accessible automata such that $L(\mathbf{A}) = L(\mathbf{B})$. We prove that $\mathbf{A}$ is isomorphic to $\mathbf{B}$. To do this, we have to conjure up an isomorphism from $\mathbf{A}$ to $\mathbf{B}$. To keep the notation simple, we shall use the 'dot' notation for both δ^* and γ^*. We shall use the following observation:

$$s_0 \cdot x \in F \quad \Leftrightarrow \quad q_0 \cdot x \in G, \tag{7.1}$$

which follows from the fact that $L(\mathbf{A}) = L(\mathbf{B})$.

Let $s \in S$. Because $\mathbf{A}$ is accessible there exists $x \in A^*$ such that $s = s_0 \cdot x$. Define

$$\theta(s) = q_0 \cdot x.$$

To show that θ is a well-defined injective function we have to prove that

$$s_0 \cdot x = s_0 \cdot y \Leftrightarrow q_0 \cdot x = q_0 \cdot y.$$

Now $\mathbf{B}$ is reduced, so it will be enough to prove that

$$s_0 \cdot x \simeq_{\mathbf{A}} s_0 \cdot y \Leftrightarrow q_0 \cdot x \simeq_{\mathbf{B}} q_0 \cdot y.$$

Now $s_0 \cdot x \simeq_{\mathbf{A}} s_0 \cdot y$ iff for all $w \in A^*$:

$$(s_0 \cdot x) \cdot w \in F \Leftrightarrow (s_0 \cdot y) \cdot w \in F.$$

By Proposition 1.5.4, this is equivalent to

$$s_0 \cdot (xw) \in F \Leftrightarrow s_0 \cdot (yw) \in F.$$

But (7.1) above this is equivalent to

$$q_0 \cdot (xw) \in G \Leftrightarrow q_0 \cdot (yw) \in G.$$

Finally $q_0 \cdot x \simeq_{\mathbf{B}} q_0 \cdot y$ by another application of Proposition 1.5.4. We have therefore proved that θ is well-defined injective function.

To show that θ is surjective, let q be an arbitrary state in **B**. By assumption, **B** is accessible and so there exists $x \in A^*$ such that $q = q_0 \cdot x$. Put $s = s_0 \cdot x$ in **A**. Then by definition $\theta(s) = q$, and so θ is surjective as required. We have therefore proved that (IM1) holds.

That (IM2) holds is immediate because $s_0 = s_0 \cdot \varepsilon$. Thus $\theta(s_0) = q_0 \cdot \varepsilon = q_0$ as required.

(IM3) holds by accessibility and (7.1).

(IM4) holds: for each $s \in S$ and $a \in A$ we have to prove that $\theta(s \cdot a) = \theta(s) \cdot a$. Let $s = s_0 \cdot x$ for some $x \in A^*$. Then $\theta(s) = q_0 \cdot x$. Thus

$$\theta(s) \cdot a = (q_0 \cdot x) \cdot a = q_0 \cdot (xa),$$

using Proposition 1.5.4. On the other hand,

$$s \cdot a = (s_0 \cdot x) \cdot a = s_0 \cdot (xa).$$

Hence by definition,

$$\theta(s \cdot a) = q_0 \cdot (xa) = \theta(s) \cdot a$$

and the result follows.

(ii) Let **A** be a reduced and accessible automaton recognising L. We prove that **A** is minimal for L. Let **B** be any automaton recognising L. Then $L = L(\mathbf{B}^{ar})$ and the number of states in $\mathbf{B}^{ar}$ is less than or equal to the number of states in **B**. But by (i), **A** and $\mathbf{B}^{ar}$ are isomorphic and so, in particular, have the same number of states. It follows that the number of states in **A** is less than or equal to the number of states in **B**. Thus **A** is a minimal automaton for L.

(iii) By Lemma 7.4.1, a minimal automaton for L is accessible and reduced. By (i), any two accessible and reduced automata recognising L are isomorphic. Thus any two minimal automata for a language are isomorphic. □

We can paraphrase the above theorem in the following way: the minimal automaton for a recognisable language is *unique up to isomorphism*. Because of this we shall often refer to *the* minimal automaton of a recognisable language.

The number of states in a minimal automaton for a recognisable language L is called the *rank* of the language L. This can be regarded as a measure of the complexity of L.

Observe that if $\mathbf{A}$ is an automaton, then $\mathbf{A}^{ar}$ and $\mathbf{A}^{ra}$ are both reduced and accessible and recognise $L(\mathbf{A})$. So in principle, we could calculate either of these two automata to find the mimimal automaton. However, it makes sense to compute $\mathbf{A}^{ar} = (\mathbf{A}^{a})^{r}$ rather than $\mathbf{A}^{ra}$. This is because calculating the reduction of an automaton is more labour intensive than calculating the accessible part. By calculating $\mathbf{A}^{a}$ first, we will in general reduce the number of states and so decrease the amount of work needed in the subsequent reduction.

Algorithm 7.4.3 (Minimal automaton) This algorithm computes the minimal automaton for a recognisable language L from any complete deterministic automaton $\mathbf{A}$ recognising L. Calculate $\mathbf{A}^{a}$, the accessible part of $\mathbf{A}$, using Algorithm 3.1.4. Next calculate the reduction of $\mathbf{A}^{a}$, using Algorithm 7.2.5. The automaton $\mathbf{A}^{ar}$ that results is the minimal automaton for L. □

Exercises 7.4

1. Find the rank of each subset of $(0+1)^2$. You should first list all the subsets; construct deterministic automata that recognise each subset; and finally, convert your automata to minimal automata.

2. Let $n \geq 2$. Define

$$L_n = \{x \in (a+b)^*: |\, x \,| \equiv 0 \pmod{n}\}.$$

Prove that the rank of L_n is n.

7.5 The method of quotients

In Section 7.4, we showed that if $L = L(\mathbf{A})$ then the minimal automaton for L is $\mathbf{A}^{ar}$. In this section, we shall construct the minimal automaton of L directly from a regular expression for L. Our method is based on a new language operation.

Let L be a language over the alphabet A and let $u \in A^*$. Define the *left quotient of* L *by* u to be

$$u^{-1}L = \{v \in A^*: uv \in L\}.$$

Similarly, define the *right quotient of* L *by* u to be

$$Lu^{-1} = \{v \in A^*: vu \in L\}.$$

The notation is intended to help you remember the meaning:

$$v \in u^{-1}L \Leftrightarrow uv \in uu^{-1}L \Leftrightarrow uv \in L,$$

because we think of u as being cancelled by u^{-1}.

Terminology In this section, I shall deal primarily with left quotients, so when I write 'quotient,' I shall always mean 'left quotient.'

Examples 7.5.1 Let A be an alphabet, $a \in A$ and L a language over A.

(1) $a^{-1}a = \varepsilon$. Remember that $a^{-1}a$ means $a^{-1}\{a\}$. By definition $u \in a^{-1}\{a\}$ iff $au \in \{a\}$. Thus $au = a$ and so $u = \varepsilon$. It follows that $a^{-1}a = \varepsilon$.

(2) $a^{-1}\varepsilon = \emptyset$. Let $u \in a^{-1}\{\varepsilon\}$. Then $au \in \{\varepsilon\}$ and so $au = \varepsilon$. However there is no string u which satisfies this condition. Consequently $a^{-1}\varepsilon = \emptyset$.

(3) $a^{-1}\emptyset = \emptyset$. This is proved by a similar argument to that in (2) above.

(4) $a^{-1}b = \emptyset$ if $b \in A$ and $b \neq a$. Let $u \in a^{-1}\{b\}$. Then $au = b$. There are no solutions to this equation and so $a^{-1}b = \emptyset$.

(5) $\varepsilon^{-1}L = L$. By definition $u \in \varepsilon^{-1}L$ iff $\varepsilon u \in L$. This just means that $u \in L$. Hence $\varepsilon^{-1}L = L$.

□

The quotients of a regular language, as we shall show, can be used to construct the minimal automaton of the language. So we shall need to develop ways of computing quotients efficiently. To do this, the following simple definition will be invaluable. Let L be any language. Define

$$\delta(L) = \begin{cases} \emptyset & \text{if } \varepsilon \notin L \\ \{\varepsilon\} & \text{if } \varepsilon \in L. \end{cases}$$

Thus $\delta(L)$ simply records the absence or presence of ε in the language. The following lemma provides the tools necessary for computing δ for any language given by means of a regular expression. The proofs are straightforward and left as exercises.[1]

Lemma 7.5.2 *Let A be an alphabet and $L, M \subseteq A^*$.*

(i) $\delta(a) = \emptyset$ *for each* $a \in A$.

(ii) $\delta(\emptyset) = \emptyset$.

(iii) $\delta(\varepsilon) = \varepsilon$.

(iv) $\delta(LM) = \delta(L) \cap \delta(M)$.

(v) $\delta(L + M) = \delta(L) + \delta(M)$.

[1]This just repeats the first part of Lemma 6.2.3.

(vi) $\delta(L^*) = \varepsilon$.

□

We now show how to compute quotients.

Proposition 7.5.3 *Let $u, v \in A^*$ and $a \in A$.*

(i) *If $L = \emptyset$ or ε then $u^{-1}(LM) = L(u^{-1}M)$.*

(ii) *If $\{L_i : i \in I\}$ is any family of languages then $u^{-1}(\sum_{i \in I} L_i) = \sum_{i \in I} u^{-1}L_i$.*

(iii) $a^{-1}(LM) = (a^{-1}L)M + \delta(L)(a^{-1}M)$.

(iv) $a^{-1}L^* = (a^{-1}L)L^*$.

(v) $(uv)^{-1}L = v^{-1}(u^{-1}L)$.

Proof (i) Straightforward.

(ii) By definition $v \in u^{-1}(\sum_{i \in I} L_i)$ iff $uv \in \sum_{i \in I} L_i$. But $uv \in \sum_{i \in I} L_i$ implies $uv \in L_i$ for some $i \in I$. Thus $v \in u^{-1}L_i$ for some $i \in I$. It follows that $v \in \sum_{i \in I} u^{-1}L_i$. The converse is proved similarly.

(iii) Write $L = \delta(L) + L_0$ where $L_0 = L \setminus \varepsilon$. Then

$$a^{-1}(LM) = a^{-1}(\delta(L)M + L_0M) = \delta(L)(a^{-1}M) + a^{-1}(L_0M),$$

using (i) and (ii). It is therefore enough to prove the result for the case where L does not contain ε. We have to prove that

$$a^{-1}(LM) = (a^{-1}L)M$$

if $\varepsilon \notin L$. Let $x \in a^{-1}(LM)$. Then $ax = lm$ where $l \in L$ and $m \in M$ and $l \neq \varepsilon$, by assumption. Thus $l = al'$ for some l'. It follows that $x = l'm$. Also $l = al' \in L$ iff $l' \in a^{-1}L$. Thus $x \in (a^{-1}L)M$. Conversely, if $x \in (a^{-1}L)M$, then $x = l'm$ for some $l' \in a^{-1}L$ and $m \in M$. But then $al' \in L$ and so $ax = (al')m \in LM$.

(iv) By definition $x \in a^{-1}L^*$ iff $ax \in L^*$. Thus $ax = u_1 \ldots u_n$ for some non-empty $u_i \in L$. Now $u_1 = au$ for some u. Hence $x = u(u_2 \ldots u_n)$, where $u \in a^{-1}L$. Thus $x \in (a^{-1}L)L^*$. Conversely, if $x \in (a^{-1}L)L^*$ then $x = u(u_2 \ldots u_n)$ for some $u \in a^{-1}L$. It follows that $au \in L$ and so $ax \in L^*$. Hence $x \in a^{-1}L^*$.

(v) By definition $x \in (uv)^{-1}L$ iff $(uv)x \in L$ iff $u(vx) \in L$ iff $vx \in u^{-1}L$ iff $x \in v^{-1}(u^{-1}L)$. Hence $(uv)^{-1}L = v^{-1}(u^{-1}L)$. □

It is important to note that in parts (iii) and (iv) above we have derived expressions for quotients by means of a *single letter only.* We shall deal with the general case in Proposition 7.5.15.

Examples 7.5.4 In the examples below, $A = \{a, b\}$.

(1) $a^{-1}A = \{\varepsilon\}$. We can write $A = a + b$. Thus

$$a^{-1}A = a^{-1}(a+b) = a^{-1}a + a^{-1}b = \varepsilon + \emptyset = \varepsilon.$$

(2) $a^{-1}A^* = A^* = b^{-1}A^*$. This is straightforward.

(3) Let x be a non-empty string that does not begin with a. Then $a^{-1}(xA^*) = \emptyset$. This is because $y \in a^{-1}(xA^*)$ iff $ay \in xA^*$. But x does not begin with a. So there is no solution for y.

(4) $a^{-1}(axA^*) = xA^*$. This is because $y \in a^{-1}(axA^*)$ iff $ay \in axA^*$. This can only be true if $y \in xA^*$.

(5) Calculate $a^{-1}(A^*abaA^*)$. We can regard A^*abaA^* as a product of two languages in a number or ways, any one of which can be chosen. We choose to regard it as A^* followed by $abaA^*$. Thus by Proposition 7.5.3(iii), we have that

$$a^{-1}(A^*abaA^*) = (a^{-1}A^*)(abaA^*) + \delta(A^*)a^{-1}(abaA^*).$$

We have already shown that $a^{-1}A^* = A^*$ and that $a^{-1}(abaA^*) = baA^*$. Thus

$$a^{-1}(A^*abaA^*) = A^*abaA^* + baA^*.$$

□

We now prove two important results.

Proposition 7.5.5

(i) *The left (respectively right) quotient of a recognisable language is recognisable.*

(ii) *A recognisable language has only a finite number of distinct left (respectively right) quotients.*

Proof (i) Let L be a recognisable language. Then $L = L(\mathbf{A})$ where $\mathbf{A} = (S, A, i, \delta, T)$ is an automaton. We prove first that every left quotient of L is recognisable. Let $u \in A^*$ and put $i' = i \cdot u$. Put $\mathbf{A}_u = (S, A, i', \delta, T)$. We claim that $L(\mathbf{A}_u) = u^{-1}L$. Let $x \in u^{-1}L$. Then $ux \in L$. Thus $i \cdot (ux) \in T$ and so $(i \cdot u) \cdot x \in T$ by Proposition 1.5.4. Hence $i' \cdot x \in T$ giving $x \in \mathbf{A}_u$. We have therefore proved that $u^{-1}L \subseteq L(\mathbf{A}_u)$. To prove the reverse inclusion let $x \in L(\mathbf{A}_u)$. Then $i' \cdot x \in T$ and so $(i \cdot u) \cdot x \in T$. By Proposition 1.5.4, this means that $i \cdot (ux) \in T$ and so $ux \in L(\mathbf{A}) = L$. Hence $x \in u^{-1}L$, as required.

To prove that the right quotient of a recognisable language is recognisable we use the above result and Proposition 3.3.1. Observe that $x \in Lu^{-1}$ iff $xu \in L$ iff $\text{rev}(xu) \in \text{rev}(L)$ iff $\text{rev}(u)\text{rev}(x) \in \text{rev}(L)$. It follows that

$$x \in Lu^{-1} \Leftrightarrow \text{rev}(x) \in \text{rev}(u)^{-1}\text{rev}(L) \Leftrightarrow x \in \text{rev}(\text{rev}(u)^{-1}\text{rev}(L)).$$

Thus

$$Lu^{-1} = \text{rev}(\text{rev}(u)^{-1}\text{rev}(L)).$$

The fact that L is recognisable implies that rev(L) is recognisable by Proposition 3.3.1. By our result above $\text{rev}(u)^{-1}\text{rev}(L)$ is recognisable, and so by Proposition 3.3.1 again, $\text{rev}(\text{rev}(u)^{-1}\text{rev}(L))$ is recognisable. Hence Lu^{-1} is recognisable.

(ii) To finish off, we have to prove that there are only finitely many left quotients. The result for right quotients then follows by the results in (i). Now the set of left quotients of L is just the set of languages $L(\mathbf{A}_s)$, where $\mathbf{A}_s = (S, A, s, \delta, T)$ and $s \in S$, and there are clearly only a finite number of these. □

We can also prove the converse of the above result.

Proposition 7.5.6 *Let L be a language with only a finite number of distinct left (respectively right) quotients. Then L is recognisable.*

Proof We need only prove the result for left quotients. We shall construct a finite automaton $\mathbf{A}_L = (S, A, i, \delta, T)$ such that $L(\mathbf{A}) = L$. Define

- $S = \{u^{-1}L\colon u \in A^*\}$, which is finite by assumption.
- $i = L = \varepsilon^{-1}L$.
- $T = \{u^{-1}L\colon \varepsilon \in u^{-1}L\}$; those quotients of L which contain ε.
- $\delta(u^{-1}L, a) = a^{-1}(u^{-1}L) = (ua)^{-1}L$, using Proposition 7.5.3(v).

By construction, $\mathbf{A}_L$ is a complete deterministic automaton. To calculate $L(\mathbf{A}_L)$ we need to determine δ^*. We claim that

$$\delta^*(u^{-1}L, x) = (ux)^{-1}L$$

for each $x \in A^*$. We leave the proof of this as an exercise.

By definition, $w \in L(\mathbf{A}_L)$ iff $\delta^*(i, w) \in T$ iff $\delta^*(L, w) \in T$. From the form of δ^* and the definition of T this is equivalent to $\varepsilon \in w^{-1}L$, which means precisely that $w \in L$. Hence $L(\mathbf{A}_L) = L$. □

Combining Propositions 7.5.5 and 7.5.6, we now have the following new characterisation of recognisable languages.

Theorem 7.5.7 *A language is recognisable if and only if it has a finite number of distinct left (respectively right) quotients.* □

The automaton $\mathbf{A}_L$ constructed from a recognisable language L in Proposition 7.5.6 is the best we can hope for.

Theorem 7.5.8 *Let L be a recognisable language. Then $\mathbf{A}_L$ is the minimal automaton of L.*

Proof By Theorem 7.4.2, it is enough to show that $\mathbf{A}_L$ is reduced and accessible. The proof that $\mathbf{A}_L$ is accessible is almost immediate: let $u^{-1}L$ be an arbitrary state in $\mathbf{A}_L$. Then $\delta^*(L,u) = u^{-1}L$ and L is the initial state and so $\mathbf{A}_L$ is accessible. To prove that $\mathbf{A}_L$ is reduced, suppose that $u^{-1}L \simeq v^{-1}L$. Then by definition, for each $x \in A^*$ we have that

$$\delta^*(u^{-1}L, x) \in T \Leftrightarrow \delta^*(v^{-1}L, x) \in T.$$

This is equivalent to saying that for each $x \in A^*$, we have that

$$\varepsilon \in (ux)^{-1}L \Leftrightarrow \varepsilon \in (vx)^{-1}L.$$

In other words, $x \in u^{-1}L \Leftrightarrow x \in v^{-1}L$. Hence $u^{-1}L = v^{-1}L$. □

We now describe an algorithm that takes as input a regular expression for a language L and produces as output the minimal automaton $\mathbf{A}_L$. This algorithm has one drawback, which we explain after Example 7.5.13.

Algorithm 7.5.9 (Method of Quotients) Given a regular expression for the recognisable language L, this algorithm constructs the minimal automaton for L. We denote the regular expression describing L also by L. We shall construct the transition tree of $\mathbf{A}_L$, the automaton defined in Proposition 7.5.6, directly from L. It is then an easy matter to construct $\mathbf{A}_L$ which is the minimum automaton by Theorem 7.5.8.

(1) The root of the tree is L. For each $a \in A$ calculate $a^{-1}L$ using Proposition 7.5.3. Join L to $a^{-1}L$ by an arrow labelled a. Any repetitions should be closed with a $\times$.

(2) Subsequently, for each non-closed vertex M calculate $a^{-1}M$ for each $a \in A$ using Proposition 7.5.3. Close repetitions using $\times$.

(3) The algorithm terminates when all leaves are closed. Mark with double circles all labels containing ε. The tree is now the transition tree of $\mathbf{A}_L$, and so $\mathbf{A}_L$ can be constructed in the usual way.

□

Example 7.5.10 Let $A = \{a, b\}$ and $L = (a+b)^*aba(a+b)^*$. We find $\mathbf{A}_L$ using the algorithm above.

(1) $\varepsilon^{-1}L = L = L_0$. By Examples 7.5.1(5).

(2) $a^{-1}L_0 = L + baA^* = L_1$. By Examples 7.5.4(5).

(3) $b^{-1}L_0 = L = L_0$, closed. By Proposition 7.5.3(iii) and Examples 7.5.4(3), and Examples 7.5.4(2).

(4) $a^{-1}L_1 = L_1$, closed. By Proposition 7.5.3(ii) and Examples 7.5.4(3).

(5) $b^{-1}L_1 = L + aA^* = L_2$. By Proposition 7.5.3(ii) and Examples 7.5.4(3) (adapted).

(6) $a^{-1}L_2 = a^{-1}L + A^* = A^* = L_3$. By Proposition 7.5.3 and Examples 7.5.4(4).

(7) $b^{-1}L_2 = L = L_0$, closed. By Proposition 7.5.4 and Examples 7.5.4(3).

(8) $a^{-1}L_3 = A^* = L_3$, closed. By Examples 7.5.4(2).

(9) $b^{-1}L_3 = A^* = L_3$, closed. By Examples 7.5.4(2).

The states of $\mathbf{A}_L$ are therefore

$$\{L_0, L_1, L_2, L_3\},$$

with L_0 as the initial state. The only quotient of L that contains ε is L_3 and so this is the terminal state. The minimal automaton for L is therefore as follows:

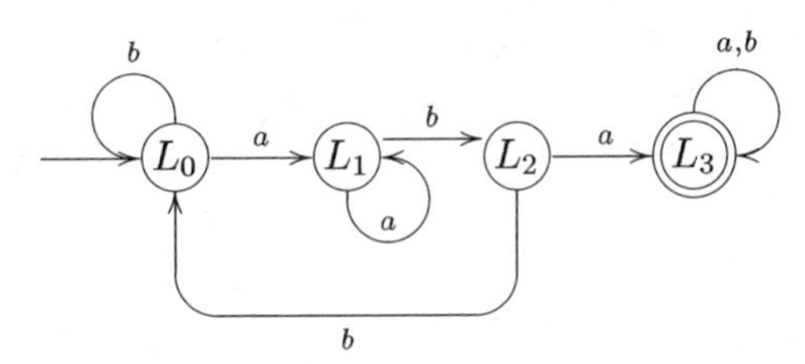

□

The Method of Quotients has an important extra feature. Regular expressions are defined using only $+$, $\cdot$, and $*$, even though we know that the complement of a regular language is regular. We define a *generalised regular expression* to be a regular expression in the usual sense except that we also allow complementation, which we denote by $'$. By De Morgan's laws, if we can do union and complementation we can do intersection. It follows that generalised regular expressions can contain arbitrary Boolean operations. Generalised regular expressions are often a much more natural way of describing recognisable languages. Here is an example.

Example 7.5.11 Let $A = \{0, 1\}$. Consider the language consisting of all strings over A that do not contain three consecutive 0's. A generalised regular expression describing this language is

$$r = [(0+1)^*000(0+1)^*]'.$$

After some thought, we can construct a regular expression describing the same language:

$$s = (1 + 01 + 001)^*(\varepsilon + 0 + 00).$$

Clearly r is a more natural way of describing this language than s. □

The problem with the complementation operation, which is the essential new ingredient distinguishing generalised regular expressions from regular expressions, is that it is only easy to handle for *deterministic* automata. Our algorithm proving one-half of Kleene's Theorem using ε-automata does not enable us to handle complementation. A similar problem arises with Algorithm 6.2.2 via 'linearisation.' However, the Method of Quotients can easily be extended to handle generalised regular expressions. The following lemma is all that is needed. The proofs are left as simple exercises.

Lemma 7.5.12 *Let A be an alphabet and L a language over A.*

(i) $\delta(L') = \begin{cases} \varepsilon & \text{if } \delta(L) = \emptyset \\ \emptyset & \text{else.} \end{cases}$

(ii) $u^{-1}(L') = (u^{-1}L)'$.

□

Example 7.5.13 We wish to construct an automaton to recognise all those strings over the alphabet $A = \{0, 1\}$, which consist of two consecutive 0's but do not end in 01. A generalised regular expression describing this language is

$$R = (A^*00A^*) \cap (A^*01)'.$$

We shall now apply the Method of Quotients to construct a deterministic automaton recognising this language. It will aid our calculations to put $P = A^*00A^*$ and $Q = A^*01$. We leave to the reader the verification of the following calculations and the construction of a corresponding automaton $\mathbf{A}$:

(1) $\varepsilon^{-1}R = R = L_0$.

(2) $0^{-1}R = (P + 0A^*) \cap (Q + 1)' = L_1$.

(3) $1^{-1}R = L_0$, closed.

(4) $0^{-1}L_1 = (Q + 1)' = L_2$.

(5) $1^{-1}L_1 = P \cap (Q + \varepsilon)'$. Here we have to be careful because $P \cap (Q + \varepsilon)' = P \cap Q' \cap \varepsilon'$. Now $\varepsilon' = A^+$ and P does not contain ε and so $P \cap \varepsilon' = P$. It follows that $P \cap (Q + \varepsilon)' = P \cap Q' = R = L_0$.

(6) $0^{-1}L_2 = L_2$, closed.

(7) $1^{-1}L_2 = (Q + \varepsilon)' = L_3$.

(8) $0^{-1}L_3 = L_2$, closed.

(9) $1^{-1}L_3 = Q' = L_4$.

(10) $0^{-1}L_4 = L_2$, closed.

(11) $1^{-1}L_4 = L_4$, closed.

□

We conclude this section by discussing the one drawback of the Method of Quotients. For the Method of Quotients to work, we have to recognise when two quotients are equal as in step (5) in Example 7.5.13 above. However, we saw in Section 5.1 that checking whether two regular expressions are equal is not always easy. If we do not recognise that two quotients are equal, then the machine we obtain will no longer be minimal. Here is another example.

Example 7.5.14 Consider the regular expression,

$$r = a^*(aa)^*.$$

We calculate $a^{-1}r = a^*(aa)^* + a(aa)^*$. This looks different from r. However,

$$a^*(aa)^* = (\varepsilon + a + a^2 + \ldots)(aa)^*,$$

and so $a(aa)^* \subseteq a^*(aa)^*$. It follows that $a^{-1}r = r$. □

Two questions are raised by this problem:

Question 1 Could Algorithm 7.5.9 fail to terminate?

Question 2 If it does terminate, what can we say about the automaton described by the transition tree?

The answer to Question 1 is 'yes' but, as long as we do even a small amount of checking, we can guarantee that the algorithm will always terminate. The answer to Question 2 is that if we fail to recognise when two quotients are the same, then we shall obtain an accessible deterministic automaton but not necessarily one that is reduced. It follows that once we have applied the Method of Quotients we should calculate the indistinguishability relation of the resulting automaton as a check. In what follows, we justify these two answers.

We say that two regular expressions are *similar* if one can be obtained from the other by using the following properties of union: idempotence, commutativity, and associativity. It is not hard to check that similarity is an equivalence relation on the set of regular expressions. Two regular expressions that are not similar are said to be *dissimilar*. If two regular expressions are similar they are certainly equal, but the equality of regular expressions does not in general imply their similarity. Proposition 7.5.16 below tells us that as long as we can check whether two regular expressions are similar or not, then we will construct a finite number of quotients. To prove this result we shall need to extend Proposition 7.5.3.

Proposition 7.5.15 *Let L and M be languages over A, and let $u \in A^*$.*

(i) $u^{-1}(LM) = (u^{-1}L)M + \sum_{u=xy, x \neq u} \delta(x^{-1}L)(y^{-1}M)$.

(ii) $u^{-1}L^* = \sum_{u=xy, x \neq u} \delta(x, y)(y^{-1}L)L^*$ *where $\delta(x, y)$ is either $\emptyset$ or ε.*

Proof (i) We prove the result by induction on the length of the string u. We have already proved in Proposition 7.5.3(iii) that the formula is correct when u has length 1. Assume the result is correct when the string has length n. We prove that this implies the formula is correct for strings of length $n+1$. Let $|v| = n+1$. Then $v = ua$ where u has length n. By Proposition 7.5.3(v), we have that

$$v^{-1}(LM) = a^{-1}(u^{-1}(LM)).$$

Thus

$$v^{-1}(LM) = a^{-1}\left((u^{-1}L)M + \sum_{u=xy, x \neq u} \delta(x^{-1}L)(y^{-1}M)\right).$$

By Proposition 7.5.3(i),(ii) and (v), this is equal to

$$(v^{-1}L)M + \delta(u^{-1}L)(a^{-1}M) + \sum_{u=xy, x \neq u} \delta(x^{-1}L)(ya)^{-1}M,$$

which is equal to the required result.

(ii) By Proposition 7.5.3(iv), the result is true when $|u| = 1$. Assume the formula is true for strings u of length n. We prove the formula is true for strings of length $n+1$. Let $v = ua$ where u has length n. Then

$$v^{-1}L^* = a^{-1}(u^{-1}L^*) = a^{-1}\left(\sum_{u=xy, x \neq u} \delta(x, y)(y^{-1}L)L^*\right),$$

which is equal to

$$\sum_{u=xy, x \neq u} \delta(x, y)((ya)^{-1}L)L^* + \delta(x, y)\delta(y^{-1}L)(a^{-1}L)L^*,$$

using various parts of Proposition 7.5.3 which in turn is of the form,

$$\sum_{v=wz, w \neq v} \delta(w, z)(z^{-1}L)L^*,$$

as required. □

In the result above, the notation $\delta(x, y)$ can take either the value $\emptyset$ or the value ε, but we do not know which in general.

Proposition 7.5.16 *Every regular expression has only a finite number of dissimilar quotients.*

Proof The proof is by induction but we have already done most of the work in Propositions 7.5.3 and 7.5.15. Suppose that L and M each have a finite number of dissimilar quotients. We prove that the same is true for $L + M$, LM, and L^*. The result is clear for $L + M$, so we need only deal with the last two cases.

To prove that LM has only finitely many dissimilar quotients we use Proposition 7.5.15(i). Let $\mathcal{L}$ be the finite set of dissimilar quotients of L, and let $\mathcal{M}$ be the finite set of dissimilar quotients of M. For each $u \in A^*$ we have that $u^{-1}(LM)$ is of the form, $L_1M + \mathsf{M}$ where $L_1 \in \mathcal{L}$ and M is a union of some elements of $\mathcal{M}$. Working with set notation is the same as assuming the associative, commutative, and idempotency laws for union. It follows that if L and M each have finitely many dissimilar quotients so too does LM.

To prove that L^* has only finitely many dissimilar quotients, we use Proposition 7.5.15(ii). For each $u \in A^*$ we have that $u^{-1}L^*$ is of the form $\mathsf{L}L^*$, where L is a a union of some elements of $\mathcal{L}$. Once again, L^* has only finitely many dissimilar quotients. □

The above result reassures us that as long as we at least recognise when two quotients are similar, then our algorithm will terminate.

It remains to explain the implications for the automaton we construct of failing to recognise the equality of two quotients. Suppose that P and Q are two quotients of L such that $P = Q$ but we fail to recognise this when constructing the transition tree used in the Method of Quotients. In the automaton we construct from the transition tree, P and Q will label two different states. These states will be indistinguishable, because for each input x we have that $\varepsilon \in x^{-1}P$ iff $\varepsilon \in x^{-1}Q$.

Exercises 7.5

1. Prove Lemma 7.5.2.

2. Complete the proof of Proposition 7.5.6.

3. Let $A = \{a, b\}$. For each of the languages below find the minimal automaton using the Method of Quotients.

(i) ab.

(ii) $(a + b)^*a$.

(iii) $(ab)^*$.

(iv) $(ab + ba)^*$.

(v) $(a+b)^*a^2(a+b)^*$.

(vi) aa^*bb^*.

(vii) $a(b^2+ab)^*b^*$.

(viii) $(a+b)^*aab(a+b)^*$.

4. Prove Lemma 7.5.12.

5. Fill in all the steps missing from Example 7.5.13.

6. Let L be a language over the alphabet A. Prove that

$$L = \delta(L) + \sum_{a \in A} a(a^{-1}L).$$

J. H. Conway [35] says that this result is "both Taylor's theorem and the mean value theorem" for the theory of quotients of languages.

7. Calculate the quotients of $\{a^n b^n : n \geq 0\}$.

7.6 Summary of Chapter 7

- *Reduction of an automaton*: From each deterministic automaton $\mathbf{A}$ we can construct an automaton $\mathbf{A}^r$ with the property that each pair of states in $\mathbf{A}$ is distinguishable and $L(\mathbf{A}^r) = L(\mathbf{A})$. The automata $\mathbf{A}^{ra}$ and $\mathbf{A}^{ar}$ are isomorphic and both are reduced and accessible.

- *Minimal automaton*: Each recognisable language L is recognised by an automaton that has the smallest number of states amongst all the automata recognising L: this is the minimal automaton for L. Such an automaton must be reduced and connected, and any reduced and connected automaton must be minimal for the language it recognises. Any two minimal automata for a language are isomorphic.

- *Method of Quotients*: The minimal automaton corresponding to the language described by a regular expression r can be constructed directly from r by calculating the quotients of r.

7.7 Remarks on Chapter 7

The minimisation algorithm, Algorithm 7.2.5, goes back to Huffman [69] and Moore [95]. The version I have presented is an informal version of the one described in [65]. An efficient algorithm can be found in [64]. The 'Method of Quotients' and, indeed, most of Section 7.5, is due to Brzozowski [19], although he points out that quotients had been used by earlier writers. Brzozowski uses the term 'derivative' instead of 'quotient' and writes $D_u L$ rather than our $u^{-1}L$. This terminology is based on the following two results (Proposition 7.5.3(ii) and (iii)):

- $D_a(L + M) = D_a L + D_a M$,
- $D_a(LM) = (D_a L)M + \delta(L)(D_a M)$,

which are analogous to the following two results in calculus:

- $\frac{d}{dx}(f + g) = \frac{d}{dx}(f) + \frac{d}{dx}(g)$,
- $\frac{d}{dx}(fg) = \frac{d}{dx}(f)g + f\frac{d}{dx}(g)$.

Conway [35] even goes so far as to have a chapter entitled "The differential calculus of events," where 'event' is an older term for 'language.'

Chapter 8

The transition monoid

In this chapter, we take the first steps in describing an algebraic method for answering questions about recognisable languages. The basic idea is simple. Consider a complete deterministic automaton. Each input string to the automaton defines a function from the set of states to itself. Although there are infinitely many input strings, there can only be a finite number of distinct functions that arise in this way because our machine has only a finite number of states. Thus the set of input strings leads to a finite set of functions defined on the set of states. This set of functions has an additional property: the composition of any two of them is again in the set. Because composition of functions is associative, we therefore obtain a set equipped with an associative binary operation. Such a set is called a 'semigroup,' and because it has only finitely many elements, the semigroup in question is a 'finite semigroup.' Thus with each automaton we can associate a finite semigroup. Now suppose we have a recognisable language. This is accepted by an essentially unique minimal automaton by Theorem 7.4.2 and, by the process above, we can associate a finite semigroup with this machine. It follows that with each recognisable language we can associate a finite semigroup. We shall see in Chapters 9–12 that this semigroup can be used to obtain important information about the language in question.

8.1 Functions on states

In this chapter, we shall be interested in the effect that input strings have on the set of states of an automaton. We begin by making clear what we mean by this. Let $\mathbf{A} = (S, A, i, \delta, T)$ be an automaton. For each $x \in A^*$ we define a function τ_x from S to S as follows:

$$s\tau_x = \delta^*(s, x),$$

where $s \in S$. The function τ_x describes the *effect* of the string x on the set of states S.

Notation Notice that we have written the function τ_x to the right of its argument rather than to the left. This is because automata process input from left to right.

The function τ_x can be explicitly described by means of its row form. To see how, we assume that an ordering of the states has been chosen. By the *row form of* τ_x we mean a table with two rows: the first consisting of the states s listed in their order and the second consisting of the corresponding values $s\tau_x$.

Example 8.1.1 Consider the following automaton **A**:

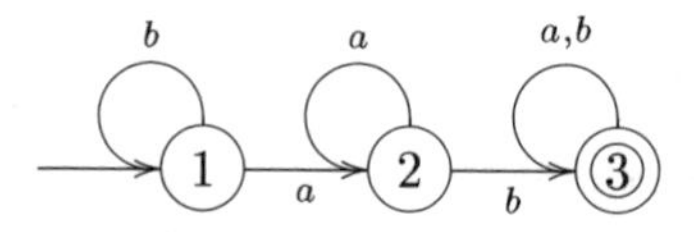

The transition table for this automaton is

	a	b
1	2	1
2	2	3
3	3	3

We have omitted the usual labelling of initial and terminal states in the transition table because they will play no role in what we want to do. Let us calculate the effect of the string ba. To do this we use the extended transition function δ^*. We have that

$$1 \cdot ba = (1 \cdot b) \cdot a = 1 \cdot a = 2,$$

$$2 \cdot ba = (2 \cdot b) \cdot a = 3 \cdot a = 3,$$

$$3 \cdot ba = (3 \cdot b) \cdot a = 3 \cdot a = 3.$$

Thus τ_{ba} is the function:

$$1\tau_{ba} = 2, \quad 2\tau_{ba} = 3, \quad 3\tau_{ba} = 3.$$

In row form, this function is represented by the table

$$\begin{pmatrix} 1 & 2 & 3 \\ 2 & 3 & 3 \end{pmatrix}.$$

□

More generally, we shall be interested in functions defined on any set to itself. Let X be a set, and let α be a function defined from X to itself. We shall write arguments on the left so that if $x \in X$ then $(x)\alpha$ is the value of α

on x. We shall usually write $x\alpha$ unless this could cause ambiguity. If β is also a function from X to itself, then we can define a new function $\alpha \circ \beta$ from X to itself as follows:

$$x(\alpha \circ \beta) = (x\alpha)\beta.$$

In other words, we evaluate α on x and then β on $x\alpha$. We call $\alpha \circ \beta$ the *composition of* α *and* β. We shall usually omit '$\circ$' and write simply $\alpha\beta$.

Example 8.1.2 Let $X = \{1, 2, 3, 4\}$ and let α and β be the two functions that map X to X defined as follows using row form:

$$\alpha = \begin{pmatrix} 1 & 2 & 3 & 4 \\ 2 & 1 & 2 & 3 \end{pmatrix} \quad \beta = \begin{pmatrix} 1 & 2 & 3 & 4 \\ 4 & 3 & 4 & 1 \end{pmatrix}.$$

If we compose the functions in accordance with our definition above we obtain the new function $\alpha\beta\colon X \to X$, which is given by

$$\alpha\beta = \begin{pmatrix} 1 & 2 & 3 & 4 \\ 3 & 4 & 3 & 4 \end{pmatrix}.$$

Let us see how this result was obtained. The composite function $\alpha\beta$ maps elements of X to elements of X. Thus the composite has the form:

$$\begin{pmatrix} 1 & 2 & 3 & 4 \\ ? & ? & ? & ? \end{pmatrix}.$$

Our job is to fill in the question marks. We begin with the first. To calculate $1(\alpha\beta)$ we have to calculate $(1\alpha)\beta$; thus we calculate the effect α has on 1 and then the effect β has on the result:

$$1 \xrightarrow{\alpha} 2 \xrightarrow{\beta} 3.$$

We can therefore fill in the first question mark:

$$\begin{pmatrix} 1 & 2 & 3 & 4 \\ 3 & ? & ? & ? \end{pmatrix}.$$

To fill in the second question mark we have to calculate $2(\alpha\beta)$; this is just

$$2 \xrightarrow{\alpha} 1 \xrightarrow{\beta} 4.$$

We can therefore fill in the second question mark:

$$\begin{pmatrix} 1 & 2 & 3 & 4 \\ 3 & 4 & ? & ? \end{pmatrix}.$$

The remaining two question marks can be calculated in the same way. The only point that has to be remembered is that **we always work from left to**

right. If we compose the functions in the opposite order we obtain the new function $\beta\alpha: S \to S$, which is given by

$$\beta\alpha = \begin{pmatrix} 1 & 2 & 3 & 4 \\ 3 & 2 & 3 & 2 \end{pmatrix}.$$

□

Let X be a set. Denote by $T(X)$ the set of all functions $\alpha: X \to X$. If $X = \{1, \ldots, n\}$ then we usually write T_n rather than $T(\{1, \ldots, n\})$. The *identity function on* X, denoted by ι, is the function from X to X defined by $x\iota = x$ for each $x \in X$. The set $T(X)$ equipped with composition of functions and the distinguished function ι is called the *full transformation monoid on* X. The word 'transformation' is also used to mean 'function'; the meaning of the word 'monoid' will be given in Section 8.4.

Proposition 8.1.3 *Let* $\alpha, \beta, \gamma \in T(X)$. *Then* $\alpha(\beta\gamma) = (\alpha\beta)\gamma$. *Furthermore,* $\iota\alpha = \alpha = \alpha\iota$. □

Proof Let $x \in X$. Then by definition,

$$x(\alpha(\beta\gamma)) = (x\alpha)(\beta\gamma) = ((x\alpha)\beta)\gamma.$$

A similar calculation shows that

$$x((\alpha\beta)\gamma) = ((x\alpha)\beta)\gamma.$$

Thus $\alpha(\beta\gamma) = (\alpha\beta)\gamma$ as required. The proof of the last assertion is left as an exercise. □

The fact that $\alpha(\beta\gamma) = (\alpha\beta)\gamma$ is similar to a property we met in Section 1.1: composition of functions, like concatenation of strings, is associative; when we compose three functions it does not matter where we put the brackets. As we shall show in Chapter 9,[1] more is true: in computing the composition of n elements $\alpha_1 \ldots \alpha_n$ of T_X it does not matter where we put the brackets. However, as Example 8.1.2 shows, the order in which we compose functions is important. If $\alpha \in T(X)$ and n is a positive integer then we shall write α^n to mean the composition of α with itself n times; if $n = 0$ then we define $\alpha^0 = \iota$.

The proof of the following, except the last claim, is just a reformulation of Proposition 1.5.4. We leave it all as an exercise.

Proposition 8.1.4 *Let* $\mathbf{A} = (S, A, i, \delta, T)$ *be an automaton. Then* $\tau_{xy} = \tau_x \tau_y$ *for all* $x, y \in A^*$. *In addition* $\tau_\varepsilon = \iota$, *the identity function on* S. □

The above result, together with the associativity of function composition, has the following consequence. Let $x = a_1 \ldots a_n$. Then $\tau_x = \tau_{a_1} \ldots \tau_{a_n}$.

[1]Specifically, Theorem 9.1.1.

Example 8.1.5 We compute τ_{aba} in the automaton of Example 8.1.1. From our result above, this is equal to $\tau_{aba} = \tau_a \tau_b \tau_a$, which is the composition of the following three functions in the given order:

$$\begin{pmatrix} 1 & 2 & 3 \\ 2 & 2 & 3 \end{pmatrix} \begin{pmatrix} 1 & 2 & 3 \\ 1 & 3 & 3 \end{pmatrix} \begin{pmatrix} 1 & 2 & 3 \\ 2 & 2 & 3 \end{pmatrix}.$$

This is simply

$$\begin{pmatrix} 1 & 2 & 3 \\ 3 & 3 & 3 \end{pmatrix}.$$

□

Let $\mathbf{A} = (S, A, i, \delta, T)$ be an automaton, and let $x, y \in A^*$. We say that *x has the same effect as y (in the automaton $\mathbf{A}$)*, written $x \equiv y$, if $\tau_x = \tau_y$. Strictly speaking we should write $\equiv_{\mathbf{A}}$ to make clear the dependence of $\equiv$ on the automaton, but we shall always use the simpler notation.

Proposition 8.1.6 *Let $\mathbf{A} = (S, A, i, \delta, T)$ be an automaton. Then the relation $\equiv$ which $\mathbf{A}$ defines on A^* is an equivalence relation that has the following additional property: if $x, y, x', y' \in A^*$ and $x \equiv x'$ and $y \equiv y'$ then $xy \equiv x'y'$.*

Proof The proof that $\equiv$ is an equivalence relation is left as an exercise.

To prove the second claim, we are given that $\tau_x = \tau_{x'}$ and $\tau_y = \tau_{y'}$. By Proposition 8.1.4,

$$\tau_{xy} = \tau_x \tau_y \text{ and } \tau_{x'y'} = \tau_{x'} \tau_{y'}.$$

Thus $\tau_{xy} = \tau_{x'y'}$ and so $xy \equiv x'y'$. □

Exercises 8.1

1. List the elements of T_2, and calculate all possible compositions.

2. If $|X| = n$, show that $|T(X)| = n^n$.

3. Consider the automaton $\mathbf{A}$ below:

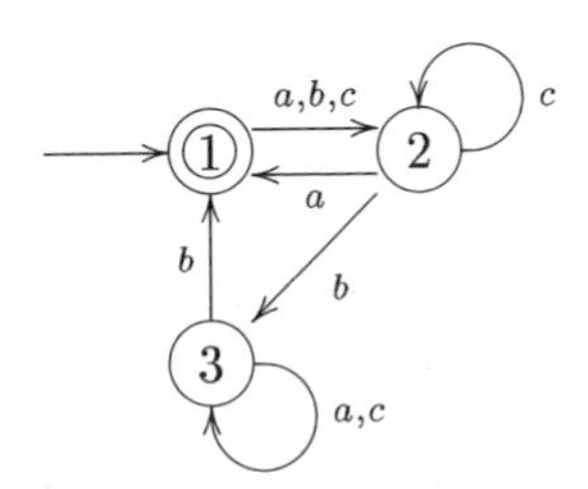

Write down τ_a, τ_b and τ_c in row form. Hence calculate the effects of a^2, b^2, b^3, and c^2.

4. Complete the proof of Proposition 8.1.3 by showing that $\iota\alpha = \alpha = \alpha\iota$.

5. Prove Proposition 8.1.4.

6. Complete the proof of Proposition 8.1.6 by showing that $\equiv$ is an equivalence relation.

8.2 The extended transition table

Let $\mathbf{A} = (S, A, i, \delta, T)$ be an automaton. As we saw in Section 8.1, each string $x \in A^*$ defines a function τ_x from the set of states of $\mathbf{A}$ to itself. We shall be interested in *all* the functions that arise in this way. In other words, the set of functions,

$$T(\mathbf{A}) = \{\tau_x \colon x \in A^*\}.$$

The question now arises of how we can find them in a systematic way. The transition table of $\mathbf{A}$ tells us how to process individual input *letters*: the columns are labelled by the input letters, the rows by the states, and the entry in row s and column a is $\delta(s, a)$. Thus the effects of the input letters can be read off from the columns of the transition table. In order to determine the effects of input *strings*, we have to use the extended transition function δ^*. In principle, we could draw up an 'extended transition table' in which the columns were labelled by the strings in A^*. Of course, this is not practical because there are infinitely many strings. However, as we shall show, all the information in the extended transition table can be presented in entirely finite terms. Before we explain how to do this in general, we give an example.

Example 8.2.1 Consider the following automaton $\mathbf{A}$:

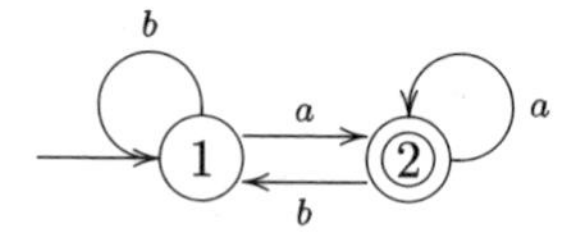

The transition table of $\mathbf{A}$ is given below:

	a	b
1	2	1
2	2	1

We wish to describe the effects of *all* the strings in A^*. We shall do this by expanding the transition table into an 'extended transition table' by adding extra columns. At the same time we shall show how to get around the problem of having infinitely many columns. The first point to note is that our extended transition table is likely to have many columns, so it makes sense to construct them vertically rather than horizontally. Thus the table above will be represented as follows:

	1	2
a	2	2
b	1	1

This has the added benefit that we can now immediately read off from this table the row form of the functions τ_a and τ_b:

$$\tau_a = \begin{pmatrix} 1 & 2 \\ 2 & 2 \end{pmatrix} \text{ and } \tau_b = \begin{pmatrix} 1 & 2 \\ 1 & 1 \end{pmatrix}.$$

The strings in A^* can be listed in the tree order:

$$\varepsilon, a, b, aa, ab, ba, bb, aaa, \ldots.$$

From the definition of the extended transition function we know that τ_ε is the identity function. We may therefore incorporate the effect of ε as follows:

	1	2
ε	1	2
a	2	2
b	1	1

Now we turn to the strings of length 2. We calculate the effect of aa by computing $\tau_a \tau_a$; this is just

$$\begin{pmatrix} 1 & 2 \\ 2 & 2 \end{pmatrix}\begin{pmatrix} 1 & 2 \\ 2 & 2 \end{pmatrix} = \begin{pmatrix} 1 & 2 \\ 2 & 2 \end{pmatrix}.$$

In the same way, we can calculate the effects of the other strings of length 2:

$$\tau_{ab} = \begin{pmatrix} 1 & 2 \\ 2 & 2 \end{pmatrix}\begin{pmatrix} 1 & 2 \\ 1 & 1 \end{pmatrix} = \begin{pmatrix} 1 & 2 \\ 1 & 1 \end{pmatrix}$$

and

$$\tau_{ba} = \begin{pmatrix} 1 & 2 \\ 1 & 1 \end{pmatrix}\begin{pmatrix} 1 & 2 \\ 2 & 2 \end{pmatrix} = \begin{pmatrix} 1 & 2 \\ 2 & 2 \end{pmatrix}$$

and

$$\tau_{bb} = \begin{pmatrix} 1 & 2 \\ 1 & 1 \end{pmatrix}\begin{pmatrix} 1 & 2 \\ 1 & 1 \end{pmatrix} = \begin{pmatrix} 1 & 2 \\ 1 & 1 \end{pmatrix}.$$

Our extended transition table now has the following form:

	1	2	
ε	1	2	
a	2	2	
b	1	1	
aa	2	2	x
ab	1	1	x
ba	2	2	x
bb	1	1	x

I have marked certain rows with an 'x.' I shall explain why below. We could continue extending our table by considering the effects of all strings of length 3; however, we do not need to. This is because each string of length 2 has the same effect on the states of **A** as a string of length at most 1; we have highlighted these rows with an 'x.' Thus we have the following:

$$a^2 \equiv a, \quad ab \equiv b, \quad ba \equiv a \text{ and } b^2 \equiv b.$$

We now have all the information needed to compute the effect of *any* input string with the minimum effort. Consider the first few strings of length 3 in the tree order:

$$\begin{aligned} aaa &= (aa)a \equiv aa \equiv a, \\ aab &= (aa)b \equiv ab \equiv b, \\ aba &= (ab)a \equiv ba \equiv a, \\ abb &= (ab)b \equiv bb \equiv b. \end{aligned}$$

These calculations are correct by Proposition 8.1.6. It should be clear that every string of length 3 has the same effect as one of the strings ε, a, b. More generally, every string in A^* has the same effect as one of these three strings. □

The above example is typical: although there are infinitely many input strings there are only finitely many effects that these input strings can have, because the automaton has only finitely many states. As a result, the extended transition table, which is potentially infinite, can in fact be represented in finite terms. It is this finite table that we shall call the *extended transition table.* The algorithm below is similar to the construction of the transition tree of an automaton, described in Section 3.1, for the following reason. If $\mathbf{A} = (S, A, i, \delta, T)$ is a deterministic automaton with n states $\{s_1, \ldots s_n\}$, construct a new automaton **B** as follows: the set of states is S^n, the initial state is $(s_1, \ldots, s_n)$, the transition function is $(s_1, \ldots, s_n) \cdot a = (s_1 \cdot a, \ldots, s_n \cdot a)$, the terminal states are chosen arbitrarily. The algorithm below just constructs the transition tree of **B**, but does so in tabular form.

Algorithm 8.2.2 (The extended transition table) Choose an order for the alphabet, and accordingly order all input strings using the tree order. Construct a table whose columns are labelled by the states of **A** in some order and whose rows will be labelled by some of the strings in A^*. In addition, certain rows will be marked with a cross.

(1) To start the algorithm off, the first row is labelled ε and contains the states of **A** in the given order. For each $a \in A$, calculate $\delta(s, a)$ for each state s; if the resulting row of states is a repeat of an earlier row, then mark it with a cross. We say that this row is 'closed.'

(2) The general procedure runs as follows: assume that we have completed the rows corresponding to strings of length n. For each string x of length n in turn, which is *not* closed, compute for each $a \in A$ the effect of xa on the states: if the resulting row of states is a repeat of an earlier row then mark it with a cross.

(3) If all rows corresponding to strings of length n are closed then the algorithm terminates.

□

Here is an example of this algorithm in action.

Example 8.2.3 Consider the automaton **A** of Example 8.1.1:

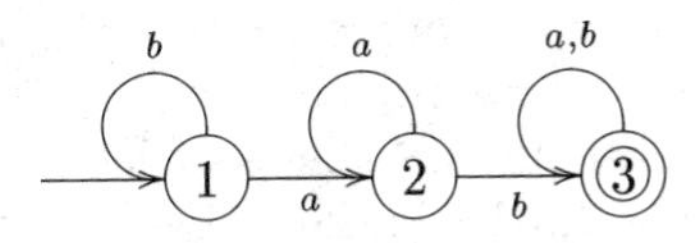

We shall calculate its extended transition table. The algorithm begins with the following three rows:

	1	2	3
ε	1	2	3
a	2	2	3
b	1	3	3

In this case, there are so far no repeats. We therefore have to calculate the effects of all the strings aa, ab, ba and bb. We obtain the following:

	1	2	3	
ε	1	2	3	
a	2	2	3	
b	1	3	3	
a^2	2	2	3	x
ab	3	3	3	
ba	2	3	3	
b^2	1	3	3	x

We see that $a^2 \equiv a$ and $b^2 \equiv b$. The next step of the algorithm considers the effect of the following strings: aba, abb and baa, bab; we do not need to consider the effects of a^2a and a^2b or of b^2a and b^2b because the rows corresponding to

a^2 and b^2 are closed. We obtain the following:

	1	2	3	
ε	1	2	3	
a	2	2	3	
b	1	3	3	
a^2	2	2	3	x
ab	3	3	3	
ba	2	3	3	
b^2	1	3	3	x
aba	3	3	3	x
ab^2	3	3	3	x
ba^2	2	3	3	x
bab	3	3	3	x

At this point our algorithm terminates because all strings of length 3 that appear in the table are closed. □

There is a more compact way of recording the information contained in the extended transition table. Let us first consider only those rows of the table that are not closed. The corresponding table is called the *table of representatives.* In the case of Example 8.2.3, this is the following table:

	1	2	3
ε	1	2	3
a	2	2	3
b	1	3	3
ab	3	3	3
ba	2	3	3

Now we turn to the closed rows. For each closed row labelled x of the extended transition, the string x has the same effect as a unique representative y. We write $x \equiv y$ and call this a *relation.* The relations arising from Example 8.2.3 are as follows:

$$\begin{aligned} a^2 &\equiv a, \\ b^2 &\equiv b, \\ aba &\equiv ab, \\ ab^2 &\equiv ab, \\ ba^2 &\equiv ba, \\ bab &\equiv ab. \end{aligned}$$

We have listed them according to the tree order of their left-hand sides. It is clear that the extended transition table contains exactly the same information as the table of representatives and the list of relations. Algorithm 8.2.2 can easily be modified to yield the table of representatives and relations directly.

Algorithm 8.2.4 (Representatives and relations) We order the alphabet, and accordingly order all input strings using the tree order. Construct a table whose columns are labelled by the states of **A** in some order and whose rows are labelled by some of the strings in A^*. We shall also construct a list of relations.

(1) To start the algorithm off, the first row is labelled ε and contains the states of **A** in the given order. For each $a \in A$ calculate $s \cdot a$ for each state s; if the resulting row of states is a repeat of an earlier row labelled y, then add $a \equiv y$ to the list of relations and erase the new row labelled a.

(2) The general procedure runs as follows: assume that we have completed the rows corresponding to strings of length n. For each string x of length n in turn that appears in the table compute for each $a \in A$ the effect of xa on the states: if the resulting row of states is a repeat of an earlier row labelled y, then add $xa \equiv y$ to the list of relations and erase the new row labelled xa.

(3) The algorithm terminates when for some n all strings x of length n that appear in the table, all the strings xa for each $a \in A$ lead to no new rows.

$\square$

Once we have the table of representatives and relations produced by this algorithm there is a simple algorithm for computing the effect of an arbitrary string over the input alphabet.

Algorithm 8.2.5 (Reduction of strings) Let **A** be an automaton and assume that we have applied Algorithm 8.2.4 to **A** to obtain a table of representatives and a list of relations. Let $w = a_1 \ldots a_n$ be a non-empty string over the input alphabet of **A**. The algorithm computes the unique representative x such that $w \equiv x$.

If w is a representative then we are done, otherwise w has a prefix u such that $u \equiv v$ is a relation. Let $w = uw'$. Then $w = uw' \equiv vw'$. By construction $|\,vw'\,| < |\,w\,|$. Repeat this procedure with w replaced by vw'. The algorithm terminates at one of the representatives x that will satisfy $x \equiv w$ by construction. $\square$

Example 8.2.6 Let us apply Algorithm 8.2.5 to the string $aababbbaa$ using the table of representatives and relations of Example 8.2.3. We obtain the following sequence of strings; prefixes used at each step are underlined and the results of applying each relation are highlighted in bold:

$$\begin{aligned} \underline{aa}babbbaa &\equiv \mathbf{a}babbbaa, \\ \underline{aba}bbbaa &\equiv \mathbf{ab}bbbaa, \\ \underline{abb}bbaa &\equiv \mathbf{ab}bbaa, \end{aligned}$$

$$\begin{aligned}
\underline{abb}baa &\equiv \mathbf{ab}baa, \\
\underline{abb}aa &\equiv \mathbf{ab}aa, \\
\underline{aba}a &\equiv \mathbf{ab}a, \\
\underline{aba} &\equiv \mathbf{ab}.
\end{aligned}$$

Thus $aababbbaa \equiv ab$. □

Algorithms 8.2.4 and 8.2.5 together provide a fast way of computing the effect of an input string on the states of an automaton: Algorithm 8.2.4 is carried out once to provide a table of representatives and a list of relations, and then we implement Algorithm 8.2.5 for each input string we are interested in. It remains for us to prove our claims.

Theorem 8.2.7 *Algorithm 8.2.4 always halts, and the table of representatives and list of relations can be used to calculate the effect of each input string in accordance with Algorithm 8.2.5.*

Proof The proof that Algorithm 8.2.4 always halts is the same as the proof that the construction of the transition tree of an automaton always halts, as explained prior to Algorithm 8.2.2.

It remains to show that we can compute the effect of any input string using Algorithm 8.2.5. We prove the following: for each $x \in A^*$ there exists a representative x' such that $x \equiv x'$ that can be found using Algorithm 8.2.5. This is trivially true for strings of length 0. Assume this is true for all strings x of length at most n. We prove it is true for all strings of length $n + 1$. Let x be a string of length $n + 1$. Then $x = ya$ where y is a string of length n and $a \in A$. By assumption, we may use Algorithm 8.2.5 to find a representative y' such that $y' \equiv y$. Thus by Proposition 8.1.6, we have that $x = ya \equiv y'a$. Now y' is a representative so according to Algorithm 8.2.4, either $y'a$ is a representative, in which case we are done, or it is equivalent to a representative y''.□

Exercises 8.2

1. Calculate the extended transition table, and find all the relations for the following automata.

(i)

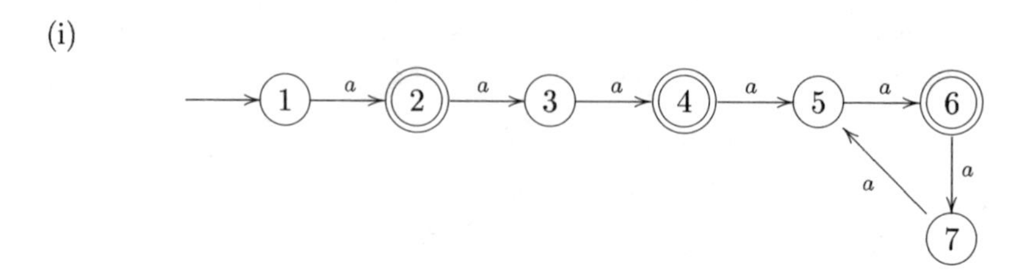

(ii)

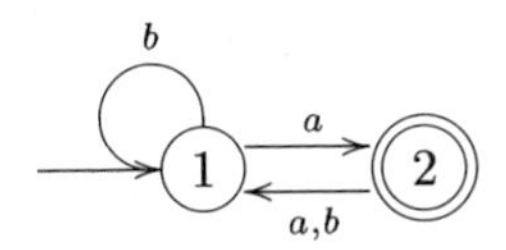

(iii)

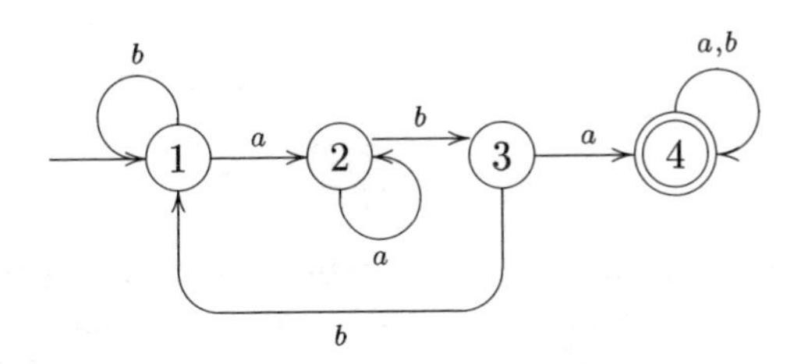

8.3 The Cayley table of an automaton

In the last two sections, we have developed ways of calculating the effects of all input strings. This is not intended to be an end unto itself: it is what we can do with them that is interesting. The first thing we shall do is discard some of the information obtained from Algorithm 8.2.4.

Example 8.3.1 Consider the table of representatives and relations for the automaton in Example 8.2.3. Our first step is to omit some of the information contained in the table of representatives: we are interested in the representatives themselves but not their effects. There are three equivalent ways of doing this.

Method 1 We represent this information as follows:

$$\langle \varepsilon, a, b, ab, ba \colon a^2 \equiv a, b^2 \equiv b, aba \equiv ab, abb \equiv ab, baa \equiv ba, bab \equiv ab \rangle;$$

before the colon we write the representatives and after the colon we write the relations. We call this a *presentation*.[2]

Method 2 The information in a presentation can also be represented diagrammatically in terms of a labelled directed graph called the *Cayley graph* of

[2]This is not the most general definition of 'presentation' in semigroup theory, but it is the only one we shall use in this book.

the automaton:

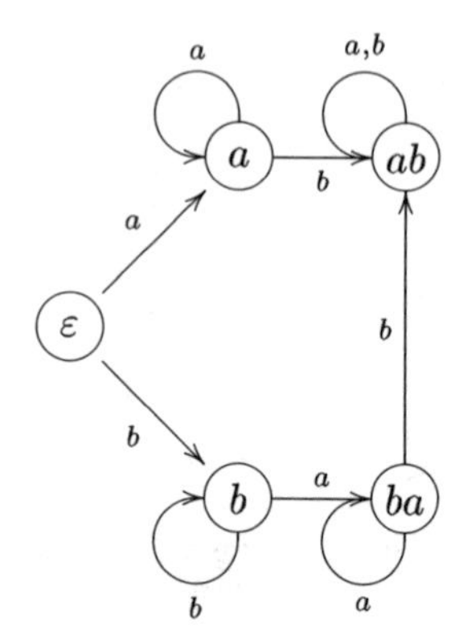

This was constructed as follows: the vertices are labelled by the representatives; for each vertex x and for each $a \in A$ we calculate the representative $y \equiv xa$ and then draw an arrow from vertex x to vertex y labelled by a.

Method 3 We have seen that Algorithm 8.2.5 enables us to calculate for each string x a representative y such that $x \equiv y$. In particular, we can apply this algorithm to the concatenation of two representative. We draw up a table, called a *Cayley table*, whose rows and columns are labelled by the representatives and where the entry in row x and column y is the unique representative z such that $xy \equiv z$.

	ε	a	b	ab	ba
ε	ε	a	b	ab	ba
a	a	a	ab	ab	ab
b	b	ba	b	ab	ba
ab	ab	ab	ab	ab	ab
ba	ba	ba	ab	ab	ab

The presentation, the Cayley graph, and the Cayley table contain the same information in different forms. Given one we can calculate the other two. □

We shall be particularly interested in the Cayley tables of automata. As we shall see there are patterns in such tables that give us information about the language accepted by the automaton.

The Cayley table of an automaton is simply a way of describing the following operation on the set of representatives: given representatives x and y then $x \circ y$ is defined to be the representative z such that $z \equiv xy$. The empty string ε is a representative and it is clear that $\varepsilon \circ x \equiv x \equiv x \circ \varepsilon$. The most important property of the operation $\circ$ is the following.

Proposition 8.3.2 *Let x, y, z be three representatives. Then*

$$x \circ (y \circ z) = (x \circ y) \circ z.$$

Proof We calculate $x \circ (y \circ z)$ first. Let $u \equiv yz$, where u is a representative, and let$v \equiv xu$, where v is a representative. Thus $v = x \circ (y \circ z)$. Now $u \equiv yz$ means $\tau_u = \tau_{yz}$ and $v \equiv xu$ means $\tau_v = \tau_{xu} = \tau_{xyz}$. In a similar fashion, if $v' = (x \circ y) \circ z$ then we can show that $\tau_{v'} = \tau_{xyz}$. It follows that $v \equiv v'$. But both v and v' are representatives and so $v = v'$, as required. □

Exercises 8.3

1. Let $A = \{a, b\}$, and consider the following languages:

(i) aA^*a.

(ii) A^*ab.

(iii) $A^*a^2A^*$.

For each language find:

(a) The minimal automaton **A**.

(b) The extended transition table of **A**.

(c) The presentation.

(d) The Cayley table.

8.4 Semigroups and monoids

The Cayley table of an automaton is an example of a semigroup. This important idea will figure prominently from now on, so in this section, we shall define what they are and give some examples.

Let X be a set. Any function $\alpha: X \times X \rightarrow X$ is called a *binary operation* on X. If $(x, y) \in X \times X$, then the value of α at (x, y) is $\alpha(x, y)$. Two examples of binary operations on the set $\mathbb{N}$ are the function that maps (a, b) to $a + b$, and the function that maps (a, b) to $a \times b$. Notice that we write the binary operation in both cases *between* the two inputs. This convention is usually adopted for all binary operations. Another point to notice is that different binary operations may be defined on the same set. For this reason, we regard a binary operation as an ordered pair: $(X, *)$ where X is the set and $*$ is the specific binary operation. So both $(\mathbb{N}, +)$ and $(\mathbb{N}, \times)$ are different binary operations on the same set.

Binary operations can be classified acording to the properties they enjoy. Let $(X, *)$ be a binary operation. We say that it is *associative* if

$$x * (y * z) = (x * y) * z$$

for all $x, y, z \in X$. A set equipped with an associative binary operation is called a *semigroup*. If $(X, *)$ is a semigroup then we often say that X is a semigroup *with respect to* the operation $*$ or *under* the operation $*$.

If $(X, *)$ is a semigroup, then an element $e \in X$ is called an *identity* for the binary operation $*$ if

$$e * x = x = x * e$$

for all $x \in X$. An element $z \in X$ is called a *zero* if

$$z * x = z = x * z$$

for all $x \in X$. The identity element does not change anything, whereas the zero makes everything into a zero.

Semigroups will usually be denoted by capitals such as $S, T, \ldots$. When dealing with general semigroup operations, for example in proofs, it is usual to denote the operation by concatenation, however when dealing with specific examples of semigroup operations we use the appropriate symbol for the binary operation.

Proposition 8.4.1 *A semigroup S has at most one identity element, and at most one zero.*

Proof Suppose that e and f are both identity elements of S. This means that $es = se = s$ and $fs = sf = a$ for all $s \in S$. We calculate the product ef in two different ways. First, e is an identity element and so $ef = f$. Second f is an identity element and so $ef = e$. Thus $e = ef = f$ and so $e = f$. Therefore if a semigroup S has an identity element e it is unique.

Now let y and z be two zeros in S. Then $ys = y = sy$ and $zs = z = sz$ for all $s \in S$. Thus $yz = y$ and $yz = z$ from which we get that $y = z$. □

A semigroup with an identity is called a *monoid*. It follows that if a semigroup has an identity it has exactly one, and so we can talk about *the* identity of a monoid. Similarly we can talk about *the* zero of a semigroup if it has one.

Let S be a semigroup, and let 1 be a symbol not in S. Put $S^1 = S \cup \{1\}$ and define a product on this set as follows: if $a, b \in S$ then ab is their product in S, and $1a = a = a1$ for all $a \in S$, and $11 = 1$. Then S^1 is a monoid with identity 1.[3]

A semigroup $(S, *)$ is said to be *finite* if S is a finite set and *infinite* otherwise. Finite semigroups $(S, *)$ can be described by means of *Cayley tables*: this is a table with rows and columns labelled by the elements of S and with the element in row a and column b being $a * b$. We first met Cayley tables in Section 8.3. A semigroup $(S, *)$ is said to be *commutative* if

$$a * b = b * a$$

for all $a, b \in S$.

[3]Readers should be aware that some authors define S^1 to be S when S is already a monoid.

Examples 8.4.2 Monoids have occurred in a number of places throughout this book.

(1) In Section 1.1, we showed that for any alphabet A, the set A^* equipped with the binary operation of concatenation is an associative binary operation with identity ε. Thus A^* is a monoid. When A consists of one letter only, the monoid A^* is commutative, but in general it is non-commutative. In all cases A^* is infinite.

(2) If L is any language, then $\varepsilon \in L^*$, and $x, y \in L^*$ implies that $xy \in L^*$ (Question 4 of Exercises 1.3). Thus L^* is a monoid.

(3) Let A be an alphabet. The pair $(\mathsf{P}(A^*), +)$ is a monoid with identity $\emptyset$ (Proposition 5.1.5).

(4) Let A be an alphabet. The pair $(\mathsf{P}(A^*), \cdot)$ is a monoid with identity $\{\varepsilon\}$ (Proposition 5.1.5).

(5) Let X be a set. Then the pair $(T(X), \circ)$ is a monoid with identity ι, the identity function on the set X, by Proposition 8.1.3.

(6) Consider the set $M_n(\mathsf{P}(A^*))$ of all $n \times n$ matrices with entries from $\mathsf{P}(A^*)$. Then with matrix multiplication as the binary operation, we have an associative operation with identity the $n \times n$ matrix all of whose entries are $\emptyset$ except the leading diagonal that instead consists entirely of ε's.

$\square$

Examples (3) and (4) can be generalised. Let S be a semigroup. Then $\mathsf{P}(S)$ is a semigroup in two different ways. First under the operation $+$ of union: $(\mathsf{P}(S), +)$ is a commutative monoid with identity $\emptyset$. Second under *product of subsets*, which is defined as follows: if $X, Y \subseteq S$ then define

$$X \cdot Y = \{xy \colon x \in X \text{ and } y \in Y\}.$$

Then $(\mathsf{P}(S), \cdot)$ is a semigroup with zero $\emptyset$. We usually denote this binary operation by concatenation. If S is a monoid then $(\mathsf{P}(S), \cdot)$ is a monoid with identity $\{1\}$. These two operations interact nicely with each other:

$$X(Y + Z) = XY + XZ \text{ and } (Y + Z)X = YX + ZX.$$

Just as we did in the case where $S = A^*$, we shall usually denote subsets of S of the form $\{s\}$ by s. We can therefore form products such as aX or Xa where $a \in S$ and $X \subseteq S$. If $X \subseteq S$ then we define $X^1 = X$ and $X^n = XX^{n-1}$ for $n \geq 1$. If S is a monoid with identity 1, we define $X^0 = \{1\}$.

To conclude this section, we return to what we did in Sections 8.1–8.3 and apply our new terminology. Let $\mathbf{A}$ be an automaton with state set S and input alphabet A. By Proposition 8.1.4, the set of functions,

$$T(\mathbf{A}) = \{\tau_x \colon x \in A^*\},$$

is closed under composition of functions and contains the identity function. Since the composition of functions is associative by Proposition 8.1.3, it follows that $T(\mathbf{A})$ is a monoid. It is called the *transition monoid* of the automaton $\mathbf{A}$. The set of representatives of an automaton forms a monoid by Proposition 8.3.2, called the *monoid of representatives*. The transition monoid and the monoid of representatives are closely related because $\tau_x \tau_y = \tau_{xy} = \tau_{x \circ y}$. We shall explain the nature of this relationship in Chapter 9.

Exercises 8.4

1. Show that S^1 really is a monoid.

2. Let S be a semigroup. Prove that both $(\mathsf{P}(S), +)$ and $(\mathsf{P}(S), \cdot)$ are semigroups. Show also that $A \cdot (B + C) = A \cdot B + A \cdot C$ for all $A, B, C \in \mathsf{P}(S)$.

3. Let $S = \{-1, 1\}$ with multiplication as product. Then S is a monoid. Draw up the Cayley table for $\mathsf{P}(S)$ with respect to product of subsets.

4. An element α of T_n is said to 'fix the letter 1' if $1\alpha = 1$. Prove that the composition of two functions that fix the letter 1 also fixes the letter 1. Deduce that the elements of T_n that fix the letter 1 form a monoid under composition. Draw up the Cayley table for those elements of T_3 that fix the letter 1.

8.5 Summary of Chapter 8

- *The transition monoid of an automaton*: This is the set of all functions on the set of states induced by input strings. It is closed under composition of functions. The elements of the transition monoid can be found in a way analogous to the calculation of the accessible states of an automaton via its transition tree.

- *Representatives and relations*: Each element of the transition monoid is induced by a string and, if we find the smallest such string in the tree order, we get a bijection between elements of the transition monoid and a finite set of strings. These are called representatives. In addition, for each representative x and each symbol a, the string xa must be equivalent to a unique representative y. We call $xa \equiv y$ a relation.

- *Semigroups and monoids*: A semigroup is a set equipped with an associative binary operation. A monoid is a semigroup with an identity.

8.6 Remarks on Chapter 8

In writing this chapter, I found inspiration by reading Chapter 4 of [88].

At this point, we come to a turning point in the book. The methods used in Chapters 1 to 8 have been elementary although sometimes quite involved, but the introduction of the transition monoid of an automaton marks the change to more advanced methods. The deeper analysis of recognisable languages would be impossible without them.

Chapter 9

The syntactic monoid

In the last chapter, we associated two monoids with an automaton: the transition monoid and the monoid of representatives. The algebraic ideas described in this chapter will enable us to prove that these two monoids are essentially the same or 'isomorphic.' We are not just interested in these monoids for their own sake, they can help us to understand the properties of recognisable languages by means of the following idea. Given a recognisable language, we can construct its minimal automaton **A**, which is, as we have seen, essentially unique. The transition monoid of **A** might therefore be expected to have some special significance, and it does. With any language whatsoever we can associate a monoid called the syntactic monoid of the language. This monoid is finite precisely when the language is recognisable, in which event the syntactic monoid is isomorphic to the transition monoid of the minimal automaton associated with the language. With each recognisable language, therefore, we can associate a finite monoid and this monoid can be explicitly calculated using the mininal automaton of the language. Encoded into the algebraic structure of this monoid is information about the language.

9.1 Introduction to semigroups

In this section, I shall describe some of the elementary properties of semigroups. The first result describes *the* fundamental property of associative binary operations.

Theorem 9.1.1 *Let S be a semigroup and let $x_1, x_2, \ldots, x_n \in S$. Then all bracketings of the n-fold product $x_1 x_2 \cdots x_n$ are equal.*

Proof The proof is by induction on the length n of the product in question. The case $n = 3$ is just the associative law that is given, so we may assume that $n \geq 4$. If $x_1, x_2, \cdots, x_n$ are elements of our semigroup S, then one particular bracketing will play an important role in our proof:

$$x_1(x_2(\cdots(x_{n-1}x_n)\cdots)).$$

We denote this by $x_1x_2\cdots x_n$. Let X denote any properly bracketed expression obtained by inserting brackets into the sequence $x_1, x_2, \cdots, x_n$. Observe that the computation of such a bracketed product involves computing $n-1$ products. This is because at each step we can only compute the product of adjacent letters x_ix_{i+1} and the result is a single letter. Thus at each step of our calculation we reduce the number of letters by one until there is only one letter left. However the expression may be bracketed, the final step in the computation will be of the form YZ, where Y and Z will each have arisen from properly bracketed expressions. In the case of Y it will involve a bracketing of some sequence $x_1, x_2, \cdots, x_r$, and for Z the sequence $x_{r+1}, x_{r+2}, \cdots x_n$ for some r such that $1 \leq r \leq n-1$. Since Y involves a product of length $r < n$, we may assume by the induction hypothesis that $Y = x_1x_2\cdots x_r = x_1(x_2\cdots x_r)$. Hence by associativity,

$$X = YZ = (x_1(x_2\cdots x_r))Z = x_1((x_2\cdots x_r)Z).$$

But $(x_2\cdots x_r)Z$ is a properly bracketed expression of length $n-1$ in $x_2, \cdots, x_n$ and so, again using the induction hypothesis, must equal $x_2x_3\cdots x_n$. This produces $X = x_1x_2\cdots x_n$, as required, to show that all possible bracketings yield the same result in the presence of associativity. □

We illustrate a special case of the above proof in the example below.

Example 9.1.2 Take $n = 5$. Then $x_1x_2x_3x_4x_5 = x_1(x_2(x_3(x_4x_5)))$. Consider the product $((x_1x_2)x_3)(x_4x_5)$ Here we have $Y = (x_1x_2)x_3$ and $Z = x_4x_5$. By associativity $Y = x_1(x_2x_3)$. Thus $YZ = (x_1(x_2x_3))(x_4x_5)$. But this is equal to $x_1((x_2x_3)(x_4x_5))$ again by associativity. By the induction hypothesis $(x_2x_3)(x_4x_5) = x_2(x_3(x_4x_5))$, and so

$$((x_1x_2)x_3)(x_4x_5) = x_1(x_2(x_3(x_4x_5))),$$

as required. □

Associativity means that computing products of elements is straightforward, because we never have to worry about how to evaluate the products as long as we maintain order. Most of the important binary operations in mathematics are associative.[1]

An important consequence of associativity is that for any element a of a semigroup we may define the nth power a^n of a as the product of a with itself n times: the result is the same, however the product is bracketed. It follows that the two *laws of exponents* apply in any semigroup: $a^{m+n} = a^m a^n$ and

[1] But not all. The vector product of two vectors in $\mathbb{R}^3$ is non-associative.

$(a^m)^n = a^{mn}$, where m and n are any positive integers. Moreover powers of the same element a commute with one another: $a^m a^n = a^n a^m$ as both products equal a^{m+n}. Indeed, for the case of monoids we can extend these laws to cover exponents involving 0 using the convention that $a^0 = 1$, where 1 is the identity.

In order to show that a binary operation is associative we have in principle to check that all possible products $(ab)c$ and $a(bc)$ are equal. However there are circumstances when showing that a binary operation is associative is easy. Let $(S, *)$ be a semigroup and let $T \subseteq S$ be any subset. If for *every* pair of elements $a, b \in T$ the product $a * b \in T$ then $*$ is also a binary operation on T and we say that T is *closed under* $*$. In addition, the pair $(T, *)$ is a semigroup in its own right because the associative law holds in S and so it must hold in T. We say that $(T, *)$ is a ***subsemigroup*** of $(S, *)$. Usually, we just say that 'T is a subsemigroup of S' as long as the binary operation in question is clear. Now let $(S, *)$ be a monoid with identity e. Let T be a subsemigroup of S and suppose in addition that $e \in T$. Then T is a monoid in its own right and we say that T is a ***submonoid*** of S. Observe that if T is a subset of a semigroup S, it is a subsemigroup precisely when $T^2 \subseteq T$. Observe also that if S is a subsemigroup of T, and T is a subsemigroup of U, then S is a subsemigroup of U.

Example 9.1.3 Here are some examples of subsemigroups and submonoids.

(1) The set of even numbers in $\mathbb{N}$ is a submonoid under addition. However the set of odd numbers is not: it contains neither the identity 0 nor is it closed under addition.

(2) The set of even numbers is a subsemigroup of $\mathbb{N}$ under multiplication because the product of two even numbers is even. However, it is not a submonoid because the multiplicative identity of $\mathbb{N}$ is 1, which is not even. The set of odd numbers is a submonoid because 1 is odd and the product of two odd numbers is odd.

(3) Let $\text{Rec}(A)$ denote the set of recognisable languages over A. Then this is a submonoid of $\mathsf{P}(A^*)$ under product of languages because if L and M are recognisable then so is LM. The identity is ε, which is recognisable. The set $\text{Rec}(A)$ is also a monoid under union because if L and M are recognisable then so is $L + M$. The identity now is $\emptyset$, which is recognisable.

(4) The transition monoid $T(\mathbf{A})$ of an automaton $\mathbf{A}$ with state set S is a submonoid of the full transformation monoid $T(S)$.

□

Submonoids of A^* played an important role in formulating Kleene's Theorem as we now show. Let S be a semigroup and let $X \subseteq S$. Define

$$X^+ = X + X^2 + X^3 + \ldots.$$

If S is a monoid with identity 1 then we can also define

$$X^* = 1 + X + X^2 + X^3 + \ldots.$$

It is easy to check that X^+ is a subsemigroup of S and that if S is a monoid then X^* is a submonoid of S.

Proposition 9.1.4 *Let S be a semigroup and let X be a subset of S. Then X^+ is the smallest subsemigroup of S containing X. If S is a monoid, then X^* is the smallest submonoid of S containing X.*

Proof We prove the first case; the second case follows almost immediately. By construction, X^+ is a subsemigroup of S containing X. Suppose that T is any subsemigroup of S containing X. Then $X^n \subseteq T^n \subseteq T$ for each $n \geq 1$. It follows that $X^+ \subseteq T$. Thus X^+ is contained in every subsemigroup of S containing X. □

Example 9.1.5 Let A be a finite alphabet. Let $L \subseteq A^*$ be any language. Then L^* is the smallest submonoid of A^* containing L. This describes the purely algebraic role of the Kleene star operation. □

Let S be a semigroup and X a subset of S. Then X^+ is called the *subsemigroup generated by* X. A subset $X \subseteq S$ is called a *generating set* if $S = X^+$; in this case, we say that S is *generated* by X. If a semigroup S can be generated by a finite generating set, it is said to be *finitely generated*. We can make the same definitions in the monoid case replacing X^+ by X^*.

Examples 9.1.6 Here are some examples of finitely generated semigroups and monoids.

(1) The monoid $(\mathbb{N}, +)$ is finitely generated; the subset consisting of 1 alone does the trick because every non-identity element can be obtained by adding 1 to itself the requisite number of times.

(2) Every finite semigroup S is finitely generated. For example, it is generated by S itself. However, it is often possible to find much smaller sets to generate finite semigroups.

(3) The monoid A^* is finitely generated by A itself when A is a finite alphabet.

□

So far, we have concentrated on the properties an individual semigroup might have, but we are also interested in comparing different semigroups. Sets can be compared using functions, and so we shall also use functions to compare semigroups. But a semigroup has more structure than the set that underlies it, so the functions we use to compare semigroups will also have to reflect that extra structure. Let $(S, *)$ and $(T, \circ)$ be semigroups. A function $\alpha: S \to T$ is called a *homomorphism* if

$$\alpha(s_1 * s_2) = \alpha(s_1) \circ \alpha(s_2).$$

Suppose S is a monoid with identity e_S, and T is a monoid with identity e_T. If $\alpha(e_S) = e_T$ then we say that α is a *monoid homomorphism*. Injective homomorphisms are often called *embeddings*.

Examples 9.1.7 Some examples of homomorphisms.

(1) Let A be an alphabet. Then A^* is a monoid under concatenation. We regard $\mathbb{N}$ as a monoid under addition. Define a function from A^* to $\mathbb{N}$ by $x \mapsto |x|$. This is a monoid homomorphism, because $|xy| = |x| + |y|$ and $|\varepsilon| = 0$.

(2) We first met monoid homomorphisms in Section 4.2 as a means of comparing languages over different alphabets.

(3) The set $S = \{\emptyset, \varepsilon\}$ is a monoid with respect to union with identity $\emptyset$. Likewise $\mathsf{P}(A^*)$ is a monoid with respect to union with identity $\emptyset$. The function $\delta: \mathsf{P}(A^*) \to S$ defined by $L \mapsto \delta(L)$ is a monoid homomorphism by Lemma 7.5.2.

(4) Let $\mathbf{A}$ be an automaton with state set S and input alphabet A. Define $\tau: A^* \to T(S)$ by $\tau(x) = \tau_x$, the effect the string x has on the set of states S. Then τ is a monoid homomorphism by Proposition 8.1.4.

□

The following properties of homomorphisms are fundamental.

Proposition 9.1.8 *Let $\alpha: S \to T$ be a homomorphism between semigroups.*

(i) *The image of α is a subsemigroup of T.*

(ii) *If T' is a subsemigroup of T then $\alpha^{-1}(T')$ is a subsemigroup of S.*

(iii) *Let $\beta: T \to U$ be a homomorphism between semigroups. Then $\beta\alpha: S \to U$ is a homomorphism.*

These results also hold when 'semigroup' is replaced by 'monoid' and 'homomorphism' by 'monoid homomorphism.'

Proof (i) Put $T' = \mathrm{im}(\alpha)$. Let $t_1, t_2 \in T'$. By assumption there exist elements $s_1, s_2 \in S$ such that $\alpha(s_1) = t_1$ and $\alpha(s_2) = t_2$. But

$$\alpha(s_1 s_2) = \alpha(s_1)\alpha(s_2) = t_1 t_2.$$

Thus $t_1 t_2 \in T'$. It follows that T' is a subsemigroup of T.

The proofs of (ii) and (iii) and the monoid case are left as exercises. □

Example 9.1.9 Let $\mathbf{A} = (S, A, i, \delta, T)$ be an automaton. Then by Example 9.1.7(4), the function $\tau \colon A^* \to T(S)$ defined by $\tau(x) = \tau_x$ is a monoid homomorphism. The image of τ is the set,

$$T(\mathbf{A}) = \{\tau_x \colon x \in A^*\},$$

which is a submonoid of $T(S)$ by Proposition 9.1.8(i). It is just the transition monoid of **A**. □

One way of comparing two semigroups is to say that despite superficial differences they are 'essentially the same.'

Example 9.1.10 The monoid a^* under concatenation has elements that are strings of a's. Whereas the monoid $(\mathbb{N}, +)$ has elements that are natural numbers. But these two monoids are essentially the same: the element a^m of a^* corresponds to the element m of $\mathbb{N}$, and the product $a^m a^n$ corresponds to the sum $m + n$. □

The mathematical way of saying that two structures of the same type are 'essentially the same' is provided by the notion of 'isomorphism.'[2] We have already met the notion of isomorphism appropriate for automata in Section 7.3. The appropriate notion for semigroups is similar. An *isomorphism of semigroups* is a bijective homomorphism whose inverse is a homomorphism. An *isomorphism of monoids* is the same as an isomorphism of semigroups except that the homomorphisms in question are monoid homomorphisms. If S and T are semigroups (or monoids) and there is an isomorphism from S to T, then we say that *S is isomorphic to T*. Isomorphic semigroups are to be regarded as essentially the same, since they have exactly the same properties as semigroups. See Question 11 of Exercises 9.1 for examples. The monoid a^* under concatenation and the monoid $\mathbb{N}$ under addition are isomorphic.

We say that we know a semigroup S *up to isomorphism* if we have a semigroup T, which is isomorphic to S. Generally speaking, we are happy to have isomorphic copies of a semigroup, and we are not usually worried about the nature of the elements.

[2]From two Greek words meaning 'same shape.'

To check that a function is an isomorphism involves three things: showing that it is a homomorphism, showing that is it a bijection, and showing that its inverse is a homomorphism. In fact, the last condition comes for free.

Proposition 9.1.11 *A bijective homomorphism is an isomorphism.*

Proof Let $\alpha\colon S \to T$ be a bijective homomorphism. We prove that α^{-1} is a homomorphism, which justifies the claim. Let $t, t' \in T$. Put $s = \alpha^{-1}(t)$ and $s' = \alpha^{-1}(t')$. Because α is a homomorphism $\alpha(ss') = \alpha(s)\alpha(s')$. Thus $ss' = \alpha^{-1}(\alpha(s)\alpha(s'))$. We therefore have $\alpha^{-1}(t)\alpha^{-1}(t') = \alpha^{-1}(tt')$ and so

$$\alpha^{-1}(tt') = \alpha^{-1}(t)\alpha^{-1}(t'),$$

as required. □

Exercises 9.1

1. Let S be a semigroup. Show explicitly that

$$(((ab)c)d)e = a(b(c(de)))$$

for all $a, b, c, d, e \in S$.

2. Let $\mathbb{R}$ be the set of all real numbers. Show that the operation of subtraction is a binary operation on $\mathbb{R}$ but that it is neither associative nor commutative.

3. Let $\mathbb{R}^+$ be the set of all positive real numbers. Let $\div$ denote the operation of division in which $a \div b = \frac{a}{b}$. Show that $\div$ is a binary operation on $\mathbb{R}^+$ that is neither associative nor commutative.

4. Consider the operation $\circ$ on $\mathbb{N}$, defined by the rule: $a \circ b = a + b + ab$. Show that $(\mathbb{N}, \circ)$ is a commutative monoid.

5. Let $\circ$ denote the binary operation on $\mathbb{R}^+$ defined by $a \circ b = (\frac{1}{a} + \frac{1}{b})^{-1}$. Show that $\circ$ is both commutative and associative so that $(\mathbb{R}^+, \circ)$ is a commutative semigroup. Is $(\mathbb{R}^+, \circ)$ a monoid?

6. Let $S = \{3x + 5y\colon x, y \in \mathbb{N}\}$. Show that S is a submonoid of $\mathbb{N}$ with respect to addition. Find an explicit description of the elements of S, in particular what is the largest element of $\mathbb{N}$ not in S?

Subsemigroups of $(\mathbb{N}, +)$ *are called* numerical semigroups. *You can find out more about them in [48], for example.*

7. Let α be the following element of T_8:

$$\alpha = \begin{pmatrix} 1 & 2 & 3 & 4 & 5 & 6 & 7 & 8 \\ 2 & 3 & 4 & 3 & 4 & 2 & 8 & 8 \end{pmatrix}.$$

Find all the elements in α^*.

8. Let S be the set of 2×2-matrices with coefficients from $\mathbb{R}$. Show explicitly that S is a monoid. Let T be the monoid $(\mathbb{R}, \times)$. Define $\delta: S \to T$ by $\delta(A) = ad - bc$ where

$$A = \begin{pmatrix} a & b \\ c & d \end{pmatrix}.$$

Show that δ is a monoid homomorphism.

9. Complete the proof of Proposition 9.1.8.

10. Let S, T and U be semigroups. Prove the following:

(i) The identity function on S is an isomorphism from S to itself.

(ii) If $\theta: S \to T$ is an isomorphism then $\theta^{-1}: T \to S$ is an isomorphism.

(iii) If $\theta: S \to T$ and $\phi: T \to U$ are both isomorphisms then $\phi\theta: S \to U$ is an isomorphism.

11. Let S and T be isomorphic semigroups.

(i) Prove that $|S| = n$ iff $|T| = n$.

(ii) Prove that S has an identity if and only if T has an identity.

(iii) Prove that S has a zero if and only if T has a zero.

(iv) Prove that S is commutative iff T is commutative.

9.2 Congruences

In Section 7.1, we showed that partitions and equivalence relations were different ways of thinking about the same thing. We now introduce another way of thinking about equivalence relations that is important in algebra.

Lemma 9.2.1 *Let $\alpha: X \to Y$ be a function between two sets. The set,*

$$Z = \{\alpha^{-1}(y) \colon y \in \mathrm{im}(\alpha)\},$$

consisting of the inverse images of each of the elements in the image of α, is a partition of X.

Proof Each $\alpha^{-1}(y)$ is non-empty, because y is chosen from the image of α. Suppose that $\alpha^{-1}(y) \cap \alpha^{-1}(y') \neq \emptyset$. Then there exists $x \in \alpha^{-1}(y) \cap \alpha^{-1}(y')$. But then $\alpha(x) = y$ and $\alpha(x) = y'$. But α is a function and so $y = y'$, which implies that $\alpha^{-1}(y) = \alpha^{-1}(y')$. Finally, if $x \in X$ then $\alpha(x) = y$ for some $y \in Y$. Hence $x \in \alpha^{-1}(y)$. We have shown that the three defining conditions for a partition of Section 7.1 hold. □

The following example illustrates this result.

Example 9.2.2 Consider the function $\alpha\colon X \to Y$ where

$$X = \{1, 2, 3, 4, 5, 6, 7, 8, 9\} \text{ and } Y = \{a, b, c, d\},$$

and where the rule is given by means of the table below:

$$\begin{pmatrix} 1 & 2 & 3 & 4 & 5 & 6 & 7 & 8 & 9 \\ a & a & c & b & b & a & c & c & b \end{pmatrix}.$$

Notice that α is not surjective, because d is not in the image of α. The partition α induces on X is

$$P = \{\{1, 2, 6\}, \{4, 5, 9\}, \{3, 7, 8\}\}.$$

□

We know from Section 7.1 that partitions and equivalence relations contain exactly the same information. We can easily write down the equivalence relation associated with a function via its partition: this relation is just

$$x \sim x' \Leftrightarrow \alpha(x) = \alpha(x').$$

The relation $\sim$ is usually denoted by $\ker(\alpha)$ and is called the *kernel of* α.

Thus each function gives rise to an equivalence relation on its domain. On the other hand, the next result tells us that we can construct surjective functions from equivalence relations.

Lemma 9.2.3 *Let ρ be an equivalence relation on the set X. Define*

$$\nu\colon X \to X/\rho$$

by mapping x to $\rho(x)$, the ρ-equivalence class that contains x. Then ν is a surjective function.

Proof The equivalence relation ρ induces a partition on the set X whose blocks are the ρ-equivalence classes. Each element of X belongs to a unique equivalence class, so that ν is a well-defined function, and because equivalence classes are non-empty, the function ν is surjective. □

The function ν is called the *natural function* associated with the equivalence relation ρ and is sometimes written $\rho^\natural$.

If we start with a function, we can construct an equivalence relation, and from that equivalence relation we can construct its natural function. We are impelled to ask how the original function and the natural function are related to each other. The following theorem answers this question.

Notation It is common to represent surjective functions thus:

$$\twoheadrightarrow$$

and injective functions so:

$$\hookrightarrow$$

Functions whose existence we assume are usually represented using solid lines, whereas functions whose existence we are trying to establish are often represented using dashed lines. We shall use these conventions whenever they help clarity.

Theorem 9.2.4 *Let* $\alpha\colon X \to Y$ *be a function. Let* $\nu\colon X \to X/\ker(\alpha)$ *be the natural function associated with the equivalence relation* $\ker(\alpha)$*. Then there is a unique function* $\theta\colon X/\ker(\alpha) \to Y$ *such that* $\theta\nu = \alpha$*. In addition,* θ *is injective.*

$$\begin{array}{ccc} X & \xrightarrow{\alpha} & Y \\ {\scriptstyle\nu}\downarrow & \nearrow{\scriptstyle\theta} & \\ X/\ker(\alpha) & & \end{array}$$

The function θ *is surjective if and only if* α *is surjective.*

Proof Denote the $\ker(\alpha)$-equivalence class containing x by $[x]$. Define θ to map $[x]$ to $\alpha(x)$. To see this is well-defined, suppose that $[x] = [x']$. Then $(x, x') \in \ker(\alpha)$ and so $\alpha(x) = \alpha(x')$. Thus the definition of $\theta([x])$ is independent of the choice of element from $[x]$ we make. Let $x \in X$. Then

$$(\theta\nu)(x) = \theta([x]) = \alpha(x).$$

Hence $\theta\nu = \alpha$. If θ' were another function such that $\theta'\nu = \alpha$, then for each $[x] \in X/\ker(\alpha)$ we would have

$$\theta'([x]) = \theta'(\nu(x)) = (\theta'\nu)(x) = \alpha(x).$$

Hence $\theta' = \theta$. Suppose that $\theta([x]) = \theta([x'])$. Then $\alpha(x) = \alpha(x')$. But then $(x, x') \in \ker(\alpha)$ and so $[x] = [x']$. It follows that θ is injective.

If α is surjective, then for each $y \in Y$ there exists $x \in X$ such that $\alpha(x) = y$. But then $\theta([x]) = y$ and so θ is also surjective. We have proved that θ is a bijection. On the other hand, if θ is surjective then so is α because $\alpha = \theta\nu$ and both functions on the right-hand side are surjective, which implies that α is surjective. □

The theorem above may seem abstract, but an example will demonstrate that the idea behind it is simple.

Example 9.2.5 Consider the function $\alpha: X \to Y$ of Example 9.2.2. For convenience, label the blocks of the partition which α induces on X as follows:

$$A = \{1, 2, 6\},\ B = \{4, 5, 9\},\ \text{and } C = \{3, 7, 8\}.$$

The function ν from X to $X/\ker(\alpha)$ is therefore given by the following table:

$$\begin{pmatrix} 1 & 2 & 3 & 4 & 5 & 6 & 7 & 8 & 9 \\ A & A & C & B & B & A & C & C & B \end{pmatrix}.$$

The unique injection from P to Y promised by Proposition 9.2.4 is therefore given by

$$\begin{pmatrix} A & B & C \\ a & b & c \end{pmatrix}.$$

□

We shall now show how to adapt Theorem 9.2.4 when the sets are semigroups and the function is a homomorphism. Let $\alpha: S \to T$ be a semigroup homomorphism. As before, we define an equivalence relation $\ker(\alpha)$ on S as follows:

$$(s, s') \in \ker(\alpha) \Leftrightarrow \alpha(s) = \alpha(s').$$

Because α is a homomorphism between semigroups this equivalence relation has an additional property.

Proposition 9.2.6 *Let $a, b, c, d \in S$ such that $(a, b) \in \ker(\alpha)$ and $(c, d) \in \ker(\alpha)$. Then $(ac, bd) \in \ker(\alpha)$.*

Proof By definition $(a, b) \in \ker(\alpha)$ means $\alpha(a) = \alpha(b)$ and $(c, d) \in \ker(\alpha)$ means $\alpha(c) = \alpha(d)$. Now α is a semigroup homomorphism and so we have the following equalities:

$$\alpha(ac) = \alpha(a)\alpha(c) = \alpha(b)\alpha(d) = \alpha(bd).$$

Thus $(ac, bd) \in \ker(\alpha)$. □

The property described in Proposition 9.2.6 is so important we single it out as the basis of a definition. Let S be a semigroup and let $\sim$ be an equivalence relation defined on S. We say that $\sim$ is a *congruence* if $a \sim b$ and $c \sim d$ implies that $ac \sim bd$. A $\sim$-equivalence class is called a *congruence class* and the $\sim$-congruence class containing s is denoted by $[s]$. Elements in the same congruence class are said to be *congruent*. The set of congruence classes is denoted $S/\sim$.

Equivalence relations are related to partitions, and congruence relations are special kinds of equivalence relations, so it is natural to ask what are the special kinds of partition that correspond to congruences? The next definition provides the answer. Let S be a semigroup. An *admissible partition on* S is a partition P of S such that if A and B are blocks of the partition, then there is a block C in the partition, necessarily unique, such that $AB \subseteq C$, where $AB = \{ab\colon a \in A, b \in B\}$.

Proposition 9.2.7 *Let S be a semigroup. The equivalence relation corresponding to an admissible partition on S is a congruence on S, and the set of congruence classes of a congruence is an admissible partition.*

Proof Let P be an admissible partition on S, and let $\sim$ be the corresponding equivalence relation. We prove that $\sim$ is a congruence. Let $a \sim b$ and $c \sim d$. Then a and b belong to the same block, A say, of P. Similarly c and d belong to the same block B. By assumption, $AB \subseteq C$ where C is a block. Now $a \in A$ and $c \in B$ so that $ac \in C$. Similarly $b \in A$ and $d \in B$ so that $bd \in C$. It follows that $ac, bd \in C$, and so $ac \sim bd$, as required.

Let ρ be a congruence on S and let $\rho(a)$ and $\rho(b)$ be two congruence classes. Let $a' \in \rho(a)$ and $b' \in \rho(b)$. Then $a' \,\rho\, a$ and $b' \,\rho\, b$. Thus $a'b' \,\rho\, ab$, because ρ is a congruence. It follows that $a'b' \in \rho(ab)$. Thus $\rho(a)\rho(b) \subseteq \rho(ab)$. So the set of congruence classes forms an admissible partition. □

Let $\sim$ be any congruence on the semigroup S. We shall now show that we can define a binary operation on the set $S/\sim$ in such a way that it becomes a semigroup. To do this, we have to find a way of multiplying two congruence classes $[s]$ and $[t]$. From the above result, we see that $[s][t]$ is a subset of the uniquely determined congruence class $[st]$. It is therefore natural to define $[s] \circ [t] = [st]$. The proof of the following is now straightforward.

Proposition 9.2.8 *Let $\sim$ be a congruence defined on the semigroup S. Define the binary operation $\circ$ on the set $S/\sim$ by $[s] \circ [t] = [st]$. Then $(S/\sim, \circ)$ is a semigroup.* □

It is worth emphasising that $[s][t]$ is usually contained in $[st]$ rather than equal to it. Nevertheless, it is usual to denote the product in $S/\sim$ by concatenation. The semigroup $S/\sim$ is called a *quotient* or *factor semigroup* of S.

Example 9.2.9 Consider the following semigroup S:

	e	a	f	b
e	e	a	f	b
a	a	e	b	f
f	f	b	f	b
b	b	f	b	f

You might like to check that this really is a semigroup. To determine whether the partition $\{\{e, f\}, \{a, b\}\}$ is admissible we proceed as follows: we try to construct the Cayley table of S/ρ; if closure fails at any point then the partition is not admissible. By definition $S/\rho = \{[e], [a]\}$. We calculate the products of the elements in S/ρ as subsets of S and determine whether the result is contained in a single block:

- $[e][e] = \{e, f\}\{e, f\} = \{ee, ef, fe, ff\} = \{e, f\} = [e]$.
- $[e][a] = \{e, f\}\{a, b\} = \{ea, eb, fa, fb\} = \{a, b\} = [a]$.
- $[a][e] = \{a, b\}\{e, f\} = \{ae, af, be, bf\} = \{a, b\} = [a]$.
- $[a][a] = \{a, b\}\{a, b\} = \{aa, ab, ba, bb\} = \{e, f\} = [e]$.

The partition is therefore admissible. In this case, the product of two congruence classes is equal to a congruence class rather than simply being contained in one. The Cayley table for S/ρ is therefore:

	$[e]$	$[a]$
$[e]$	$[e]$	$[a]$
$[a]$	$[a]$	$[e]$

□

Let $\sim$ be a congruence on S. There is a function $\nu\colon S \to S/\sim$, which maps s to $[s]$. This function is defined between two semigroups, so it is natural to ask if it is a homomorphism.

Lemma 9.2.10 *Let S be a semigroup and $\sim$ a congruence on S. Then the function $\nu\colon S \to S/\sim$ defined by $s \mapsto [s]$ is a semigroup homomorphism.*

Proof By definition $\nu(ab) = [ab] = [a][b] = \nu(a)\nu(b)$. □

The function ν above is called the *natural homomorphism* associated with the congruence $\sim$. The following theorem extends Theorem 9.2.4 to semigroups and homomorphisms.

Theorem 9.2.11 (First Isomorphism Theorem) *Let $\alpha\colon S \to T$ be a semigroup homomorphism. Then there is a unique homomorphism $\theta\colon S/\ker(\alpha) \to T$ such that $\theta\nu = \alpha$. In addition, θ is injective. The homomorphism θ is surjective if and only if α is surjective. If α is surjective, then T is isomorphic to $S/\ker(\alpha)$.*

Proof All the set-theoretic results are as in Theorem 9.2.4, it only remains to prove that ν and θ are homomorphisms: we proved that the former was a homomorphism in Lemma 9.2.10. To prove that θ is a homomorphism, let $[x], [x'] \in X/\ker(\alpha)$. Then

$$\theta([x][x']) = \theta([xx']) = \alpha(xx') = \alpha(x)\alpha(x') = \theta([x])\theta([x']).$$

□

The result above tells us that any homomorphic image of a semigroup S can be constructed, up to isomorphism, by means of a congruence on S.

Proposition 9.2.12 *Let S be a semigroup and let ρ and σ be congruences on S such that $\rho \subseteq \sigma$. Then there is a surjective homomorphism $\alpha\colon S/\rho \to S/\sigma$ such that $\alpha\rho^\natural = \sigma^\natural$.*

Proof The diagram below shows the homomorphisms we are given as solid arrows and the homomorphism we wish to construct is a dashed arrow:

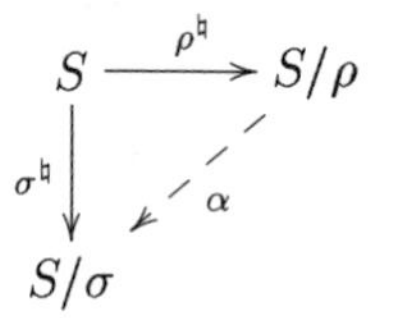

$$\begin{array}{ccc} S & \xrightarrow{\rho^\natural} & S/\rho \\ {\scriptstyle \sigma^\natural}\downarrow & \swarrow{\scriptstyle \alpha} & \\ S/\sigma & & \end{array}$$

Define $\alpha\colon S/\rho \to S/\sigma$ by $\rho(s) \mapsto \sigma(s)$. Observe that if $\rho(s) = \rho(s')$ then $s\rho s'$ and so $s\sigma s'$, which gives $\sigma(s) = \sigma(s')$. Thus the function α is well-defined. It is a homomorphism because $\alpha(\rho(s)\rho(t)) = \alpha(\rho(st)) = \sigma(st)$, whereas $\alpha(\rho(s))\alpha(\rho(t)) = \sigma(s)\sigma(t) = \sigma(st)$. It is clear that α is surjective. Finally,

$$(\alpha\rho^\natural)(s) = \alpha(\rho^\natural(s)) = \alpha(\rho(s)) = \sigma(s) = \sigma^\natural(s),$$

so that $\alpha\rho^\natural = \sigma^\natural$. □

Dealing with semigroups described in terms of congruence classes can be inconvenient. However, there is a simple way around this problem. Let $\sim$ be a congruence on the semigroup S. Choose one element from each congruence class in any way whatsoever. Denote the set of these elements by T. The key property of T is that for each $s \in S$ there exists a unique $t \in T$ such that $s \sim t$. We call T a set of *congruence representatives* for $\sim$. Define a binary operation $\circ$ on T in the following way. Let $t_1, t_2 \in T$. Then t_1t_2 is some element in S. By assumption t_1t_2 is congruent to a unique element t_3 of T. Define $t_1 \circ t_2 = t_3$.

Proposition 9.2.13 *$(T, \circ)$ is a semigroup isomorphic to $S/\sim$.*

Proof We show first that $\circ$ is an associative binary operation. Let $a, b, c \in T$. Let $a \circ b = u$ and $(a \circ b) \circ c = u \circ c = v$. Let $b \circ c = x$ and $a \circ (b \circ c) = a \circ x = y$. We have to prove that $v = y$. We use the fact that $\sim$ is a congruence. Now $ab \sim u$ and $uc \sim v$. Thus $abc \sim uc \sim v$. Similarly $bc \sim x$ and $ax \sim y$ and so $abc \sim ax \sim y$. Thus $v \sim y$. But $v, y \in T$ and so by assumption $v = y$, as required.

Define θ: $S/\sim \rightarrow T$ by $\theta([s]) = t$ where $t \in T$ and $t \sim s$. This is a well-defined function, for if $[s] = [s']$ then $s \sim s'$ and so $s' \sim t$. Thus $\theta([s']) = t$.

Suppose that $\theta([s]) = \theta([s'])$. Then there is $t \in T$ such that $s' \sim t \sim s$ and so $s' \sim s$. It follows that $[s] = [s']$. Thus θ is injective. Let $t \in T$. Then $\theta([t]) = t$, and so θ is surjective. We have proved that θ is a bijection.

To show that θ is a homomorphism, let $[a], [b] \in S/\sim$. Let $u \sim a$ and $v \sim b$ where $u, v \in T$. Then $uv \sim ab$, and $uv \sim u \circ v \in T$. Thus $u \circ v \sim ab$. Then

$$\theta([a][b]) = \theta([ab]) = u \circ v$$

and

$$\theta([a]) \circ \theta([b]) = u \circ v.$$

It follows that

$$\theta([a][b]) = \theta([a])\theta([b])$$

and so θ is a bijective homomorphism, and consequently an isomorphism. □

The example below illustrates the above result.

Example 9.2.14 We work with the monoid $(\mathbb{Z}, +)$. Define $\equiv$ on $\mathbb{Z}$ as follows:

$$m \equiv n \Leftrightarrow 5 \mid (m - n).$$

It is an easy exercise to check that $\equiv$ is an equivalence relation. It is also a congruence. To see why, let $a, b, c, d \in \mathbb{Z}$ and suppose that $a \equiv c$ and $b \equiv d$. Then $a - c = 5p$ and $b - d = 5q$ for some integers p and q. Thus $(a + b) - (c + d) = 5(p + q)$. Hence $a + b \equiv c + d$, which shows that $\equiv$ is a congruence. There are 5 congruence classes:

- $\{\ldots, -10, -5, 0, 5, 10, \ldots\}$,
- $\{\ldots, -9, -4, 1, 6, \ldots\}$,
- $\{\ldots, -8, -3, 2, 7, \ldots\}$,
- $\{\ldots, -7, -2, 3, 8, \ldots\}$,
- $\{\ldots, -6, -1, 4, 9, \ldots\}$.

The binary operation on $\mathbb{Z}/\equiv$ is simple but unwieldy. For example,

$$\{\ldots, -8, -3, 2, 7, \ldots\} + \{\ldots - 7, -2, 3, 8, \ldots\}$$

is defined to be the $\equiv$-congruence class containing, for example, $2 + 3 = 5$. Thus

$$\{\ldots, -8, -3, 2, 7, \ldots\} + \{\ldots - 7, -2, 3, 8, \ldots\} = \{\ldots, -10, -5, 0, 5, 10, \ldots\}.$$

An obvious set of congruence representatives in this case is $\{0,1,2,3,4\}$. The Cayley table of this set with the binary operation defined according to Proposition 9.2.13 is

+	0	1	2	3	4
0	0	1	2	3	4
1	1	2	3	4	0
2	2	3	4	0	1
3	3	4	0	1	2
4	4	0	1	2	3

□

The above example can be extended to any modulus $n \geq 2$ defined on $\mathbb{Z}$ both for addition and for multiplication. In this way, both $(\mathbb{Z}_n, +)$ and $(\mathbb{Z}_n, \times)$ are monoids.

Exercises 9.2

1. Let $n \geq 2$ be a fixed integer. Define $\alpha\colon \mathbb{N} \to \{0, 1, \ldots, n-1\}$ by $\alpha(a) = r$ where r is the remainder $0 \leq r < n$ when a is divided by n. Describe the corresponding equivalence relation.

2. Define the function $\alpha\colon \mathbb{R}^2 \to \mathbb{R}$ by $\alpha(x, y) = x - y$. Describe the corresponding equivalence relation.

3. Use the semigroup of Example 9.2.9 to show that the equivalence relations $\sim$ below are admissible.

(i) $\{\{e, a\}, \{f, b\}\}$.

(ii) $\{\{e\}, \{a\}, \{f, b\}\}$.

In each case, draw up the Cayley table of $S/\sim$.[3]

4. Let $\sim$ be an equivalence relation on a semigroup S satisfying the following two conditions:

(LC) $a \sim b$ and $c \in S$ implies $ca \sim cb$.

(RC) $a \sim b$ and $c \in S$ implies $ac \sim bc$.

Prove that $\sim$ is a congruence. Conversely prove that every congruence satisfies (LC) and (RC).

An equivalence relation on a semigroup is called a left congruence *if it satisfies condition (LC), and is called a* right congruence *if it satisfies (RC). Thus*

[3]Example 9.2.9 and this question are taken from [67].

a congruence is an equivalence relation that is both a left and a right congruence.

5. Let $n \geq 2$ be an integer. Consider the monoid $(\mathbb{Z}, +)$. Define

$$a \equiv b \pmod{n} \Leftrightarrow n \mid (a - b).$$

(i) Prove that $\equiv$ is a congruence on $\mathbb{Z}$ with respect to addition.

(ii) Show that $\mathbb{Z}_n = \{0, 1, \ldots, n-1\}$ is a set of congruence representatives.

(iii) Draw up the Cayley table for $(\mathbb{Z}_6, +)$.

Repeat the above question for the monoid $(\mathbb{Z}, \times)$. In the case of (iii), draw up the Cayley table for $(\mathbb{Z}_6, \times)$.

6. Let ρ be an equivalence relation on a semigroup S. Define $\bar{\rho}$ as follows: $a \, \bar{\rho} \, b$ iff $xay \, \rho \, xby$ for all $x, y \in S^1$. Show that $\bar{\rho}$ is a congruence on S contained in ρ, and that if σ is any congruence on S contained in ρ then $\sigma \subseteq \bar{\rho}$.

The congruence $\bar{\rho}$ is the one that best approximates the equivalence relation ρ. There is an important application of this result. Let A and B be two finite alphabets. Let $f \colon A^+ \to B$ be any function. Then f induces an equivalence relation $\ker(f)$ *on A^+ by Lemma 9.2.1. The largest congruence contained in* $\ker(f)$ *is denoted $\equiv_f$. It follows that $A^+/\equiv_f$ is a semigroup. This semigroup is the starting point for John Rhodes' theory of sequential functions. See Chapter 5 of [5] for a discussion of this theory.*

A special case of this construction occurs when $B = \{0, 1\}$. The function f is then determined by the subset of A^+ consisting of all those strings $x \in A^+$ such that $f(x) = 1$. Such strings form a language $L \subseteq A^+$. It is easy to check that the congruence $\equiv_f$ is defined by $x \equiv_f y$ iff $uxv \in L \Leftrightarrow uyv$ for all $u, v \in A^*$. The semigroup $A^+/\equiv_f$ is called the syntactic semigroup *of the language L.*

9.3 The transition monoid of an automaton

We now have all the ideas necessary to understand Chapter 8.

Let $\mathbf{A} = (S, A, i, \delta, T)$ be an automaton. We have seen that the function $\tau \colon A^* \to T(S)$ defined by $\tau(x) = \tau_x$ is a monoid homomorphism whose image is the transition monoid $T(\mathbf{A})$ of $\mathbf{A}$. We now look at the congruence that τ induces on A^*. This is nothing other than the relation $\equiv$ defined in Section 8.1. It follows that $A^*/\equiv$ is isomorphic to the image of τ by Theorem 9.2.11. Thus $A^*/\equiv$ is isomorphic to the transition monoid of $\mathbf{A}$.

The monoid $A^*/\equiv$ can also be described by picking representatives of the congruence classes as in Proposition 9.2.13. There are many ways of doing this, but let us do so by picking from the class $[x]$ the string y that is smallest in the tree order. The monoid we obtain is isomorphic to $A^*/\equiv$, and is nothing other than the monoid of representatives as described in Chapter 8.

If two semigroups are each isomorphic to a third, they are isomorphic to each other (by Question 10 of Exercises 9.1). We conclude that the transition monoid and the monoid of representatives are isomorphic to each other. We summarise what we have found in the following theorem.

Theorem 9.3.1 *Let* $\mathbf{A} = (S, A, i, \delta, T)$ *be an automaton and let* $\tau \colon A^* \to T(S)$ *be the map associating with* x *the function* τ_x *it induces on* S. *The kernel of* τ *is the congruence* $\equiv$. *The semigroup* $A^*/\equiv$ *is isomorphic to the transition monoid of* $\mathbf{A}$; *it is also isomorphic to the monoid of representatives of* $\equiv$. □

Example 9.3.2 We return to the automaton $\mathbf{A}$ of Example 8.2.1:

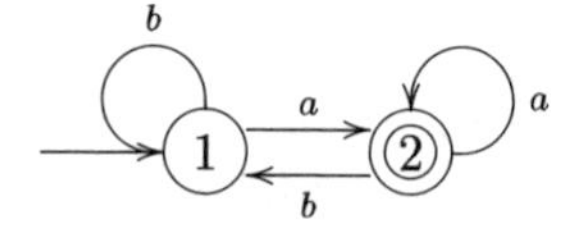

The transition monoid consists of the following elements:

$$T(\mathbf{A}) = \left\{ \begin{pmatrix} 1 & 2 \\ 1 & 2 \end{pmatrix}, \begin{pmatrix} 1 & 2 \\ 2 & 2 \end{pmatrix}, \begin{pmatrix} 1 & 2 \\ 1 & 1 \end{pmatrix} \right\},$$

and the monoid of representatives consists of the elements $\{\varepsilon, a, b\}$. The Cayley table of the transition monoid is

	$\begin{pmatrix} 1 & 2 \\ 1 & 2 \end{pmatrix}$	$\begin{pmatrix} 1 & 2 \\ 2 & 2 \end{pmatrix}$	$\begin{pmatrix} 1 & 2 \\ 1 & 1 \end{pmatrix}$
$\begin{pmatrix} 1 & 2 \\ 1 & 2 \end{pmatrix}$	$\begin{pmatrix} 1 & 2 \\ 1 & 2 \end{pmatrix}$	$\begin{pmatrix} 1 & 2 \\ 2 & 2 \end{pmatrix}$	$\begin{pmatrix} 1 & 2 \\ 1 & 1 \end{pmatrix}$
$\begin{pmatrix} 1 & 2 \\ 2 & 2 \end{pmatrix}$	$\begin{pmatrix} 1 & 2 \\ 2 & 2 \end{pmatrix}$	$\begin{pmatrix} 1 & 2 \\ 1 & 1 \end{pmatrix}$	$\begin{pmatrix} 1 & 2 \\ 1 & 1 \end{pmatrix}$
$\begin{pmatrix} 1 & 2 \\ 1 & 1 \end{pmatrix}$	$\begin{pmatrix} 1 & 2 \\ 1 & 1 \end{pmatrix}$	$\begin{pmatrix} 1 & 2 \\ 2 & 2 \end{pmatrix}$	$\begin{pmatrix} 1 & 2 \\ 1 & 1 \end{pmatrix}$

The Cayley table for the monoid of representatives is

	ε	a	b
ε	ε	a	b
a	a	b	b
b	b	a	b

The isomorphism between these two monoids is determined by

$$\varepsilon \mapsto \begin{pmatrix} 1 & 2 \\ 1 & 2 \end{pmatrix} \quad a \mapsto \begin{pmatrix} 1 & 2 \\ 2 & 2 \end{pmatrix} \quad b \mapsto \begin{pmatrix} 1 & 2 \\ 1 & 1 \end{pmatrix},$$

respectively. □

Exercises 9.3

1. Let $\mathbf{A} = (Q, A, q_0, \delta, F)$ and $\mathbf{B} = (S, A, s_0, \gamma, G)$ be automata. Let $\theta: \mathbf{A} \to \mathbf{B}$ be an isomorphism. Prove that the transition monoids $T(\mathbf{A})$ and $T(\mathbf{B})$ are isomorphic.

9.4 The syntactic monoid of a language

In this section, we shall show how to construct a monoid from *any* language. This monoid is finite if and only if the language is recognisable, in which case it is isomorphic to the transition monoid of the minimal automaton of the language.

Theorem 9.4.1 *Let L be a recognisable language, whose minimal automaton is $\mathbf{A}$ with set of states S. Let $\tau: A^* \to T(S)$ be the function taking x to the function τ_x it induces on S. Then the kernel of τ is given by the following: $(x, y) \in \ker(\tau)$ if and only if for all $u, v \in A^*$ we have that*

$$uxv \in L \Leftrightarrow uyv \in L.$$

Proof Let $\mathbf{A} = (S, A, i, \delta, T)$. Suppose that $x, y \in A^*$ satisfy the condition that

$$uxv \in L \Leftrightarrow uyv \in L$$

for all $u, v \in A^*$. We prove that $\tau_x = \tau_y$. Let $s \in S$ be an arbitrary state. We have to show that $s \cdot x = s \cdot y$. We shall do this by proving that $s \cdot x$ and $s \cdot y$ are indistinguishable. Equality of $s \cdot x$ and $s \cdot y$ will then follow from the fact that $\mathbf{A}$ is minimal and so reduced. By assumption, $\mathbf{A}$ is minimal and so accessible. It follows that there exists $u \in A^*$ such that $i \cdot u = s$. Suppose that $(s \cdot x) \cdot v \in T$. Then $i \cdot (uxv) \in L$ and so $uxv \in L$. By assumption, $uyv \in L$. Thus $i \cdot (uyv) \in T$ and so $(s \cdot y) \cdot v \in T$. We have shown that

$$(s \cdot x) \cdot v \in T \Rightarrow (s \cdot y) \cdot v \in T.$$

The reverse implication can be proved similarly. We have therefore proved that $s \cdot x \simeq s \cdot y$. Hence $s \cdot x = s \cdot y$.

Now suppose that $(x, y) \in \ker(\tau)$. By definition this means that $\tau_x = \tau_y$. For any $u, v \in A^*$ we have that

$$\tau_{uxv} = \tau_u \tau_x \tau_v = \tau_{uyv}.$$

Suppose that $uxy \in L$. Then $i \cdot (uxv) \in T$. But $\tau_{uxv} = \tau_{uyv}$ and so $i \cdot (uxv) = i \cdot (uyv)$. Hence $i \cdot (uyv) \in T$, and so $uyv \in L$. We have proved that

$$uxy \in L \Rightarrow uxy \in L.$$

The reverse implication is proved similarly. □

The above theorem motivates the following definition. Let L be an *arbitrary* language over the alphabet A. Define the relation σ_L on A^* by $(x, y) \in \sigma_L$ if and only if

$$uxv \in L \Leftrightarrow uyv \in L$$

for all $u, v \in A^*$. Using this definition, we see that Theorem 9.4.1 says that $\ker(\tau) = \sigma_L$ when L is recognisable. The following is left as an exercise.

Proposition 9.4.2 *Let L be a language over the alphabet A. The relation σ_L is a congruence on A^* with the property that L is a union of some of the σ_L-classes.* □

Let L be a language over the alphabet A. We call σ_L the *syntactic congruence of* L. The monoid $M(L) = A^*/\sigma_L$ is called the *syntactic monoid* of L.

The definition of the syntactic congruence can be rendered more intuitive. One of the features of natural languages is that words can be assigned to different grammatical categories, where words belonging to the same category behave in the same way in the language. For example, in English we classify words as being nouns, verbs, adjectives, adverbs, and so forth. To a first approximation a word is recognised as a noun if it occurs in the same sorts of places in sentences as other nouns; likewise for verbs, adjectives, etc. Let $L \subseteq A^*$ be a language and let $x \in A^*$. A *context* for x is a pair of strings $u, v \in A^*$ such that $uxv \in L$. The set of all contexts of x is denoted by $C(x)$. Using this notation, we see that $(x, y) \in \sigma_L$ means precisely that $C(x) = C(y)$. In other words, x and y occur as factors of elements of L in exactly the same places. In terms of natural languages, x and y belong to the same 'grammatical category.'

Theorem 9.4.3 *Let L be any language over the alphabet A. The syntactic monoid of L is finite if and only if L is recognisable. If L is recognisable then the syntactic monoid is isomorphic to the transition monoid of the minimal automaton for L.*

Proof Suppose that L is recognisable with minimal automaton $\mathbf{A}$. By Theorem 9.4.1, the kernel of τ is σ_L. Thus A^*/σ_L is the same as $A^*/\equiv$, and so by Theorem 9.3.1, the monoid A^*/σ_L is isomorphic to the transition monoid of $\mathbf{A}$. In particular, A^*/σ_L is finite.

Suppose that A^*/σ_L is finite. We prove that L is recognisable. Put $Q = A^*/\sigma_L$ and denote the σ_L-congruence class containing x by $[x]$. Put $i = [\varepsilon]$ and $T = \{[x]: x \in L\}$. Let $a \in A$. Define $[x] \cdot a = [xa]$. This is well-defined, because if $(x, x') \in \sigma_L$ then $(xa, x'a) \in \sigma_L$ because σ_L is a congruence. Thus the definition of $[x] \cdot a$ is independent of the choice of x. It follows that

$\mathbf{A} = (Q, A, i, \cdot, T)$ is a well-defined automaton. It is easy to check that if u is an arbitrary string in A^* and $[x]$ an arbitrary state in Q then $[x] \cdot u = [xu]$. To calculate $L(\mathbf{A})$, observe that $y \in L(\mathbf{A})$ if and only if $i \cdot y = [\varepsilon] \cdot y \in L$. That is $[y] \in L$. But L is a union of σ_L-classes and so $[y] \in L$ if and only if $y \in L$. It follows that $y \in L(\mathbf{A})$ if and only if $y \in L$. Hence we have proved that $L(\mathbf{A}) = L$, and so L is recognisable. □

It is worth stating explicitly how to calculate the syntactic monoid of a recognisable language.

Algorithm 9.4.4 Let L be a recognisable language. If $L = L(r)$, where r is a regular expression, then we can find the minimal automaton $\mathbf{A}$ of L using the Method of Quotients. Alternatively, if $L = L(\mathbf{B})$ where $\mathbf{B}$ is an automaton, then we can replace $\mathbf{B}$ by $\mathbf{A} = \mathbf{B}^{ar}$, which is again the minimal automaton for L. In either case, we can then calculate the transition monoid of $\mathbf{A}$ using either Algorithm 8.2.2, thereby obtaining explicitly the set of all functions induced by input strings to $\mathbf{A}$, or Algorithms 8.2.4 and 8.2.5, thereby obtaining an isomorphic copy of the transition monoid in terms of the monoid of representatives. In either case, the monoid obtained is isomorphic to the syntactic monoid. □

Describing the syntactic monoid of a recognisable language by means of a Cayley table is only practical when it has a small number of elements. In general, the monoid is better described by means of its representatives and relations.

Example 9.4.5 Let $L = (ab + ba)^*$. It is easy to check that the minimal automaton for L is the machine $\mathbf{A}$ below.

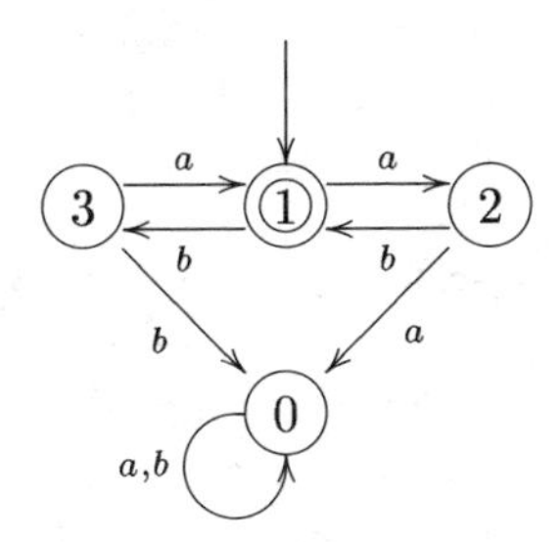

The table of representatives is

	0	1	2	3
ε	0	1	2	3
a	0	2	0	1
b	0	3	1	0
a^2	0	0	0	2
ab	0	1	0	3
ba	0	1	2	0
b^2	0	0	3	0
a^3	0	0	0	0
a^2b	0	0	0	1
ab^2	0	3	0	3
ba^2	0	2	0	0
b^2a	0	0	1	0
a^2b^2	0	0	0	3
ab^2a	0	1	0	0
b^2a^2	0	0	2	0

The set of relations is

(1) $aba \equiv a$
(2) $bab \equiv b$
(3) $b^3 \equiv a^3$
(4) $a^4 \equiv a^3$
(5) $a^3b \equiv a^3$
(6) $a^2ba \equiv a^2$
(7) $ab^3 \equiv a^3$
(8) $ba^3 \equiv a^3$
(9) $ba^2b \equiv ab^2a$
(10) $b^2ab \equiv b^2$
(11) $a^2b^2a \equiv a^2b$
(12) $a^2b^3 \equiv a^3$
(13) $ab^2a^2 \equiv ba^2$
(14) $ab^2ab \equiv ab^2$
(15) $b^2a^3 \equiv a^3$
(16) $b^2a^2b \equiv b^2a$

Many of these relations are either superfluous or can be simplified using the fact that $\equiv$ is a congruence. Relations (4), (5) and (8) tell us that a^3 is the zero for the semigroup. I shall signal this by writing $a^3 \equiv 0$ and removing relations (4), (5) and (8). Relation (3) tells us that $b^3 \equiv 0$. The relations (7), (12) and (15) are now superfluous. Notice that (6) follows from (1), because $a^2ba = a(aba) \equiv a^2$; and (10) follows from (2), because $b^2ab = b(bab) \equiv b^2$. Relation (14) follows from (2), because $ab^2ab = ab(bab) \equiv ab^2$; and relation (16) follows from (9) and (2), because

$$b^2a^2b = b(ba^2b) \equiv b(ab^2a) = (bab)ba \equiv b^2a.$$

We therefore end up with the following set of relations:

(1) $a^3 \equiv 0 \equiv b^3$.

(2) $aba \equiv a$.

(3) $bab \equiv b$.

(4) $ba^2b \equiv ab^2a$.

(5) $a^2b^2a \equiv a^2b$.

(6) $ab^2a^2 \equiv ba^2$.

These provide a more succinct description of the syntactic monoid of L, as against the Cayley table, which has $15^2 = 225$ entries. □

Exercises 9.4

1. Prove Proposition 9.4.2.

2. Compute the Cayley tables of the syntactic monoids of the following recognisable languages over $A = \{a, b\}$. Use as congruence representatives the strings arising from Algorithm 8.2.4.

(i) $(a + b)^*$.

(ii) $(a + b)^+$.

(iii) $(ab)^*$.

(iv) $(a^2)^*$.

(v) aba.

(vi) $(a + b)^*ab(a + b)^*$.

(vii) $(a + b)^*aba(a + b)^*$.

(viii) $ababa$.

9.5 Summary of Chapter 9

- *Semigroup theory*: Semigroups and monoids form the algebraic component of automata theory. Fundamental ideas concerned with semigroups are: subsemigroups, homomorphisms, and isomorphisms. Each homomorphism of a semigroup is associated with a congruence and the image of the homomorphism is isomorphic to the quotient or factor semigroup constructed from the congruence.

- *The syntactic monoid*: With each language we can construct a monoid, called the syntactic monoid. This monoid is finite if and only if the language is recognisable. In this case, the syntactic monoid is isomorphic to the transition monoid of the minimal automaton of the language.

9.6 Remarks on Chapter 9

Theorem 9.2.11 is the 'First Isomorphism Theorem'. There are two others in common use. The *Second Isomorphism Theorem* runs as follows: let ρ be a congruence on a semigroup S, and let T be a subsemigroup of S. Then $\rho' = \rho \cap (T \times T)$ is a congruence on T, so we can form the quotient semigroup T/ρ'. On the other hand,

$$T' = \{s \in S\colon s\,\rho\,t \text{ for some } t \in T\}$$

is a subsemigroup of S containing T. Observe that if $a \in T'$ and $a\,\rho\,b$, then $b \in T'$. We can therefore form the quotient semigroup T'/ρ. The Second Isomorphism Theorem states that T/ρ' is isomorphic to T'/ρ. The *Third Isomorphism Theorem* is essentially Proposition 9.2.12: since S/σ is a homomorphic image of S/ρ there is a congruence, τ say, such that $(S/\rho)/\tau$ is isomorphic to S/σ. The congruence τ is usually denoted σ/τ.

Our characterisation in Theorem 9.4.3 of recognisable languages in terms of the syntactic congruence is the one most useful to us, but many books give another characterisation, which is worth describing here. Let $\mathbf{A} = (Q, A, i, \delta, T)$ be an *accessible* automaton. Define the relation ρ on A^* by $(x, y) \in \rho_{\mathbf{A}}$ iff $i \cdot x = i \cdot y$. It is easy to check that $\rho_{\mathbf{A}}$ is an equivalence relation on A^*, where each equivalence class is of the form $\{x \in A^*\colon i \cdot x = q\}$ for some $q \in Q$. In addition, $\rho_{\mathbf{A}}$ is a right congruence, in the sense of Question 4 of Exercises 9.2, and if $(x, y) \in \rho$, then $x \in L$ iff $y \in L$. We now use these properties as the basis of a definition. A *Myhill-Nerode relation for the language* $L \subseteq A^*$ is any right congruence on A^* with the property that if $(x, y) \in \rho$, then $x \in L$ iff $y \in L$. Thus $\rho_{\mathbf{A}}$ is a Myhill-Nerode relation with a finite number of right congruence classes. Now let $L \subseteq A^*$ be an arbitrary language. Define ρ_L on A^* by $(x, y) \in \rho_L$ iff $xv \in L \Leftrightarrow yv \in L$ for all $v \in A^*$. Then it can be checked that ρ_L is a Myhill-Nerode relation for L, and that if ρ is any Myhill-Nerode relation for L, then $\rho \subseteq \rho_L$.

The Myhill-Nerode Theorem states that the following are equivalent:

(i) A language L over the alphabet A is recognisable.

(ii) There is a Myhill-Nerode relation for L with finitely many right congruence classes.

(iii) The relation ρ_L has only finitely many right congruence classes.

I shall sketch the proof.

(i) $\Rightarrow$ (ii). If L is recognisable, then it is recognised by some accessible automaton $\mathbf{A}$. From the above, $\rho_{\mathbf{A}}$ is a Myhill-Nerode relation for L with finitely many right congruence classes.

(ii) $\Rightarrow$ (iii). If ρ is some Myhill-Nerode relation for L with finitely many right congruence classes, then from $\rho \subseteq \rho_L$, we deduce that ρ_L has only finitely many right congruence classes.

(iii) $\Rightarrow$ (i). From the relation ρ_L, we can construct an accessible automaton L, whose states are the right congruence classes and so, by assumption, there really are only finitely many states.

The details can be found in Lectures 15 and 16 of [75]. According to [65], the proof of this result is due to Nerode [98], with Myhill [97] being responsible for that part of Theorem 9.4.3 that states that the syntactic congruence of a language has only finitely many congruence classes if and only if the language in question is recognisable.

The right congruence ρ_L, and the congruence σ_L are special cases of the 'principal congruences' (see page 175 of [77]) that can be defined for any subset L of any semigroup S; see Proposition 10.3.2, for example.

The syntactic monoid of a language surfaced first in [114] and their theory was systematically worked out in [87].

Chapter 10

Algebraic language theory

In Chapter 9, we proved that a language is recognisable precisely when its syntactic monoid is finite. When the language is recognisable, the syntactic monoid can easily by calculated as the transition monoid of the minimal automaton of the language. The goal of algebraic language theory is to answer questions about recognisable languages by studying their syntactic monoids. To do this, we first have to learn more about finite semigroups. This is the subject of Section 10.1. To demonstrate that the algebraic approach is viable, we show in Section 10.2 that many of the results about recognisable languages proved in Chapters 3 and 4 using automata can easily be proved using finite monoids. The real vindication of this approach, however, will come in Chapter 11, when we prove a new result about recognisable languages using the syntactic monoid. The results of this chapter lead us to look for a correspondence between recognisable languages and finite monoids. In Section 10.3, we prove that this correspondence cannot be interpreted in a naive way. The correct correspondence is described in Chapter 12.

10.1 Finite semigroups

An element e of a semigroup is said to be *idempotent* if $e^2 = e$. Zero elements and identities are idempotents, but there are idempotents that are neither.

Example 10.1.1 Let R_n be any non-empty finite set with n elements. Define a binary operation on R_n by $ab = b$ for all $a, b \in R_n$. Then R_n is a semigroup. For $n \geq 2$, every element of R_n is idempotent but none is an identity nor a zero. □

The following result is a direct consequence of finiteness.

Theorem 10.1.2 *Every element in a finite semigroup has a non-zero power that is idempotent, and so every finite semigroup contains an idempotent.*

Proof Let S be a semigroup with n elements and let $s \in S$. Consider the list $s, s^2, \ldots, s^{n+1}$. Because S contains n elements, there must exist at least two powers of s, say s^i and s^j, where $j > i$, such that $s^i = s^j$. Let $j = i + k$ for some $k \geq 1$, so that $s^j = s^i s^k$. Now $s^i = s^i s^k = s^i s^k s^k = s^{i+2k}$. By induction,

$$s^i = s^{i+qk}$$

for all $q \geq 0$. The numbers $i, i+1, \ldots i + (k-1)$ are k consecutive numbers. A little thought will convince you that these numbers must each be congruent modulo k to exactly one of the numbers $0, 1, \ldots, k-1$ in some order. Let p be in the range $0 \leq p \leq k-1$ such that $i + p \equiv 0 \pmod{k}$. Then $i + p = qk$ for some q. The element s^{i+p} is idempotent because

$$(s^{i+p})^2 = s^{2i+2p} = s^p s^{i+(i+p)} = s^p s^{i+qk} = s^p s^i = s^{i+p}.$$

□

A finite semigroup might well have neither identity nor zero, but it will always have idempotents. Finiteness is essential in the above theorem. For example, the infinite semigroup A^+ does not have any idempotents.

Amongst the subsemigroups of a semigroup, there are some with a stronger property. Let S be a semigroup. A subset I of S is said to be a *right ideal* if $IS \subseteq I$, a *left ideal* if $SI \subseteq I$, and a *two-sided ideal* if it is ambidextrous. An ideal (left, right two-sided) I in S is said to be *proper* if it is not equal to S. If $a \in S$ then aS^1, S^1a, and S^1aS^1 are, respectively, the smallest right, the smallest left, and the smallest two-sided ideals containing a.[1] Ideals arise in the following way. In ordinary arithmetic, we are often interested in whether one number divides another. For example, we define even numbers to be those divisible by 2 and prime numbers as those numbers (apart from 0 and 1) having the smallest possible number of divisors: namely, 1 and themselves. We can similarly talk about divisibility in semigroups. However, we come up against an important difference: in general, the multiplication in a semigroup is not commutative. For that reason, we have to study three different kinds of divisibility. In the definition below, $a \in S^1b$ is just a way of saying that either $a = b$ or $a \in Sb$. Let S be a semigroup, and let $a, b \in S$. If $a \in bS^1$, then we write $a \leq_R b$. If $a \leq_R b$ and $a \neq b$, then we write $a <_R b$. If $a \leq_R b$ and $b \leq_R a$, then we write $a \mathcal{R} b$. We may dually define $\leq_L$, $<_L$ and $\mathcal{L}$. Define $a \leq_J b$ if $a \in S^1bS^1$. We define $<_J$ and $\mathcal{J}$ in the obvious ways. Finally, put $\mathcal{H} = \mathcal{L} \cap \mathcal{R}$. When S is a monoid, we can clearly use S itself rather than S^1 in the definitions above.

Example 10.1.3 Consider the monoid A^*. Then $x \leq_R y$ means precisely that y is a prefix of x; $x <_R y$ means that y is a proper prefix of x. In this case, $\mathcal{R}$ is the equality relation (why?). By the same token, $x \leq_L y$ means that y is a suffix of x, and $x \leq_J y$ means that y is a factor of x. □

[1] The curious terminology 'ideal' goes back to their origins in ring theory. See [46].

The proof of the following is left as an exercise.

Lemma 10.1.4 *Each of the relations $\leq_R$, $\leq_L$ and $\leq_J$ is reflexive and transitive.* $\square$

Relations that are both reflexive and transitive are called *preorders*. Such relations have the following important properties.

Lemma 10.1.5 *Let $\preceq$ be a preorder on the set X. Define $x \equiv y$ iff $x \preceq y$ and $y \preceq x$. Then $\equiv$ is an equivalence relation on X.*

Denote by $X/\equiv$ the set of equivalence classes and $[x]$ the equivalence class containing x. Define $[x] \leq [y]$ iff $x \preceq y$. Then $\leq$ is a partial order on $X/\equiv$.

Proof It is easy to check that $\equiv$ is an equivalence relation. We show that $\leq$ is well-defined. Let $x' \in [x]$ and $y' \in [y]$. Then $x' \preceq x \preceq y \preceq y'$ and so $x' \preceq y'$ by transitivity. It is left as an exercise to check that $\leq$ is a partial order. $\square$

It follows from Lemmas 10.1.4 and 10.1.5 that $\mathcal{L}$, $\mathcal{R}$, and $\mathcal{J}$ are equivalence relations on any semigroup S. The relation $\mathcal{H}$, which is the intersection of two equivalence relations, is itself an equivalence relation (by Question 3 of Exercises 7.1). They are called *Green's relations.* Observe that $\mathcal{L}, \mathcal{R} \subseteq \mathcal{J}$. This means that each $\mathcal{J}$-class is a union of some of the $\mathcal{L}$-classes and some of the $\mathcal{R}$-classes (again by Question 3 of Exercises 7.1). In addition, $S/\mathcal{J}$ is a partially ordered set by Lemma 10.1.5. There is one piece of idiosyncratic notation concerning Green's relations that is common in semigroup books. If $\mathcal{K}$ is any one of Green's relation, then the $\mathcal{K}$-class containing the element s is denoted K_s.

Theorem 10.1.6 *Let S be a finite semigroup. The following are equivalent.*

(i) $a \,\mathcal{J}\, b$.

(ii) $a \,\mathcal{L}\, c \,\mathcal{R}\, b$ *for some* $c \in S$.

(iii) $a \,\mathcal{R}\, c \,\mathcal{L}\, b$ *for some* $c \in S$.

Proof (i) $\Rightarrow$ (ii). Suppose that $a \,\mathcal{J}\, b$. Then by definition,

$$xay = b \text{ and } ubv = a$$

for some $x, y, u, v \in S^1$. Notice that

$$a = ubv = u(xay)v = (ux)a(yv).$$

It follows that $a = (ux)^m a(yv)^m$ for all $m \geq 1$. Similarly $b = (xu)^m b(vy)^m$. Because S is finite there is a positive integer p such that $(ux)^p = e$, an idempotent by Theorem 10.1.2. We have that $a = ea(yv)^p$ and so

$$ea = e^2 a(yv)^p = ea(yv)^p = a.$$

It follows that $a = ea = ((ux)^{p-1}u)xa$. Hence $a\,\mathcal{L}\,xa$. Similarly, there is a positive integer q such that $(vy)^q$ is an idempotent. Now

$$xa = x(ux)^{q+1}a(yv)^{q+1} = (xu)^{q+1}xay(vy)^q v,$$

by simply rewriting $x(ux)^{q+1}$ and $(yv)^{q+1}$. But

$$(xu)^{q+1}xay(vy)^q v = (xu)^{q+1}b(vy)^q(vy)^q v$$

because $(vy)^q$ is an idempotent. But then,

$$(xu)^{q+1}b(vy)^q(vy)^q v = (xu)^{q+1}b(vy)^{q+1}\cdot(vy)^{q-1}v = b(vy)^{q-1}v.$$

Thus $b\cdot(vy)^{q-1}v = xa$. But we also have $xa\cdot y = b$, so we have proved that $xa\,\mathcal{R}\,b$.

(ii) $\Rightarrow$ (iii). The proof of this implication does not require finiteness. Suppose that $a\,\mathcal{L}\,c\,\mathcal{R}\,b$ for some $c \in S$. By definition there are $x, y, u, v \in S^1$ such that

$$xa = c, \quad yc = a, \quad cu = b, \quad bv = c.$$

Put $d = ycu$. We prove that $a\,\mathcal{R}\,d$ and $d\,\mathcal{L}\,b$. To prove the first equivalence,

$$au = (yc)u = d,$$

and

$$dv = (ycu)v = y(cu)v = ybv = y(bv) = yc = a.$$

To prove the second equivalence,

$$yb = y(cu) = d,$$

and

$$xd = x(ycu) = x(yc)u = xau = (xa)u = cu = b.$$

(iii) $\Rightarrow$ (i). Straightforward. □

The result above is the basis of a pictorial device for representing the divisibility properties of finite semigroups. I shall describe how this is constructed in stages. First, draw the Hasse diagram of the partially ordered set $S/\mathcal{J}$. Next each vertex of the diagram is replaced by the corresponding $\mathcal{J}$-class. Then each $\mathcal{J}$-class is divided into rows, where each row is an $\mathcal{R}$-equivalence class. Finally, each $\mathcal{J}$-class is divided into columns, where each column is an $\mathcal{L}$-class. Theorem 10.1.6 tells us that each row and each column in a $\mathcal{J}$-class intersect in an $\mathcal{H}$-class. The resulting diagram is called an *eggbox diagram*.

Example 10.1.7 What follows is the Cayley table of a monoid S.

	0	e	f	1	a	b	c
0	0	0	0	0	0	0	0
e	0	e	0	e	0	b	b
f	0	0	f	f	a	0	a
1	0	e	f	1	a	b	c
a	0	a	0	a	0	f	f
b	0	0	b	b	e	0	e
c	0	a	b	c	e	f	1

We shall calculate Green's relations for this monoid, and display them by means of an eggbox diagram. Observe that because S is a monoid, Green's relations can be defined in terms of S rather than S^1. To calculate $x \mathcal{R} y$ we have to show that $xS = yS$. The set xS is simply the set of elements occurring in the row of the Cayley table belonging to x. The $\mathcal{R}$-classes are therefore:

$$\{0\}, \{e, b\}, \{f, a\}, \{1, c\}.$$

To show that $x \mathcal{L} y$, we have to show that $Sx = Sy$. The set Sx is simply the set of elements occurring in the column belonging to x. The $\mathcal{L}$-classes are therefore:

$$\{0\}, \{e, a\}, \{f, b\}, \{1, c\}.$$

To show that $x \mathcal{J} y$ we have to show that $SxS = SyS$. We can use our previous calculations since $SxS = Sx \cdot S$. The $\mathcal{J}$-classes are therefore:

$$\{1, c\}, \{e, f, a, b\}, \{0\}.$$

The partially ordered set $S/\mathcal{J}$ can therefore be represented by the following Hasse diagram:

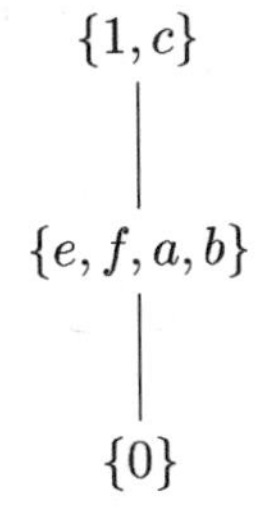

The eggbox diagram is therefore:

$1, c$

e	b
a	f

0

□

There is an important technique for combining semigroups to make new ones. Let $(S, \circ)$ and $(T, *)$ be semigroups. On the set $S \times T$ define a binary operation as follows: if $(s,t), (s',t') \in S \times T$, then

$$(s,t)(s',t') = (s \circ s', t * t').$$

Define functions $\pi_1, \pi_2 \colon S \times T \to S$ by $\pi_1(s,t) = s$ and $\pi_2(s,t) = t$. The proof of the following is left as an exercise.

Lemma 10.1.8 *With the definition above, if S and T are semigroups, then $S \times T$ is a semigroup, and π_1 and π_2 are semigroup homomorphisms. If S and T are both monoids, then $S \times T$ is a monoid, and π_1 and π_2 are monoid homomorphisms.* □

The construction above can be generalised as follows. Let

$$S_1, \ldots, S_n$$

be semigroups (respectively monoids). Then

$$S_1 \times \ldots \times S_n$$

is a semigroup (respectively monoid) under *pointwise* multiplication: the product of $(a_1, \ldots, a_n)$ and $(b_1, \ldots, b_n)$ is defined to be $(a_1 b_1, \ldots, a_n b_n)$. The function,

$$\pi_i \colon S_1 \times \ldots \times S_n \to S_i \text{ defined by } (s_1, \ldots, s_n) \mapsto s_i,$$

where $1 \leq i \leq n$, is a homomorphism (respectively, monoid homomorphism), called the ith *projection homomorphism*. The semigroup $S_1 \times \ldots \times S_n$ is called the *product* of the semigroups S_i. We shall sometimes write this product as $\prod_{i=1}^{n} S_i$. An element of $\prod_{i=1}^{n} S_i$ is an n-tuple, which we shall write as (s_i), where $1 \leq i \leq n$.

Example 10.1.9 Let S_1, S_3, S_3 be three semigroups. We can first form the product $S_1 \times S_2$ and then we can form the product $(S_1 \times S_2) \times S_3$. Similarly, we can form the product $S_1 \times (S_2 \times S_3)$. The semigroups $(S_1 \times S_2) \times S_3$ and $S_1 \times (S_2 \times S_3)$ are isomorphic where the isomorphism takes $((s_1, s_2), s_3)$ to $(s_1, (s_2, s_3))$. We leave the simple verification as an exercise. It is also easy to check that both $(S_1 \times S_2) \times S_3$ and $S_1 \times (S_2 \times S_3)$ are isomorphic to $S_1 \times S_2 \times S_3$. This result can be generalied to deal with any product of semigroups. □

The following combines subsemigroups, products, and homomorphic images and is left as an exercise.

Lemma 10.1.10 *Let $S_1, \ldots, S_n$ and $T_1, \ldots, T_n$ be semigroups.*

(i) *Let S_i be a subsemigroup of T_i for each $1 \leq i \leq n$. Then $S_1 \times \ldots \times S_n$ is a subsemigroup of $T_1 \times \ldots \times T_n$.*

(ii) *Let $\alpha_i \colon S_i \to T_i$ be a homomorphism for each $1 \leq i \leq n$. Then*

$$\alpha \colon S_1 \times \ldots \times S_n \to T_1 \times \ldots \times T_n,$$

defined by

$$\alpha(s_1, \ldots, s_n) = (\alpha_1(s_1), \ldots, \alpha_n(s_n)),$$

is a homomorphism. □

Exercises 10.1

1. Let R_n be a set with n elements and product defined by $ab = b$. Show that R_n is a semigroup in which every element is idempotent.

2. Show that α is an idempotent in T_n if and only if α restricted to $\text{im}(\alpha)$ is the identity function.

3. Prove Lemma 10.1.4.

4. Complete the proof of Lemma 10.1.5.

5. Show that in a commutative semigroup $\mathcal{L} = \mathcal{R} = \mathcal{J}$.

6. Let e and f be idempotents in a semigroup S. Prove that

(i) $e \,\mathcal{L}\, f \Leftrightarrow ef = e$ and $fe = f$.

(ii) $e \,\mathcal{R}\, f \Leftrightarrow ef = f$ and $fe = e$.

7. Let X be a finite, non-empty set. Show that in the full transformation monoid $T(X)$ the following hold:

(i) $\alpha \,\mathcal{L}\, \beta \Leftrightarrow \text{im}(\alpha) = \text{im}(\beta)$.

(ii) $\alpha \,\mathcal{R}\, \beta \Leftrightarrow \ker(\alpha) = \ker(\beta)$.

(iii) $\alpha \,\mathcal{J}\, \beta \Leftrightarrow |\,\text{im}(\alpha)\,| = |\,\text{im}(\beta)\,|$.

Hence construct the eggbox diagrams of T_2 and T_3.

8. Prove Lemma 10.1.8.

9. Let S_1, S_2, S_3 be semigroups. Prove that $S_1 \times (S_2 \times S_3)$ is isomorphic to $(S_1 \times S_2) \times S_3$.

10. Prove Lemma 10.1.10.

10.2 Recognisability by a monoid

The syntactic monoid of a recognisable language can be constructed from the minimal automaton of the language. But just as languages can be recognised by automata that are not minimal so they can also be recognised by monoids that are not their syntactic monoids. To make this precise we need to develop a little more semigroup theory.

Theorem 10.2.1 *Let A be a finite alphabet.*

(i) *Let S be a semigroup and let $\alpha: A \to S$ be a function. Then there is a unique homomorphism $\overline{\alpha}: A^+ \to S$ such that $\overline{\alpha}(a) = \alpha(a)$ for each $a \in A$.*

(ii) *Let S be a monoid and let $\alpha: A \to S$ be a function. Then there is a unique monoid homomorphism $\overline{\alpha}: A^* \to S$ such that $\overline{\alpha}(a) = \alpha(a)$ for each $a \in A$.*

Proof We prove (i), and then the proof of (ii) is immediate. Let $x = a_1 \dots a_n \in A^+$. Define

$$\overline{\alpha}(x) = \alpha(a_1) \dots \alpha(a_n).$$

It is easy to check that $\overline{\alpha}$ is a homomorphism, and by construction it has the correct property. □

The above theorem has an important consequence.

Proposition 10.2.2

(i) *Let S be a finitely generated semigroup. Then S is a homomorphic image of A^+ for some finite alphabet A.*

(ii) *Let S be a finitely generated monoid. Then S is a monoid homomorphic image of A^* for some finite alphabet A.*

Proof We prove the semigroup case. The monoid case is then immediate. Since S is a finitely generated semigroup, there is a finite subset X of S such that $S = X^+$. Let A be any set that has the same number of elements as X. This means that there is a bijection from A to X. Thus each element of A corresponds to a unique element of X. By Theorem 10.2.1, there is a homomorphism α from A^+ to S, which maps each element of A to the corresponding element of X. By Proposition 9.1.8, the image of α is a subsemigroup of S

containing X. Thus the image of α contains $X^+ = S$. Hence α is surjective.□

Because of Proposition 10.2.2, the semigroups A^+ and the monoids A^* play a special role in the theory of semigroups. Loosely speaking, they form the templates from which all (finitely generated) semigroups and monoids can be built by means of homomorphic images.

Let A be a finite alphabet. The semigroup A^+ is called the *free semigroup on the set* A because of the property described in Theorem 10.2.1. Similarly, we say that A^* is the *free monoid on the set* A.

There is a further consequence of Theorem 10.2.1, which we shall use later in this chapter.

Proposition 10.2.3 *Let α: $A^+ \to S$ be a homomorphism, and let β: $T \to S$ be a surjective homomorphism. Then there is a homomorphism γ: $A^+ \to T$ such that $\beta\gamma = \alpha$. The same result holds when A^+ is replaced by A^* and all the homomorphisms are monoid homomorphisms.*

Proof In the diagram below, the solid arrows represent the homomorphisms we have been given and the dashed arrow is the one we have to define.

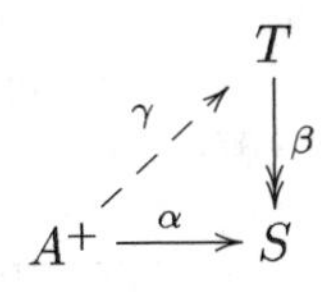

Because β is surjective, $\beta^{-1}(s)$ is non-empty for each $s \in S$ the set. In particular, for each $a \in A$ the set $\beta^{-1}(\alpha(a))$ is non-empty. Choose $t_a \in \beta^{-1}(\alpha(a))$. We have therefore defined a function from A to T given by $a \mapsto t_a$. By Theorem 10.2.1, this can be extended to a unique homomorphism γ: $A^+ \to T$ such that $\gamma(a) = t_a$ for each $a \in A$. It follows that $\beta(\gamma(a)) = \alpha(a)$ for each $a \in A$. Thus the homomorphisms $\beta\gamma$ and α agree on the generating set A, and so they must agree on the whole of A^+. It follows that $\beta\gamma = \alpha$, as required. □

To motivate what follows, we isolate an important property of the syntactic monoid of a language.

Lemma 10.2.4 *Let L be a language over the alphabet A, let $M(L)$ be its syntactic monoid, and let ν: $A^* \to M(L)$ be the homomorphism that maps a string x to $\sigma_L(x)$. Let $P = \nu(L)$, a subset of the syntactic monoid. Then $L = \nu^{-1}(P) = (\nu^{-1}\nu)(L)$.*

Proof Clearly $L \subseteq \nu^{-1}(P)$. We prove that equality holds. Let $x \in \nu^{-1}(P)$. Then there exists $y \in L$ such that $\nu(x) = \nu(y)$ and so $x \, \sigma_L \, y$. But by Proposition 9.4.2, the language L is a union of σ_L-classes. Thus $x \in L$, as required.□

The result above motivates the following definition. Let A be a finite alphabet, M a monoid, and α: $A^* \to M$ a monoid homomorphism. Let $L \subseteq A^*$.

We say that α *recognises* L if there is a subset $P \subseteq M$ such that $L = \alpha^{-1}(P)$. Usually, we shall just say that the monoid M itself *recognises* the language L. In particular, the syntactic monoid $M(L)$ recognises L by Lemma 10.2.4. We say that a language L is *recognisable (by a monoid)* if it is recognised by a *finite* monoid. Our first task is to show that this new type of recognisability agrees with our usual definition via automata.

Theorem 10.2.5 *A language is recognisable by a monoid if and only if it is recognisable in the automata-theoretic sense.*

Proof Let $\mathbf{A} = (Q, A, i, \delta, T)$ be an automaton such that $L(\mathbf{A}) = L$, and let $T(\mathbf{A})$ be the transition monoid of $\mathbf{A}$. Denote by $\tau\colon A^* \to T(\mathbf{A})$ the function associating a string with the effect of the string on the set of states Q. Thus $\tau(x) = \tau_x$. Put $P = \tau(L)$, the image of the language in the transition monoid. Clearly $L \subseteq \tau^{-1}(P)$. We show that in fact equality holds. Let $x \in \tau^{-1}(P)$. Then by definition there exists $y \in L$ such that $\tau(x) = \tau(y)$. But $y \in L(\mathbf{A})$ and so $x \in L(\mathbf{A})$ because x and y have the same effect on the states of $\mathbf{A}$. Hence $x \in L$ as required.

To prove the converse, suppose that L is recognised by a finite monoid. Specifically, there is a monoid homomorphism $\alpha\colon A^* \to M$ and a subset $P \subseteq M$ such that $L = \alpha^{-1}(P)$. We shall prove that L is recognised by a finite automaton. Define $\mathbf{A} = (M, A, 1, \delta, P)$, where 1 is the identity of M and $\delta(m, a) = m\alpha(a)$. It is clear that $\mathbf{A}$ is a finite automaton. If u is a string in A^* then $\delta^*(m, u) = m\alpha(u)$. We now determine $L(\mathbf{A})$. We have that $x \in L(\mathbf{A})$ iff $1\alpha(x) \in P$ iff $x \in \alpha^{-1}(P) = L$. Hence $L(\mathbf{A}) = L$. □

A regular language may be recognised by many different monoids. The syntactic monoid occupies a privileged position that can be characterised using a definition combining subsemigroups with homomorphic images. Let M and N be semigroups. We say that N *divides* M if N is a homomorphic image of a subsemigroup of M; that is, there is a subsemigroup $M' \subseteq M$ and a surjective homomorphism $\alpha\colon M' \to N$. If M and N are both monoids, we say that N divides M if N is a monoid homomorphic image of a submonoid of M.

Theorem 10.2.6 *Let L be a language over the alphabet A. The monoid M recognises L if and only if $M(L)$ divides M where $M(L)$ is the syntactic monoid of L.*

Proof Suppose that M is a monoid that recognises L. Then by definition there is a monoid homomorphism $\alpha\colon A^* \to M$ and a subset $B \subseteq M$ such that $L = \alpha^{-1}(B)$. We show first that $\ker(\alpha) \subseteq \sigma_L$. Let $\alpha(x) = \alpha(y)$ and suppose that $uxv \in L$ for some $u, v \in A^*$. Then $\alpha(uxv) \in B$ and so $\alpha(uyv) \in B$, which implies that $uyv \in L$. A similar argument shows that $uyv \in L$ implies that $uxv \in L$. We have therefore proved that $(x, y) \in \ker(\alpha)$ implies $(x, y) \in \sigma_L$. Let $M' = \operatorname{im}(\alpha)$. By Proposition 9.2.12, there is a surjective homomorphism from M' to $M(L)$. It follows that $M(L)$ divides M, as required.

Suppose that $M(L)$ divides M. We prove that M recognises L. Because $M(L)$ divides M, there is a submonoid N of M and a surjective homomorphism β: $N \to M(L)$. Let γ: $A^* \to M(L)$ be the natural homomorphism determined by the congruence σ_L. Let α: $N \to M$ be the embedding function that takes an element of N and maps it to the same element inside M. We therefore have the following diagram of monoid homomorphisms indicated using solid arrows:

$$\begin{array}{ccc} M & & A^* \\ \uparrow \alpha & \swarrow_{\delta} & \downarrow \gamma \\ N & \xrightarrow[\beta]{} & M(L) \end{array}$$

By Proposition 10.2.3, there is a monoid homomorphism δ: $A^* \to N$ such that $\beta\delta = \gamma$ indicated in the diagram by the dashed arrow. We therefore have a monoid homomorphism $\zeta = \alpha\delta$: $A^* \to M$. Put $P = (\alpha\beta^{-1}\gamma)(L) \subseteq M$. We claim that $\zeta^{-1}(P) = L$. To see this, we calculate as follows:

$$\zeta^{-1}(P) = (\alpha\delta)^{-1}(P) = (\delta^{-1}\alpha^{-1}\alpha\beta^{-1}\gamma)(L) = (\delta^{-1}\beta^{-1}\gamma)(L) = (\gamma^{-1}\gamma)(L) = L,$$

where we use the fact that $\alpha^{-1}\alpha$ is the identity function on N. Thus the monoid M recognises the language L. □

Examples 10.2.7 Here are two examples of languages recognised by monoids.

(1) Let S be the monoid with the following Cayley table:

	1	s	t
1	1	s	t
s	s	s	s
t	t	t	t

Let $A = \{a, b\}$ and define a monoid homomorphism α: $A^* \to S$ by $\alpha(a) = s$ and $\alpha(b) = t$. Then $\alpha^{-1}(s) = a(a + b)^*$. This result can either be verified directly, or using the construction given in the proof of Theorem 10.2.5. The automaton corresponding to the monoid homomorphism α and subset $\{s\}$ of the monoid S is

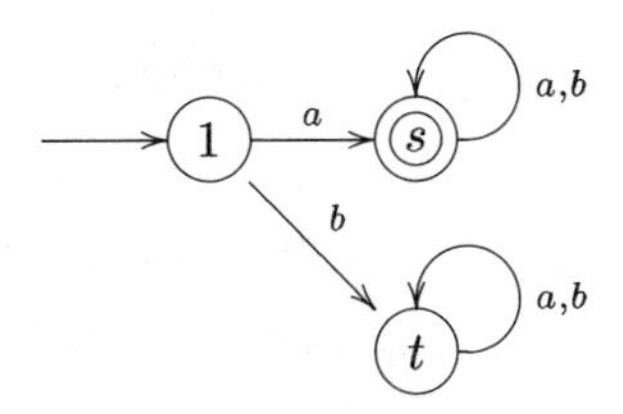

The language recognised by this machine is, by construction, $\alpha^{-1}(s)$.

(2) Let $A = \{a, b, c\}$. Then $(\mathsf{P}(A), +)$ is a commutative monoid with identity $\emptyset$. Define $\beta\colon A^* \to \mathsf{P}(A)$ by

$$\beta(x) \text{ is equal to the set of letters occurring in } x.$$

This is a monoid homomorphism because $\beta(\varepsilon) = \emptyset$, and $\beta(xy) = \beta(x) + \beta(y)$. Let $B = \{a, b\} \in \mathsf{P}(A)$, a single element of the monoid. Then $\beta^{-1}(B)$ consists of all those strings in $(a+b+c)^*$ that contain both a and b but not c. Thus $\beta^{-1}(B) = B^*aB^*bB^* + B^*bB^*aB^*$. This can also be verified by constructing the automaton from the monoid homomorphism β and subset $\{\{a, b\}\}$ of the monoid $\mathsf{P}(A)$ using the construction given in the proof of Theorem 10.2.5.

□

Some of the results we have proved about recognisable languages using finite automata can easily be proved using finite monoids.

Proposition 10.2.8 *Let A be an alphabet.*

(i) *Let $L \subseteq A^*$. If M recognises L, then M recognises L'.*

(ii) *Let $L_1, L_2 \subseteq A^*$. If M_1 recognises L_1 and M_2 recognises L_1, then $M_1 \times M_2$ recognises $L_1 \cap L_2$ and $L_1 + L_2$.*

(iii) *Let $L \subseteq A^*$ and $u \in A^*$. If M recognises L, then M recognises $u^{-1}L$ and Lu^{-1}.*

(iv) *Let M recognise $L \subseteq A^*$ and let $\alpha\colon B^* \to A^*$ be a monoid homomorphism. Then M recognises $\alpha^{-1}(L)$.*

Proof (i) Let $\alpha\colon A^* \to M$ and $P \subseteq M$ be such that $L = \alpha^{-1}(P)$. It is easy to check that $L' = \alpha^{-1}(M \setminus P)$.

(ii) Let $\alpha_1\colon A^* \to M_1$ and $\alpha_2\colon A^* \to M_2$ and $P_1 \subseteq M_1$ and $P_2 \subseteq M_2$ where $L_1 = \alpha_1^{-1}(P_1)$ and $L_2 = \alpha_2^{-1}(P_2)$. Define $\beta\colon A^* \to M_1 \times M_2$ by

$$\beta(x) = (\alpha_1(x), \alpha_2(x)).$$

This is a monoid homomorphism. It is easy to check that

$$L_1 \cap L_2 = \beta^{-1}(P_1 \times P_2)$$

and

$$L_1 + L_2 = \beta^{-1}((P_1 \times M_2) \cup (M_1 \times P_2)).$$

(iii) Let M recognise L. Then there is a monoid homomorphism $\alpha\colon A^* \to M$ and a subset $P \subseteq M$ such that $L = \alpha^{-1}(P)$. We prove that M recognises $u^{-1}L$. The proof of the other case is left as an exercise. Let $\alpha(u) = b$. Put

$Q = \{a \in M\colon ba \in P\}$. We prove that $\alpha^{-1}(Q) = u^{-1}L$. Let $x \in \alpha^{-1}(Q)$. Then $\alpha(x) \in Q$. Thus $b\alpha(x) \in P$; that is $\alpha(ux) = \alpha(u)\alpha(x) \in P$. Thus $ux \in \alpha^{-1}(P)$ and so $x \in u^{-1}L$. To prove the reverse inclusion, let $x \in u^{-1}L$. Then $ux \in L$, and so $\alpha(ux) \in P$. Thus $\alpha(u)\alpha(x) \in P$, which is just $b\alpha(x) \in P$. Thus $\alpha(x) \in Q$ and so $x \in \alpha^{-1}(Q)$.

(iv) Let M recognise L. Then there is a monoid homomorphism $\beta\colon A^* \to M$ and a subset $P \subseteq M$ such that $L = \beta^{-1}(P)$. Let $\alpha\colon B^* \to A^*$ be a monoid homomorphism. Then $\beta\alpha\colon B^* \to M$ is a monoid homomorphism and $(\beta\alpha)^{-1}(P) = \alpha^{-1}(L)$. Thus M recognises $\alpha^{-1}(L)$. □

In Proposition 3.3.4, we proved that if L and M are both recognisable, then LM is recognisable. To conclude this section, we shall prove the algebraic version of this result. To do this, we shall need a new way of combining two monoids. Let M and N be monoids and let $X \subseteq M \times N$. For each $m \in M$ and $n \in N$ define

$$mX = \{(mx, y)\colon (x, y) \in X\} \text{ and } Xn = \{(x, yn)\colon (x, y) \in X\}.$$

Define $M \diamond N$ to consist of the following set of 2×2-matrices:

$$\begin{pmatrix} m & X \\ \emptyset & n \end{pmatrix},$$

where $m \in M$, $n \in N$ and $X \subseteq M \times N$. We define the product of two such matrices as follows:

$$\begin{pmatrix} m_1 & X_1 \\ \emptyset & n_1 \end{pmatrix}\begin{pmatrix} m_2 & X_2 \\ \emptyset & n_2 \end{pmatrix} = \begin{pmatrix} m_1m_2 & m_1X_2 + X_1n_2 \\ \emptyset & n_1n_2 \end{pmatrix}.$$

Proposition 10.2.9 *Let M and N be monoids. Then $M \diamond N$ is a monoid.*

Proof We leave it as an exercise to check that the multiplication is associative and that

$$\begin{pmatrix} 1 & \emptyset \\ \emptyset & 1 \end{pmatrix}$$

is the identity, where the entry in row 1 and column 1 is the identity of M, and the entry in row 2 and column 2 is the identity of N. □

The monoid $M \diamond N$ is called the *Schützenberger product of M and N*. We shall usually denote the elements of $M \diamond N$ by triples (m, X, n), with the above matrix form being a useful mnemonic.

The proof of our main result depends on some calculations involving prefixes. If x is a string, we denote by Prefix(x) the set of prefixes of x. If u is a prefix of x, then we can write $x = uv$ for some string v. We denote v by $u^{-1}x$. The proof of the following is left as an exercise.

Lemma 10.2.10 *Let $x, y \in A^*$. Then*

$$\text{Prefix}(xy) = \text{Prefix}(x) + x\text{Prefix}(y).$$

□

We now have our main result.

Proposition 10.2.11 *Let L_1 and L_2 be languages over the alphabet A. Let M_1 and M_2 be monoids that recognise L_1 and L_2, respectively. Then there is a semigroup homomorphism β: $A^* \to M_1 \diamond M_2$ and a subset Q of $M_1 \diamond M_2$ such that $L_1L_2 = \beta^{-1}(L_1L_2)$.*

In particular, L_1L_2 is recognised by a monoid that is a subsemigroup of $M_1 \diamond M_2$.

Proof Let α_1: $A^* \to M_1$ and α_2: $A^* \to M_2$ be monoid homomorphisms such that and $L_1 = \alpha_1^{-1}(P_1)$ and $L_2 = \alpha_2^{-1}(P_2)$. Define β: $A^* \to M_1 \diamond M_2$ by

$$\beta(x) = (\alpha_1(x), X(x), \alpha_2(x)),$$

where

$$X(x) = \{(\alpha_1(u), \alpha_2(u^{-1}x)) \colon u \in \text{Prefix}(x)\}.$$

We shall prove that β is a monoid homomorphism. To do this, it is enough to prove that

$$X(xy) = \alpha_1(x)X(y) + X(x)\alpha_2(y)$$

for all $x, y \in A^*$, since this condition implies that $\beta(xy) = \beta(x)\beta(y)$. By definition,

$$X(xy) = \{(\alpha_1(u), \alpha_2(u^{-1}(xy))) \colon u \in \text{Prefix}(xy)\}.$$

By Lemma 10.2.10, this is equal to

$$\{(\alpha_1(u), \alpha_2(u^{-1}(xy))) \colon u \in \text{Prefix}(x)\} + \{(\alpha_1(u), \alpha_2(u^{-1}(xy))) \colon u \in x\text{Prefix}(y)\},$$

which simplifies to

$$\{(\alpha_1(u), \alpha_2(u^{-1}x)\alpha_2(y)) \colon u \in \text{Prefix}(x)\} + \{(\alpha_1(x)\alpha_1(v), \alpha_2(v^{-1}y)) \colon v \in \text{Prefix}(y)\}.$$

This is just

$$X(xy) = X(x)\alpha_2(y) + \alpha_1(x)X(y) = \alpha_1(x)X(y) + X(x)\alpha_2(y),$$

as required.

We shall now show that if we put $Q = \beta(L_1L_2)$ then $L_1L_2 = \beta^{-1}(Q)$. It is enough to show that $\beta^{-1}(Q) \subseteq L_1L_2$. Let $x \in \beta^{-1}(Q)$. Then $\beta(x) = \beta(l_1l_2)$ for some $l_1 \in L_1$ and $l_2 \in L_2$. Thus

$$\beta(x) = (\alpha_1(x), X(x), \alpha_2(x)) = (\alpha_1(l_1l_2), X(l_1l_2), \alpha_2(l_1l_2)).$$

Now $(\alpha_1(l_1), \alpha_2(l_2)) \in X(l_1l_2) = X(x)$. It follows that we can write $x = uv$, where $\alpha_1(u) = \alpha_1(l_1) \in P_1$ and $\alpha_2(v) = \alpha_2(l_2) \in P_2$. Thus $u \in L_1$ and $v \in L_2$ and so $x \in L_1L_2$ as required.

The final claim follows from the fact that the image of β is a monoid, S say, because A^* is a monoid. The monoid S clearly recognises L_1L_2. □

Exercises 10.2

1. Let $\theta \colon \mathbb{N} \to \mathbb{Z}_4$ be the function defined by $\theta(n) = r$, where $r \in \mathbb{Z}_4$ and $n \equiv r \pmod 4$. Calculate $\theta^{-1}(2)$ and $\theta^{-1}(\theta(\{5, 10, 15\}))$.

2. Let T_2 be the full transformation monoid on the set $\{1, 2\}$. Define elements σ and τ of T_2 as follows:

$$\sigma = \begin{pmatrix} 1 & 2 \\ 2 & 1 \end{pmatrix} \quad \tau = \begin{pmatrix} 1 & 2 \\ 1 & 1 \end{pmatrix}.$$

Show that $T_2 = \{\sigma, \tau\}^*$. Let $A = \{a, b\}$ and define $\theta \colon A^* \to T_2$ by $\theta(a) = \sigma$ and $\theta(b) = \tau$. Thus θ is a surjective monoid homomorphism. Calculate $\theta^{-1}(\gamma)$, where

$$\gamma = \begin{pmatrix} 1 & 2 \\ 2 & 2 \end{pmatrix}.$$

3. Complete the proof of Proposition 10.2.8.

4. Prove Lemma 10.2.10.

5. Let $G = \{-1, 1\}$ under multiplication. How many elements does $G \diamond G$ contain? How many idempotents?

10.3 Two counterexamples

The results of this chapter have been leading us toward some kind of correspondence between finite monoids and recognisable languages. The two counterexamples we describe in this section will therefore appear to be a little discouraging from the point of view of setting up an 'algebraic theory of recognisable languages.' However, in Chapter 12 we shall show that in fact these results simply indicate that the algebraic theory is more subtle than might be supposed.

Our first counterexample shows that different languages can have the same syntactic monoid. Recall that R_n is the semigroup with n elements and product defined by $ab = b$ for all $a, b \in R_n$, and R_n^1 is the semigroup R_n with an identity adjoined.

Proposition 10.3.1 *Let A be any alphabet with at least two elements. Let $B \subseteq A$ be any proper, non-empty subset. Then the syntactic monoid of A^*B is R_2^1.*

Proof It is left as an exercise to check that the automaton below is a minimal automaton for the language A^*B and to check that its transition monoid is R_2^1.

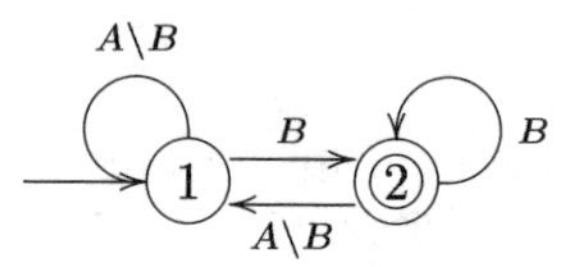

□

Our next counterexample is more hard-won. First we need to extend the definition of the syntactic congruence from subsets of free monoids to subsets of arbitrary ones. Let S be a monoid and let $L \subseteq S$ be an arbitrary subset. Define the relation σ_L on S as follows: $(a, b) \in \sigma_L$ iff $xay \in L \Leftrightarrow xby \in L$ for all $x, y \in S$.

Proposition 10.3.2 *The relation σ_L is a congruence on S such that L is a union of some of the σ_L-classes. If σ is any congruence on S such that L is a union of some of the σ-classes then $\sigma \subseteq \sigma_L$.*

Proof Denote the identity of S by 1. We leave it as an exercise to show that σ_L is a congruence on S. Let $a \in L$ and suppose that $(a, b) \in \sigma_L$. Then $1a1 \in L$ and so $1b1 \in L$, which gives $b \in L$. Thus $\sigma_L(a) \subseteq L$ for each $a \in L$, which proves that L is a union of some of the σ_L-classes.

Now let σ be an arbitrary congruence on S with the property that L is a union of some of the σ-classes. Let $(a, b) \in \sigma$. Suppose that $xay \in L$. Then $(xay, xby) \in \sigma$ because σ is a congruence. But $xay \in L$ implies that $xby \in L$ because L is a union of σ-classes. By symmetry $xby \in L$ implies that $xay \in L$. We have therefore proved that $(a, b) \in \sigma_L$. □

A subset L of a monoid S is said to be *disjunctive* if the congruence σ_L is the equality relation.

Proposition 10.3.3 *A finite monoid is the syntactic monoid of a recognisable language if and only if it contains a disjunctive subset.*

Proof Let $L \subseteq A^*$ be a recognisable language with syntactic monoid $M(L)$, and let $\phi\colon A^* \to M(L)$ be the natural homomorphism associated with σ_L. Put

$P = \phi(L)$, a subset of $M(L)$. Recall that $L = \phi^{-1}(P)$. We shall prove that P is a disjunctive subset of $M(L)$. Let $a, b \in M(L)$ and suppose that for all $c, d \in M(L)$ we have that

$$cad \in P \Leftrightarrow cbd \in P.$$

We shall prove that $a = b$. Let $\phi(u) = a$ and $\phi(v) = b$. We shall prove that $(u, v) \in \sigma_L$. Suppose that $xuy \in L$. Then $\phi(x)a\phi(y) \in P$ and so $\phi(x)b\phi(y) \in P$. Thus $\phi(xvy) \in P$. It follows that $xvy \in L$. By symmetry, $xvy \in L$ implies $xuy \in L$. Thus we have shown that $(u, v) \in \sigma_L$, and so $a = b$ as required.

Now let M be a finite monoid containing a disjunctive subset P. Let $\theta: A^* \to M$ be a surjective homomorphism for some alphabet A, such a homomorphism exists because M is finite and so we can apply Proposition 10.2.2. Put $L = \theta^{-1}(P)$. Let $\phi: A^* \to M(L)$ be the natural homomorphism associated with the congruence σ_L. From the proof of Theorem 10.2.6, there is a surjective homomorphism $\alpha: M \to M(L)$ such that $\alpha\theta = \phi$. We shall prove that α is injective. This implies that M is isomorphic to $M(L)$, which will prove the result. Suppose that $\alpha(a) = \alpha(b)$. Let $\theta(u) = a$ and $\theta(v) = b$. Then $\phi(u) = \phi(v)$ and so $(u, v) \in \sigma_L$. Suppose that $cad \in P$ for some $c, d \in M$. Let $\theta(x) = c$ and $\theta(y) = d$. Then $\theta(xuy) \in P$, so that $xuy \in L$. But $(u, v) \in \sigma_L$ so that $xvy \in L$, which implies that $cbd \in P$. Thus $cad \in P$ implies that $cbd \in P$. By symmetry we can prove the converse and so $(a, b) \in \sigma_P$. But P is a disjunctive subset and so $a = b$, as required. □

We have characterised when a finite monoid is a syntactic monoid. All we need now for our counterexample is a finite semigroup that does not contain a disjunctive subset.

Proposition 10.3.4 *The monoid R_n^1 contains no disjunctive subset for $n > 2$. In particular, R_n^1 is not a syntactic monoid for $n > 2$.*

Proof Let $n > 2$. We prove that R_n^1 has no disjunctive subsets. Let $P \subseteq R_n^1$. We can assume that P is non-empty because when P is empty the congruence σ_P is the universal congruence. Two elements are σ_P-related precisely when they have the same contexts. So we shall calculate the possible contexts of elements of R_n^1:

- Let $a \in P$ where $a \neq 1$. Then $xay \in P$ iff $(x, y) \in (R_n^1 \times P) \cup (R_n^1 \times \{1\})$.
- Let $a \in R_n \setminus P$. Then $xay \in P$ iff $(x, y) \in R_n^1 \times P \setminus \{1\}$.
- Let $a = 1$. Then $xy \in P$ iff $(x, y) \in (R_n^1 \times P) \cup (P \times \{1\})$.

It follows that σ_P can have at most three congruence classes, whereas R_n^1 has at least four elements when $n > 2$. □

Exercises 10.3

1. Complete the proof of Proposition 10.3.1.

2. Complete the proof of Proposition 10.3.2.

10.4 Summary of Chapter 10

- *Finite semigroups*: Every finite semigroup contains idempotents. The mutual divisibility properties of finite semigroups can be displayed by eggbox diagrams. Products of semigroups are a way of constructing new semigroups from old.

- *Recognisability by a monoid*: We can define what it means for a monoid to recognise a language as the algebraic analogue of what it means for an automaton to recognise a language. The syntactic monoid plays the role of the minimal automaton in the sense that it divides every monoid recognising a given language. Some of the standard properties of recognisable languages can be proved using monoids. The algebraic proof that the product of two recognisable languages is recognisable requires the introduction of a new way of combining semigroups called the Schützenberger product.

10.5 Remarks on Chapter 10

This book is not primarily about semigroups, so if you want to learn more about them, then [68] or [32] are good places to start. Two classes of semigroups have proved themselves to be particularly important: finite semigroups and inverse semigroups. As the results of this chapter show, finite semigroups are closely related to recognisable languages. Inverse semigroups, on the other hand, are related to notions of symmetry. See [78] for a cornucopia of information about inverse semigroups.

The 1965 paper by Schützenberger [116] uses the notion of a language being recognised by a finite monoid; moreover, he did something with this construction of considerable interest. See Chapter 11. This landmark paper is also the source of the Schützenberger product of monoids. You can find out more about this construction in [83], for example. I learnt about the results of Section 10.3 from [77].

Chapter 11

Star-free languages

Kleene's Theorem was the first major result on finite automata. The second deep theorem on finite automata, due to Schützenberger, is the subject of this chapter. The set of recognisable languages is closed under union, product, and Kleene star, and it is also closed under all the other Boolean operations. Suppose we retain all the other operations but discard Kleene star, what can we say about the languages that can then be described? Such languages are said to be 'star-free.' The goal of this chapter is to characterise star-free languages, and the main tool we shall use to obtain our characterisation will be the syntactic monoid of the language. Our characterisation provides conclusive proof of the importance of the syntactic monoid in studying recognisable languages.

11.1 Introduction

Before delving into the theory, I would like to use this section to introduce the problem we are going to solve and some of the ideas we shall use to solve it. I begin with the definition of the class of languages we shall be interested in.

Let A be a finite alphabet. A *star-free (generalised regular) expression (over A)* is defined as follows:

(SF1) $\emptyset$, ε and a, where $a \in A$, are star-free generalised regular expresssions.

(SF2) If s and t are star-free generalised regular expressions so are $s + t$, $s \cap t$, s', and $s \cdot t$.

(SF3) Every star-free regular expression arises by a finite number of applications of the rules (SF1) and (SF2).

The symbols $+$, $\cap$, $'$, and $\cdot$ are called the *operators*. As usual we denote $\cdot$ by concatenation. Each star-free generalised regular expression s describes a language that we denote by $L(s)$, just as in the case of regular expressions, except that we also have the symbol $'$ that denotes the complement of a subset of A^*. The languages $\emptyset$, ε, and a, where $a \in A$, will be called 'basic languages.' We say that a language L over an alphabet A is *star-free* if there is a star-free generalised regular expression s over A such that $L = L(s)$. In other words, a language is star-free if it can be constructed from the basic languages using only the Boolean operations and the product but not the Kleene star. To help understand this definition, we begin with an example.

Example 11.1.1 Is the language $(ab)^*$ star-free? On the face of it, the answer might appear to be 'no,' but this would be incorrect. To show that a language is star-free we have to show that it can be written without using the Kleene star, and to show that it is not star-free we have to show that this is impossible. Observe that a string in $(ab)^*$ does not contain the strings aa and bb as factors, and it cannot begin with a b nor end with an a. In fact, this precisely describes the strings in $(ab)^*$. Thus

$$(ab)^* = A^* \setminus (bA^* + A^*a + A^*aaA^* + A^*bbA^*).$$

If X and Y are sets, then we can rewrite $X \setminus Y$ as $X \cap Y'$. Clearly we can rewrite A^* as $\emptyset'$. If we make these changes in our expression above, we see that $(ab)^*$ *is* star-free. □

This example shows that determining whether a recognisable language can be described by a star-free regular expression may be no easy matter. To gain some insight into this problem, therefore, we shall begin by restricting our attention first to the simplest class of recognisable languages: those over one-letter alphabets. Let $A = \{a\}$. We know from Theorem 5.2.2 that a recognisable language over such an alphabet has the form $X + Y(a^p)^*$, where X and Y are finite languages. Our analysis of this language will depend on whether $p = 0, 1$ or $p > 1$. If $p = 0$, then $X + Y(a^p)^* = X + Y$ is just a finite language, which is evidently star-free. Its minimal automaton will be similar to that of the next case. If $p = 1$, then we can write this language as $X + Y\emptyset'$, which shows that it is star-free. The corresponding minimal automaton consists of a set of $m + 1$ states labelled $0, \ldots, m$ with the state 0 as the initial state. The transition monoid of this automaton consists of the function mapping 0 to 0 if there is only one state, or in the case of more than one state, all powers of the function that maps i to $i + 1$ for $i = 0, \ldots, m - 1$ and that maps m to m.

Example 11.1.2 Consider the language $a + a^3 + a^4a^*$. The minimal automaton for this language is

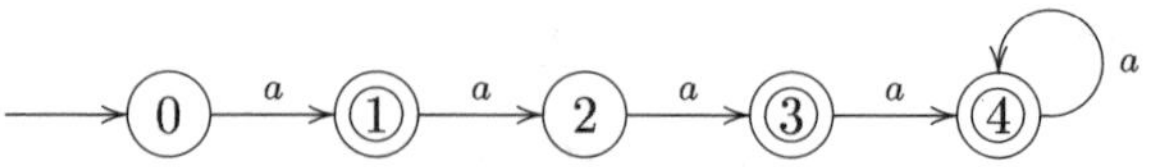

Observe that

$$\tau_a = \begin{pmatrix} 0 & 1 & 2 & 3 & 4 \\ 1 & 2 & 3 & 4 & 4 \end{pmatrix},$$

where τ is the function defined in Section 8.1 that describes the effect of each input string on the set of states. □

Now suppose that $p \geq 2$, and concentrate first on the second summand. It is a union of languages of the form $a^q(a^p)^*$: all strings of a's of length q modulo p. We assume that $q < p$: the case where $q \geq p$ will be treated in Example 11.1.4. We now look at the corresponding minimal automaton for this language. This consists of p states arranged in a circle, numbered $0, 1, \ldots, p-1$, with the state 0 as the initial state and the state q as the terminal state. The transition monoid of this automaton consists of all powers of the function that maps i to $i+1$ for $i = 0, 1, \ldots, p-2$ and that maps $p-1$ to 0. This function has an additional property: it is a bijective function on the set of states or, to use a more common term, a 'permutation' of the states.

Example 11.1.3 Consider the language $a^3(a^4)^*$. The minimal automaton for this language is

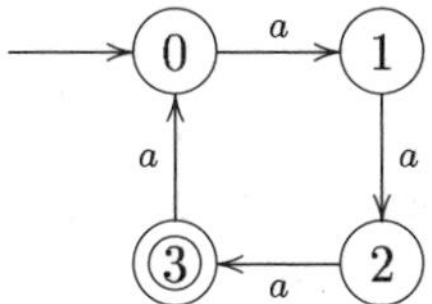

We have that

$$\tau_a = \begin{pmatrix} 0 & 1 & 2 & 3 \\ 1 & 2 & 3 & 0 \end{pmatrix},$$

which is a permutation. □

We now consider the case where $p > 1$, $q \geq p$ and where there is also an X term. Then the transition monoid again consists of all powers of a single function and this function combines features of both the above cases. However, for $p \geq 2$ at least two of the states are permutated amongst themselves. This is best illustrated by means of a concrete example.

Example 11.1.4 Consider the language $a + a^3 + a^5(a^3)^*$. Its minimal automaton is

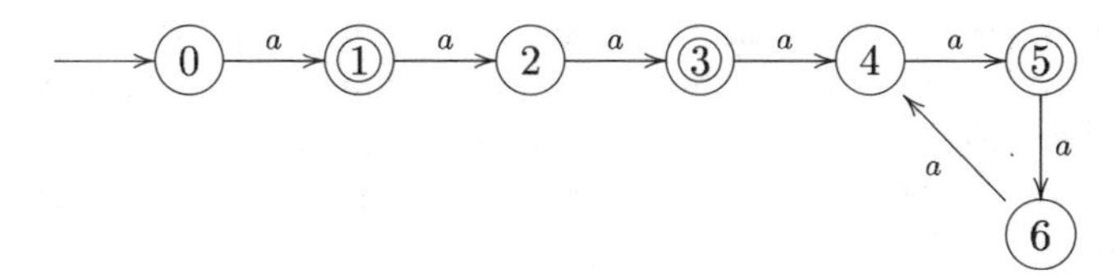

We have that

$$\tau_a = \begin{pmatrix} 0 & 1 & 2 & 3 & 4 & 5 & 6 \\ 1 & 2 & 3 & 4 & 5 & 6 & 4 \end{pmatrix}.$$

We see that the states $\{4, 5, 6\}$ are permuted amongst themselves by this function. □

These examples lead us to formulate the following *conjecture*:

> over a one-letter alphabet, the star-free languages are characterised by the property that the transition monoids of their minimal automata contain no element that induces a permutation on any subset of the set of states containing at least two elements.

We shall prove that this conjecture is correct, and that it also holds for arbitrary alphabets.

Exercises 11.1

1. Construct the Cayley tables of the transition monoids in Examples 11.1.2, 11.1.3, and 11.1.4; use the representation of elements by means of powers of a.

11.2 Groups

We begin with a definition that formalises the one we encountered informally in Section 11.1. A bijective function in the full transformation monoid $T(X)$ is called a *permutation* of X. We denote the set of permutations in $T(X)$ by $S(X)$; the letter 'S' stands for 'symmetry.' If $X = \{1, \ldots, n\}$ then we denote $S(X)$ by S_n. When an element of $T(X)$ is written in two-row form, it is easy to recognise when it represents a permutation: the bottom row is simply the top row written in a different order. Our discussion in Section 11.1 has highlighted the importance of permutations in $T(X)$. There is another way of thinking about them that we shall also need. Remember that, as usual, arguments of elements of $T(X)$ are written on the left.

Proposition 11.2.1 *Let α: $X \to X$ be a function. Then α is a bijection if and only if there is a function β: $X \to X$ such that $\alpha\beta = 1_X = \beta\alpha$, the identity function on X.*

Proof Let β: $X \to X$ be a function such that $\alpha\beta = 1_X = \beta\alpha$. We show that α is a bijection. Suppose that $x\alpha = x'\alpha$. Then $(x\alpha)\beta = (x'\alpha)\beta$, and so $x(\alpha\beta) = x'(\alpha\beta)$. But $\alpha\beta$ is the identity function on X. Thus $x = x'$, and α is injective. Now suppose that $y \in X$. Then $y(\beta\alpha) = y$ because $\beta\alpha$ is the identity on X. Thus $(y\beta)\alpha = y$, which shows that α is surjective. We have shown that α is a bijection.

Now suppose that α is a bijection. Let $y \in X$. Because α is a bijection there exists a unique element $x \in X$ such that $x\alpha = y$. Define $y\beta = x$. This is a well-defined function. By definition $y(\beta\alpha) = y$, which means that $\beta\alpha = 1_X$. Suppose that $x\alpha = y$. By definition $y\beta = x$ and so $x(\alpha\beta) = x$, which means that $\alpha\beta = 1_X$. □

The characterisation of bijections we found above is the basis for the following abstract definition. Let S be a monoid with identity 1. An element $s \in S$ is said to be *invertible* if there is an element $t \in S$ such that

$$st = 1 = ts.$$

The element t is called an *inverse* of s. Proposition 11.2.1 therefore shows that the invertible elements in $T(X)$ are precisely the permutations.

Lemma 11.2.2 *Let S be a monoid.*

(i) *If an element of a monoid has an inverse, then it has a unique inverse.*

(ii) *If a and b are both invertible with inverses a' and b', respectively, then ab is invertible with inverse $b'a'$.*

Proof (i) Let a have inverses b and b'. Then $1 = ab$ and so

$$b' = b'(ab) = (b'a)b = 1b = b.$$

Hence $b = b'$.

(ii) It is enough to show that

$$(b'a')ab = 1 = ab(b'a'),$$

which is straightforward. □

Because of the above result we may speak of *the* inverse of an invertible element.

Notation We denote the inverse of the element s, if it has one, by s^{-1}. However, if the operation is denoted by $+$, it is usual to denote the inverse by $-s$ and the identity by 0 rather than 1. This has no other basis than to make equations 'look right.'

A *group* is a monoid in which every element is invertible. A *trivial group* is a group containing only the identity element. The *order* of a finite group is simply the number of elements in the group. Let G be a group and let H be a submonoid. We say that H is a *subgroup* of G if, in addition, $h^{-1} \in H$ for each $h \in H$. It follows that H is a group in its own right. If S is a semigroup and G is a subsemigroup of S that is a group, then we say that G is a *group in S*.

Lemma 11.2.3 *Let G be a group.*

(i) $(g^{-1})^{-1} = g$ *for each* $g \in G$.

(ii) *If* $g_1, \ldots, g_n \in G$ *then* $(g_1 \ldots g_n)^{-1} = g_n^{-1} \ldots g_1^{-1}$.

Proof (i) Simply observe that g^{-1} is *an* inverse of g and so it is *the* inverse of g by Lemma 11.2.2(i).

(ii) We use induction. The case $n = 2$ is proved in Lemma 11.2.2(ii) and is our base case. Our induction hypothesis is that the result is true for n. We prove it for $n + 1$. Observe that $(g_1 \dots g_n g_{n+1})^{-1} = ((g_1 \dots g_n)g_{n+1})^{-1}$. This is equal to $g_{n+1}^{-1}(g_1 \dots g_n)^{-1}$ using the base case. The result now follows from the induction hypothesis. □

Denote the set of invertible elements in a monoid S by $U(S)$.[1] Then $U(S)$ is a group in S, by Lemma 11.2.2(ii), called the *group of invertible elements of* S. A semigroup S is a group precisely when $S = U(S)$.

Examples 11.2.4 Here are some examples of groups of invertible elements.

(1) $(\mathbb{Z}, +)$, $(\mathbb{Q}, +)$ and $(\mathbb{R}, +)$ are all groups.

(2) $(\mathbb{Z}_n, +)$ is a group. This is because for each $a \in \mathbb{Z}_n$ there exists $b \in \mathbb{Z}_n$ such that $a + b \equiv 0 \pmod{n}$.

(3) The group of invertible elements in $(\mathbb{Z}, \times)$ is $\{-1, 1\}$.

(4) The groups of invertible elements in $(\mathbb{Q}, \times)$ and $(\mathbb{R}, \times)$ are $\mathbb{Q} \setminus \{0\}$ and $\mathbb{R} \setminus \{0\}$, respectively.

(5) The group of invertible elements in $T(X)$ is $S(X)$, called the *symmetric group*.

□

Groups are special kinds of semigroups, and so we can look at homomorphisms between groups. These have extra properties.

Lemma 11.2.5 *Let α: $G \to H$ be a homomorphism between groups. Then α is a monoid homomorphism and $\alpha(g^{-1}) = \alpha(g)^{-1}$. In particular, the image of α is a subgroup of H.*

Proof We begin by showing that α is a monoid homomorphism. Denote both identities in G and H by 1. We leave it as an exercise to prove that the only idempotent in a group is the identity. But $\alpha(1)^2 = \alpha(1)$ in H, consequently $\alpha(1)$ is an idempotent and so is the identity of H.

Let $g \in G$. Then $gg^{-1} = 1 = g^{-1}g$. Thus $\alpha(g)\alpha(g^{-1}) = 1 = \alpha(g^{-1})\alpha(g)$. But these two equations imply that $\alpha(g^{-1})$ is the inverse of $\alpha(g)$ in H and so by the uniqueness of inverses we have that $\alpha(g^{-1}) = \alpha(g)^{-1}$. We say that group homomorphisms 'preserve inverses.'

We already know by Proposition 9.1.8, that the image of α is a submonoid of H. The fact that inverses are preserved by α shows that the image of α is

[1] Invertible elements are sometimes called *units*, which explains the notation.

actually a group and so a subgroup of H. □

It follows that an *isomorphism of groups* is simply a bijective homomorphism.

Example 11.2.6 We have already noted that $(\mathbb{R}, +)$ is a group. We denote the set of positive real numbers by $\mathbb{R}^+$. It is an easy exercise to check that $(\mathbb{R}^+, \times)$ is a group. The (natural) logarithm function $x \mapsto \ln(x)$ is a homomorphism from $(\mathbb{R}^+, \times)$ to $(\mathbb{R}, +)$. This is because $\ln(ab) = \ln(a) + \ln(b)$. There is an inverse homomorphism $x \mapsto \exp(x)$. Thus the two groups are isomorphic. This isomorphism is historically important because it converts hard calculations involving multiplication into easy calculations involving addition. For generations log-tables, and their mechanical analogues slide-rules, were the only tools available for easing the burden of calculation. □

Congruences can be defined on groups just as they can on arbitrary monoids. However, in the case of groups congruences can be handled much more easily. We shall only need the result below, but see Question 10 of Exercises 11.2 for the general result.

Proposition 11.2.7 *Let $\theta\colon G \to H$ be a homomorphism between groups. Put*

$$K = \{g \in G\colon \theta(g) = 1\}.$$

Then K is a subgroup of G. Furthermore, it is trivial if and only if θ is injective.

Proof The subset K contains the identity of G because homomorphisms between groups are automatically monoid homomorphisms by Lemma 11.2.5. Let $a, b \in K$. Then $\theta(ab) = \theta(a)\theta(b) = 1$. It follows that $ab \in K$. Suppose that $a \in K$. Then $\theta(a^{-1}) = \theta(a)^{-1} = 1$ by Lemma 11.2.5. It follows that $a^{-1} \in K$. Thus K is a subgroup of G. If θ is injective, then clearly K will be trivial. We prove the converse. Suppose that K is trivial and $\theta(a) = \theta(b)$. Then $\theta(a)\theta(b)^{-1} = 1$ and so $\theta(ab^{-1}) = 1$. Thus $ab^{-1} \in K$. By assumption $ab^{-1} = 1$, from which it follows that $a = b$, and so θ is injective. □

The result above tells us that to check whether or not a homomorphism between groups is injective, it is enough to consider the inverse image of the identity: this is trivial if and only if the homomorphism is injective.

Let G be a group. Then we can construct the monoid $\mathsf{P}(G)$ with respect to subset multiplication. In this case we have an additional operation. If X is a subset of a group G define

$$X^{-1} = \{x^{-1}\colon x \in X\}.$$

Now define

$$\langle X \rangle = (X \cup X^{-1})^*.$$

In a monoid S, we defined a^n for $n \in \mathbb{N}$. In a group G, we define $a^{-n} = (a^{-1})^n$ when $n \geq 1$. Thus in a group a^n is defined for all $n \in \mathbb{Z}$.

Proposition 11.2.8 *Let G be a group, and $X \subseteq G$. Then $\langle X \rangle$ is a subgroup of G, the smallest subgroup of G containing X.*

If $X = \{a\}$, then $\langle a \rangle = \{a^n \colon n \in \mathbb{Z}\}$.

Proof Clearly $\langle X \rangle$ is a submonoid of G containing X. It is actually a subgroup of G by Lemma 11.2.3. Clearly, any subgroup of G containing X must also contain X^{-1}, and so it must contain $X \cup X^{-1}$. But a subgroup is also a submonoid, and so a subgroup containing X must therefore also contain $(X \cup X^{-1})^*$. It follows that $\langle X \rangle$ is the smallest subgroup of G containing the set X.

To prove the final claim, note that any subgroup of G containing a must contain all the elements a^n where $n \in \mathbb{Z}$. But the set $\{a^n \colon n \in \mathbb{Z}\}$ is also a subgroup of G. The result now follows. □

Let G be a group. For $X \subseteq G$ we call $\langle X \rangle$ the subgroup of G *generated by* X. A group G is said to be *cyclic* if $G = \langle a \rangle$ for some $a \in G$.

Notation It is important to remember that for $n \geq 1$ the notation a^n means 'a composed with itself n times using the group operation.' Thus if the group operation is denoted by $+$, we usually write na rather than a^n, where na means 'a added to itself n times.' If $n < 0$ then na means $(-n)(-a)$ where we use $-a$ to denote the inverse of a.

Examples 11.2.9 We so far have two classes of examples of cyclic groups.

(1) The group $(\mathbb{Z}, +)$ is an infinite cyclic group. The elements of $\langle 1 \rangle$ are of the form $m1 = m$ where $m \in \mathbb{Z}$, and so the result is clear.

(2) For $n > 1$ the group $(\mathbb{Z}_n, +)$ is cyclic of order n. The elements of $\langle 1 \rangle$ contain $m1 = m$ for $m \geq 0$ modulo n and so will contain all elements of $\mathbb{Z}_n$.

□

The following theorem shows that the examples (1) and (2) above constitute a complete list of all cyclic groups.

Theorem 11.2.10 *Each infinite cyclic group is isomorphic to $(\mathbb{Z}, +)$, and each finite cyclic group is isomorphic to $(\mathbb{Z}_n, +)$ for some n.*

Proof Let G be a cyclic group. Then $G = \langle a \rangle$ for some $a \in G$. Observe first that if two distinct powers of a are equal then G must be finite. For suppose that $a^i = a^j$ for some $i < j$. Then $1 = a^{j-i}$. By taking inverses of both sides of this equation if necessary, it follows that there is a positive integer n such that $a^n = 1$. Choose $n > 0$ as small as possible such that $a^n = 1$. Let $i \in \mathbb{Z}$ be arbitrary. Then $i = nq + r$ where $0 \leq r < n$ by the Remainder Theorem. But then

$$a^i = a^{nq+r} = (a^n)^q a^r = a^r.$$

Thus

$$a^i = a^r \text{ where } r \equiv i \pmod{n} \text{ and } r \in \mathbb{Z}_n.$$

It follows that

$$G = \{a^i\colon 0 \leq i \leq n-1\}$$

and so G is finite.

Suppose that G is infinite. Then by the above result distinct powers of a are distinct. Therefore the function $\alpha\colon G \to \mathbb{Z}$ given by $\alpha(a^i) = i$ is well-defined. It is immediate that α is a bijection. It is a homomorphism because

$$\alpha(a^i)\alpha(a^j) = i + j = \alpha(a^{i+j}).$$

We have shown that G is isomorphic to $\mathbb{Z}$ under addition.

Now suppose that G is finite. Then not all powers of a can be distinct. Let n be the smallest positive integer such that $a^n = 1$. Then $G = \{a^0, \ldots, a^{n-1}\}$ by our result above. We prove that G is isomorphic to $(\mathbb{Z}_n, +)$. Define $\alpha\colon G \to \mathbb{Z}_n$ by $\alpha(a^i) = i$. This is clearly a bijection. It is a homomorphism because $a^i a^j = a^{i+j}$ and $a^{i+j} = a^k$ where $k \in \mathbb{Z}_n$ and $k \equiv i + j \pmod{n}$. But in $(\mathbb{Z}_n, +)$, we have that $i + j = k$. Thus G is isomorphic to $(\mathbb{Z}_n, +)$. □

Let S be a semigroup, and let $e \in S$ be an idempotent. It is easy to check that

$$eSe = \{ese\colon s \in S\}$$

is a subsemigroup of S, which is also a monoid in its own right with identity e. We call eSe a *local monoid in S*.

Proposition 11.2.11 *Let S be a semigroup.*

(i) *For each idempotent e in S, the group $U(eSe) = H_e$, the $\mathcal{H}$-class containing the idempotent e.*

(ii) *Let G be a group in S with identity e. Then $G \subseteq U(eSe)$.*

Proof (i) Let $a \in U(eSe)$. Then there exists $b \in eSe$ such that $ab = e = ba$. Now $ae = a$ and $ba = e$ imply that $e\,\mathcal{L}\,a$, and $ea = a$ and $ab = e$ imply that $e\,\mathcal{R}\,a$. It follows that $e\,\mathcal{H}\,a$.

Now let $e\,\mathcal{H}\,a$. We prove that $a \in U(eSe)$. By definition there exist $x, y, u, v \in S^1$ such that

$$xa = e, \quad ye = a, \quad au = e, \quad ev = a.$$

First,

$$ae = (ye)e = ye = a \text{ and } ea = e(ev) = ev = a.$$

Thus $a \in eSe$. Observe that

$$a(eue) = (ae)ue = aue = ee = e$$

and

$$(exe)a = ex(ea) = exa = ee = e.$$

Put $u' = eue$ and $x' = exe$. Then

$$u' = eu' = (x'a)u' = x'(au') = x'e = x'.$$

Thus $u'a = e = au'$ and $u' \in eSe$, and so $a \in U(eSe)$.

(ii) If $g \in G$ then $ge = g = eg$ since e is the identity of G. Thus $g = ege$. It follows that $G \subseteq eSe$. The next assertion is immediate because each element of G is invertible in eSe. □

The following definition now makes sense. Let S be a semigroup and let $e \in S$ be an idempotent. Then the group $U(eSe) = H_e$ is called a *maximal group in S around the idempotent e*.

Example 11.2.12 Let $S = M_2(\mathbb{R})$ be the monoid of 2×2-matrices under multiplication with entries from $\mathbb{R}$. The identity of this monoid is the 2×2 identity matrix, and the zero is the 2×2 zero matrix. The matrix,

$$e = \begin{pmatrix} 1 & 0 \\ 0 & 0 \end{pmatrix},$$

is an idempotent that is neither the identity nor the zero. The subsemigroup eSe consists of all matrices of the form,

$$\begin{pmatrix} a & 0 \\ 0 & 0 \end{pmatrix},$$

where $a \in \mathbb{R}$, whereas $U(eSe)$ consists of all such matrices where $a \neq 0$. □

Lemma 11.2.13 *Let α: $S \to T$ be a semigroup homomorphism.*

(i) *If e is an idempotent in S then $\alpha(e)$ is an idempotent in T.*

(ii) *If α is restricted to eSe then the image is contained in $\alpha(e)T\alpha(e)$.*

(iii) *If α is restricted to H_e then the image is contained in $H_{\alpha(e)}$.*

Proof (i) Let e be an idempotent in S. Then $\alpha(e) = \alpha(e^2) = \alpha(e)^2$. Thus $\alpha(e)$ is an idempotent in T.

(ii) Let $a \in eSe$. Then $ea = a = ae$ and so

$$\alpha(e)\alpha(a) = \alpha(a) = \alpha(a)\alpha(e),$$

which implies that $\alpha(a) \in \alpha(e)T\alpha(e)$.

(iii) Let $a \in H_e$. By (ii) above, we have that $\alpha(a) \in \alpha(e)T\alpha(e)$. By assumption, there exists $b \in H_e$ such that $ab = e = ba$ and so $\alpha(a)\alpha(b) = \alpha(e) = \alpha(b)\alpha(a)$, and so $\alpha(a)$ is invertible in the monoid $\alpha(e)T\alpha(e)$ with inverse $\alpha(b)$. Hence $\alpha(a) \in H_{\alpha(e)}$. $\square$

In general, eSe is not itself a group. However, it is useful to have a criterion when it is.

Proposition 11.2.14 *Let S be a semigroup and let e be an idempotent in S. Then eSe is a group if and only if the following property holds: if I is any non-empty left ideal and $I \subseteq S^1e$, then $I = S^1e$.*

Proof Let eSe be a group. Suppose that $I \subseteq S^1e$ is a non-empty left ideal. Let $b \in I$. Then $b \in S^1e$. Thus $b = ue$ for some $u \in S^1$. It follows that $eb = eue \in eSe$. Put $c = eb$. Thus $c \in I$ because I is a left ideal, and $c \in eSe$ by construction. By assumption eSe is a group. So there exists $v \in eSe$ such that $e = vc$. Thus $e \in I$ because I is a left ideal. It follows that $S^1e \subseteq I$ and so $I = S^1e$, as required.

To prove the converse, assume that S^1e contains no proper non-empty left ideals. We prove that eSe is a group. Let $a \in eSe$. Then $a = ae$. It follows that $S^1a = S^1ae \subseteq S^1e$. But S^1a is a non-empty left ideal. Thus by assumption $S^1a = S^1e$. It follows that there exists $b \in S^1$ such that $e = ba$. But $e = (ebe)a$. Put $b' = ebe$. Then $b' \in eSe$ and $b'a = e$. Now $S^1b' = S^1b'e \subseteq S^1e$. Thus $S^1b' = S^1e$. It follows that there exists $d \in S^1$ such that $db' = e$. But $e = (ede)b'$. Put $d' = ede$. Then $d' \in eSe$ and $d'b' = e$. We therefore have that $a = ea = (d'b')a = d'(b'a) = d'e = d'$. Therefore $b'a = e = ab'$ and so $a \in U(eSe)$. It follows that $eSe = U(eSe)$. $\square$

Let e be an idempotent in a semigroup S. We say that S^1e is a *minimal left ideal* if $I \subseteq S^1e$, where I is a non-empty left ideal, implies that $I = S^1e$. Thus the above result tells us that eSe is a group precisely when S^1e is a minimal left ideal.

To finish off this section we shall look at two important examples of groups in semigroups: groups in transition monoids and groups in Schützenberger products.

We usually present the syntactic monoid as a submonoid of the full transformation monoid on the set of states, so it is useful to know how groups in transition monoids can be identified. The following proposition gives us some information; the important result is (iii) below.

Proposition 11.2.15 *Let S be a submonoid of a full transformation monoid $T(X)$ where X is a finite set.*

(i) *Let G be a non-trivial group in S. Then there exists a subset $Y \subseteq X$ such that the restrictions of the elements in G to Y are permutations that form a group isomorphic to G.*

(ii) *Suppose there exists a subset $Y \subseteq X$ and an element $s \in S$ such that s restricted to Y is a non-trivial permutation. Let H be the subgroup of $S(Y)$ generated by all such permutations. Then there is a group G in S such that H is a homomorphic image of G.*

(iii) *The monoid S contains a non-trivial group if and only if there is an element $s \in S$ and a subset Y of X with at least two elements such that s induces a non-trivial permutation on Y.*

Proof In this proof, remember that elements of S are functions that have arguments written on the left, so that if $s \in S$ and $x \in X$ then $xs \in X$.

(i) By Proposition 11.2.11, $G \subseteq U(eSe)$ for some idempotent $e \in S$. Put $Y = Xe$, the image of X under the function e. For each $a \in G$ define the function f_a as follows: its domain is Y, its codomain is Y, and for each $y \in Y$ we define $yf_a = ya$. The function f_a will be well-defined if we show that its codomain really is Y, but this is true because for each $y \in Y$ we have that

$$(yf_a)e = y(ae) = ya = yf_a.$$

Put

$$K = \{f_a\colon a \in G\}.$$

Observe that f_e is the identity function on Y, because if $y \in Y$ there exists $x \in X$ such that $y = xe$ and so

$$yf_e = ye = (xe)e = xe^2 = xe = y.$$

The function f_a is a bijection of Y because there is an element $b \in G$ such that $ab = e = ba$. Observe that $f_a f_b = f_{ab}$ for all $a, b \in G$. It follows that K is a subgroup of $S(Y)$. Define a function from G to K by $a \mapsto f_a$. It is immediate that it is a homomorphism. We prove that it is a bijection. By construction it is surjective, so we need only prove that it is injective. Let $a, b \in G$ be such that $f_a = f_b$. Then $yf_a = yf_b$ for all $y \in Y$. Let $x \in X$ be arbitrary. Then $xa = x(ea) = (xe)f_a$, where $xe \in Y$. Similarly $xb = (xe)f_b$. But $(xe)f_a = (xe)f_b$ by assumption, and so $xa = xb$ for all $x \in X$. It follows that $a = b$, as required.

(ii) Let T be the subset of S consisting of all those elements that induce a permutation of Y. By assumption, this set contains more than just the identity function. It is evident that T is closed under composition of functions and so is a submonoid of S. Because T is finite it contains idempotents by

Theorem 10.1.2. Choose $e \in T$ an idempotent such that for every idempotent $f \in T$ we have that $|Te| \leq |Tf|$. We claim that Te is a minimal left ideal in T. Let $I \subseteq Te$ be a non-empty left ideal in T. Then I is in particular a finite subsemigroup and so contains an idempotent f. Because I is a left ideal of T we have that $Tf \subseteq I$. But then $Tf \subseteq Te$. It follows that $Tf = Te$ and so $I = Te$. By Proposition 11.2.14, we have that $G = eTe$ is a group in T and so in S. Now let H be the subgroup of $S(Y)$ generated by the permutations induced by elements of S on the set Y. Define $\alpha: G \to H$ by $\alpha(a) = f_a$ where f_a is the permutation on Y induced by a. This is clearly a surjective homomorphism.

(iii) If G is a non-trivial group in S, then the claim follows by (i). Conversely, suppose that the conditions of the claim hold. Then the result follows by (ii). □

The result in (iii) above is important, because it enables us to connect our conjecture of Section 11.1 with the definitions of this section. Specifically the presence or absence of groups in a transition monoid is the same as the presence or absence of elements inducing non-trivial permutations on subsets of the set of states. The conjecture at the end of Section 11.1 can now be rephrased as follows: a language over a one-letter alphabet is star-free iff its syntactic monoid contains no groups.

Finally, we characterise groups in Schützenberger products.

Proposition 11.2.16 *Let M and N be monoids. Then every group in $M \diamond N$ is isomorphic to a group in $M \times N$.*

Proof The function $\theta: M \diamond N \to M \times N$ given by $(m, X, n) \mapsto (m, n)$ is a homomorphism. Let (e, E, f) be an idempotent in $M \diamond N$. Then e is an idempotent in M, f is an idempotent in N, and $eE + Ef = E$. It follows by Lemma 11.2.13 that θ maps $H_{(e,E,f)}$ to $H_{(e,f)}$. We use Proposition 11.2.7 to prove that θ is injective. Suppose there is an element in $H_{(e,E,f)}$ that maps to (e, f). Such an element is of the form (e, X, f), where

$$eE + Xf = X = eX + Ef.$$

We prove that (e, X, f) must be (e, E, f). By assumption, (e, X, f) is invertible in $H_{(e,E,f)}$ so there exists an element (e, X', f) such that

$$eX' + Xf = E = eX + X'f.$$

Now

$$X = X + X = (eX + Ef) + (eE + Xf) = (eX + Xf) + (eE + Ef),$$

but

$$E = eE + Ef$$

and so $E \subseteq X$. Also,

$$E = eE + Ef = eE + (eX'f + Xf) = (eE + Xf) + eX'f,$$

but

$$X = eE + Xf$$

and so $X \subseteq E$. Hence $E = X$, as required. □

Exercises 11.2

1. Draw up Cayley tables for S_2 and S_3.

2. Prove that if G and H are groups then $G \times H$ is a group. Show that the product of the groups $(\mathbb{Z}_2, +)$ and $(\mathbb{Z}_4, +)$ is not cyclic.

3. Find all the maximal subgroups in T_3.

4. Let $(\mathbb{Z}_n, \times)$ be the monoid of integers modulo n under multiplication. Prove that an element is invertible if and only if it is coprime to n.

5. Let M be a monoid. For each $a \in M$ define a function $f_a \in T(M)$ by $(x)f_a = xa$ for each $x \in M$. Let $N = \{f_a\colon a \in M\}$. Show that N is a submonoid of $T(M)$ and that M is isomorphic to N.

6. Let G be a group. For each $a \in G$ define a function $f_a \in T(G)$ by $(x)f_a = xa$ for each $x \in G$. Show that f_a is a permutation of G. Deduce that $H = \{f_a\colon a \in G\}$ is a subgroup of $S(G)$. Show that G is isomorphic to H.

7. Show that the only idempotent in a group is the identity. Now let S be a finite monoid with exactly one idempotent. Prove that S is a group.

The second result is not true for infinite monoids. For example, A^* is a monoid with exactly one idempotent, but it is not a group.

8. Let $\alpha\colon S \to G$ be a homomorphism from the finite semigroup S onto the finite group G. Show that there is a group H in S such that $\alpha(H) = G$.

9. Let S and T be semigroups. Show that each maximal group in $S \times T$ is of the form $G \times H$, where G is a maximal group in S, and H is a maximal group in T.

10. Let ρ be a congruence on a group G, and put $N = \rho(1)$.

(i) Prove that N is a subgroup of G and that $g^{-1}Ng = N$ for each $g \in G$.

(ii) Prove that the admissible partition associated with ρ has as blocks the subsets of the form Ng where $g \in G$.

Now let N be any subgroup of G such that $g^{-1}Ng = N$ for all $g \in N$. Prove that $P = \{Ng: g \in G\}$ is an admissible partition of G.

Subgroups N satisfying the condition above are said to be normal. *The results of this question show that each congruence on a group determines and is determined by a normal subgroup of G. This means that in group theory we need only work with normal subgroups rather than with the congruences they determine.*

11. Let S be a semigroup that contains an element e such that $se = s$ for each $s \in S$. In addition, suppose that for each element $s \in S$ there exists $t \in S$ such that $st = e$. Prove that S is a group with identity e.

11.3 Aperiodic semigroups

At the end of Section 11.1, we highlighted the absence or presence of groups in a syntactic monoid as being crucial in determining whether the corresponding language was star-free or not. In Section 11.2, we investigated groups in semigroups in general. In this section, we shall characterise those semigroups all of whose groups are trivial. The key step in our characterisation will be an investigation of the sorts of semigroups that can be syntactic monoids of recognisable languages over one-letter alphabets.

Theorem 11.3.1 *Let S be a finite semigroup, and let $a \in S$ be an arbitrary element. Then there are positive integers m and r such that the elements $a^1, \ldots, a^m, \ldots, a^{m+r-1}$ are distinct, but $a^{m+r} = a^m$. Furthermore, the elements $C_a = \{a^m, \ldots, a^{m+r-1}\}$ form a cyclic group of order r in S.*

Proof Consider the subsemigroup $a^+ = \{a, a^2, \ldots\}$. Because S is finite there must be a power of a, say a^i, which is repeated later on, say $a^i = a^j$ where $i \neq j$. Thus the set,

$$\{i \in \mathbb{N} \setminus \{0\}: a^i = a^j \text{ and } i \neq j\},$$

is non-empty, and so has a smallest element which we denote by m. The set,

$$\{j \in \mathbb{N} \setminus \{0\}: a^m = a^{m+j}\},$$

is therefore non-empty, and so has a smallest element which we denote by r. By construction the elements $a^1, \ldots, a^{m+r-1}$ are distinct. Put

$$C_a = \{a^m, \ldots, a^{m+r-1}\}.$$

From $a^m = a^{m+r}$ and induction, we deduce that

$$a^{m+nr} = a^m$$

for all $n \in \mathbb{N}$. Let $s \geq m$ be any number. By the Remainder Theorem, we may write

$$s - m = qr + u \text{ where } u < r.$$

Thus

$$a^s = a^{m+qr+u} = a^{m+qr}a^u = a^m a^u = a^{m+u} \in C_a.$$

It follows that C_a is a subsemigroup of S. We show that C_a is a cyclic group in S, and to do this we shall use the group $(\mathbb{Z}_r, +)$. Define $\theta \colon C_a \to \mathbb{Z}_r$ by $\theta(a^{m+i}) = i'$, where $i' \in \mathbb{Z}_r$ and $i' \equiv m + i \pmod r$. The numbers $m, \ldots, m + (r-1)$ are r consecutive numbers and so are congruent modulo r to the numbers $0, \ldots, r-1$ in some order. Thus θ is a bijection. We show that it is a homomorphism. Let $0 \leq i, j \leq r-1$. Then

$$\theta(a^{m+i}) = i' \text{ and } \theta(a^{m+j}) = j',$$

where

$$i' \equiv m + i \pmod r \text{ and } j' \equiv m + j \pmod r.$$

It follows that

$$\theta(a^{m+i})\theta(a^{m+j}) \equiv i' + j' \pmod r.$$

Let $m + i + j = qr + u$ where $u < r$. Then

$$a^{2m+i+j} = a^{m+(m+i+j)} = a^{m+qr+u} = a^{m+qr}a^u = a^m a^u = a^{m+u}.$$

Thus

$$a^{m+i}a^{m+j} = a^{2m+i+j} = a^{m+u}.$$

It follows that

$$\theta(a^{m+i}a^{m+j}) = \theta(a^{m+u}) \equiv m + u \pmod r.$$

But

$$2m + i + j \equiv m + u \pmod r$$

and

$$2m + i + j \equiv i' + j' \pmod r.$$

Thus $i' + j' \equiv m + u \pmod r$. Hence θ is a homomorphism, and so an isomorphism of semigroups. But a semigroup isomorphic to a group is itself a group. Thus C_a is a group isomorphic to $\mathbb{Z}_r$ under addition. □

Let S be a finite semigroup and $a \in S$. Then the number m is called the *index of* a and the number r is called the *period of* a.

We can now return to the question of when a semigroup contains only trivial groups.

Theorem 11.3.2 *Let S be a finite semigroup. Then the following are equivalent:*

(i) *The $\mathcal{H}$ relation is equality.*

(ii) *The maximal subgroups in S are trivial.*

(iii) *For each element $a \in S$ the period of a is 1.*

(iv) *There exists $n > 0$ such that $a^n = a^{n+1}$ for all $a \in S$.*

Proof (i) $\Rightarrow$ (ii). Immediate from Proposition 11.2.11.

(ii) $\Rightarrow$ (iii). Suppose that the maximal subgroups of S are trivial. Then for each $a \in S$ the group C_a, defined in the proof of Theorem 11.3.1, must be trivial. Hence each element has period 1.

(iii) $\Rightarrow$ (iv). Suppose that each element of S has period 1. Let $n = |S|$ and let $a \in S$. We claim that $a^n = a^{n+1}$. Observe that the $n+1$ elements $a, a^2, \ldots, a^{n+1}$ cannot all be distinct. Thus the index of a is at most n. If the index of a is n then the result is immediate. If the index is $m < n$, then $a^n \in C_a = \{a^m\}$. Thus $a^n = a^{n+1}$.

(iv) $\Rightarrow$ (i). Let $x \,\mathcal{H}\, y$. Then there exist $a, b, c, d \in S^1$ such that

$$ax = y, \quad by = x, \quad xc = y, \quad yd = x.$$

We can therefore write $x = axd$ and so $x = a^n x d^n$ for all $n \geq 1$. By assumption, we can choose n such that $a^n = a^{n+1}$. Then $ax = a(a^n x d^n) = a^{n+1} x d^n = a^n x d^n = x$. Thus $ax = x$. But $ax = y$ as well. We have proved that $x = y$. □

A semigroup S is *aperiodic* if every group in S is trivial. If the semigroup is a subsemigroup of a full transformation monoid, we can use Proposition 11.2.15 to determine if it is aperiodic. Here is an alternative algorithm using Theorem 11.3.2.

Algorithm 11.3.3 Let S be a finite semigroup.

(1) Choose $a \in S$ and calculate the subsemigroup a^+.

(2) There are now several cases:

- If the period of a is not 1 then S is not aperiodic and the algorithm terminates.
- If the period of a is 1 and $S = a^+$ then S is aperiodic and the algorithm terminates.
- If the period of a is 1 and $S \neq a^+$, then we replace S by $S \setminus a^+$ and repeat (1).

□

A detailed analysis of the problem of determining whether a monoid is aperiodic or not is the subject of [26]. Our next result summarises some important properties of aperiodic monoids.

Theorem 11.3.4 *Let S and T be monoids.*

(i) *If S is aperiodic and T is a submonoid of S then T is aperiodic.*

(ii) *If S is aperiodic and T is a homomorphic image of S then T is aperiodic.*

(iii) *If S is aperiodic and T is a monoid dividing S then T is aperiodic.*

(iv) *If S and T are aperiodic then $S \times T$ is aperiodic.*

(v) *If S and T are aperiodic then $S \diamond T$ is aperiodic.*

Proof (i) By assumption S contains only trivial groups. It is immediate that T can only contain trivial groups.

(ii) Let θ: $S \to T$ be a surjective monoid homomorphism. Let $t \in T$. Then there exists $s \in S$ such that $\theta(s) = t$. By assumption, there exists $n > 0$ such that $s^n = s^{n+1}$. But $\theta(s^n) = t^n$ and $\theta(s^{n+1}) = t^{n+1}$. Thus $t^n = t^{n+1}$. It follows that T is aperiodic by Theorem 11.3.2.

(iii) This is immediate by (i) and (ii) above.

(iv) A group in $S \times T$ is of the form $G \times H$ where G is a group in S and H is a group in T by Question 9 of Exercises 11.2. Thus $S \times T$ is aperiodic if S and T are.

(v) This follows from Proposition 11.2.16 and Question 9 of Exercises 11.2. □

Exercises 11.3

1. Find the index and period of each element in T_3.

11.4 Schützenberger's theorem

In this section, we shall characterise star-free languages by means of their syntactic monoids. The proof of the theorem is in two parts: first, if L is star-free, then its syntactic monoid $M(L)$ is aperiodic; second, if L is a recognisable language with an aperiodic syntactic monoid $M(L)$, then L is star-free. The proof of the first half is now easy.

Theorem 11.4.1 *The syntactic monoid of a star-free language is aperiodic.*

Proof Let A be a finite alphabet. The star-free languages are constructed from the basic languages $\emptyset$, ε and a, where $a \in A$, using the Boolean operations and product of languages. It is easy to check that the syntactic monoids of the basic languages are aperiodic. Our induction hypothesis is that any language described by a star-free, generalised regular expression with at most n operators has an aperiodic syntactic monoid. Let r be a star-free generalised regular expression with $n+1$ operators. Then $r = s'$ or $r = s+t$ or $r = s \cap t$ or $r = s \cdot t$. Each of the star-free expressions s and t has at most n star-free operators. Thus the syntactic monoids of $L(s)$ and $L(t)$, M and N, respectively, are both aperiodic. We now prove in each case that the syntactic monoid of $L(r)$ is aperiodic. If $r = s'$ then $L(s)$ recognises s' by Proposition 10.2.8(i); in fact, in this case $L(s)$ is also the syntactic monoid of s'. Thus $L(r)$ has a syntactic monoid that is aperiodic. If $r = s + t$ or $r = s \cap t$ then $L(r)$ is recognised by $M \times N$. But $M \times N$ is aperiodic by Theorem 11.3.4(iv), and the syntactic monoid of $L(r)$ divides $M \times N$ by Theorem 10.2.6, consequently by Theorem 11.3.4(iii), the syntactic monoid of $L(r)$ is aperiodic. Finally, if $r = s \cdot t$ then $L(r)$ is recognised by a monoid S, which is a subsemigroup of $M \diamond N$ by Proposition 10.2.11. But $M \diamond N$ is aperiodic by Theorem 11.3.4(v), and so S is aperiodic by Theorem 11.3.4(i). The syntactic monoid of $L(r)$ divides S by Theorem 10.2.6, consequently by Theorem 11.3.4(iii), the syntactic monoid of $L(r)$ is aperiodic. It follows that in all cases the syntactic monoid of $L(r)$ is aperiodic. □

The result above is already useful.

Example 11.4.2 Let $A = \{a\}$. A recognisable language L over A has the form $L = X + Y(a^p)^*$ where X and Y are finite languages. We claim that L is star-free if and only if $p = 0, 1$. If $p = 0, 1$, then we established that L is star-free in Section 11.1. We shall now prove that if $p \geq 2$ then L is not star-free. To do this, we shall prove that the syntactic monoid of L is not aperiodic, which gives the required result by Theorem 11.4.1. As we saw in the proof of Theorem 5.2.2, the minimal automaton $\mathbf{A}$ for L consists of a stem, and a cycle of p states $r_1, \ldots, r_p$. It follows that the letter a induces a non-trivial permutation on the set $\{r_1, \ldots, r_p\}$. Consequently, the transition monoid of $\mathbf{A}$ contains a non-trivial group by Proposition 11.2.15(iii). But the transition monoid of the minimal automaton of L is isomorphic to the syntactic monoid of L by Theorem 9.4.3. It follows that the syntactic monoid of L is not aperiodic, and so L is not star-free by Theorem 11.4.1.

From this result, it follows that the language $(aa)^*$ is not star-free. This should be contrasted with the language $(ab)^*$, which we explicitly showed to be star-free in Example 11.1.1. □

The proof of the converse, that an aperiodic syntactic monoid implies a star-free language, is much harder in the case of a general alphabet and we shall need a number of preparatory steps.

Lemma 11.4.3 (Cancellation property) *Let M be an aperiodic monoid. Suppose $p, q, r \in M$ are such that $q = pqr$. Then $q = pq = qr$.*

Proof The monoid M is aperiodic, and so by Theorem 11.3.2 there is a positive integer n such that $a^n = a^{n+1}$ for each $a \in M$. From $q = pqr$ we get by induction that $q = p^n q r^n$. Thus $q = p^{n+1} q r^n = p(p^n q r^n) = pq$. A similar argument shows that $q = qr$. □

Lemma 11.4.4 *Let M be a monoid, and $a \in M$. Put*

$$W(a) = \{b \in M \colon a \notin MbM\}.$$

Then $W(a)$ is the largest ideal in M that does not contain a.

Proof Suppose that $W(a)$ is empty. We prove that every non-empty ideal of M contains a. Let I be a non-empty ideal of M and let $b \in I$. Then $MbM \subseteq I$. If $a \notin MbM$ then $b \in W(a)$, but this contradicts the assumption that $W(a)$ is empty. Thus $a \in MbM$ and so $a \in I$.

Suppose that $W(a)$ is non-empty. Let $b \in W(a)$ and let $m \in M$. We show that $mb \in W(a)$. If $a \in M(mb)M$, then from $M(mb)M \subseteq MbM$ we get that $a \in MbM$, which is a contradiction. Thus $mb \in W(a)$. We may similarly show that $bm \in W(a)$, and it follows that $W(a)$ is an ideal. We now prove that $W(a)$ is the largest ideal in M not containing a. Let I be a non-empty ideal of M that does not contain a. Let $b \in I$. Then $MbM \subseteq I$, because I is an ideal. Since I does not contain a neither can MbM. It follows that $b \in W(a)$. Hence $I \subseteq W(a)$. □

We now have the following crucial property of aperiodic monoids.

Proposition 11.4.5 *Let M be an aperiodic monoid. Then*

$$\{a\} = (aM \cap Ma) \setminus W(a)$$

for each $a \in M$.

Proof From Lemma 11.4.4, the element a belongs to the complement of $W(a)$ in M, and it is evident that $a \in aM \cap Ma$. Thus the left-hand side is always contained in the right-hand side. To prove the reverse containment, let $b \in (aM \cap Ma) \setminus W(a)$. Then $b = ap = qa$ and $a = ubv$ for some $p, q, u, v \in M$. Then $a = ubv = (uq)av$, and so by the cancellation property $a = uqa = av$. Thus $a = ub = uap$ and so by the cancellation property $a = ua = ap$. Hence $a = b$, as required. □

Suppose that L is a language whose syntactic monoid is aperiodic. In particular, there is a monoid homomorphism $\phi\colon A^* \to M$, where M is aperiodic and a subset $P \subseteq M$ such that $L = \phi^{-1}(P)$. We can therefore write $L = \sum_{a \in P} \phi^{-1}(a)$. So to show that L is star-free it is enough to prove that languages of the form $\phi^{-1}(a)$ are star-free, and this will be the approach we adopt.

Let M be a monoid and let $a \in M$. By the *rank of* a, denoted $\mathbf{r}(a)$, we mean the number $|MaM|$. If two elements a and b of M belong to the same $\mathcal{J}$-class, then they must have the same rank, because by definition $a \mathcal{J} b$ means that $MaM = MbM$. Since MaM is a subset of M, the largest the rank can be is $|M|$. This is reached by the identity because $M1M = M^2 = M$. In an aperiodic monoid, the identity is the only element reaching maximum rank, as we now show.

Lemma 11.4.6 *Let M be an aperiodic monoid. If $a \in M$ is such that $\mathbf{r}(a) = |M|$, then $a = 1$.*

Proof From $|MaM| = |M|$, and $MaM \subseteq M$ we have that $M = MaM$ because M is finite. Thus there are $p, q \in M$ such that $1 = paq$. From $1 = (pa)1q$ and the cancellation property for aperiodic monoids, we have that $1 = pa = q$. And from $1 = p1a$ we get $1 = p = a$ by the cancellation property.□

The next result tells us that for the element of maximum rank, the inverse image will be star-free.

Lemma 11.4.7 *Let $\phi: A^* \to M$ be a monoid homomorphism to an aperiodic monoid. Then $\phi^{-1}(1)$ is a star-free subset of A^*.*

Proof Define $W = \{a \in A: \phi(a) \neq 1\}$. We claim that

$$\phi^{-1}(1) = A^* \setminus A^*WA^*,$$

which is clearly star-free.

Let $a \in W$ and let $u, v \in A^*$. Suppose that $\phi(uav) = 1$, so that

$$1 = \phi(u)\phi(a)1\phi(v).$$

By the cancellation property for aperiodic monoids, we have that

$$1 = \phi(u)\phi(a) = \phi(v).$$

By the cancellation property applied to $1 = \phi(u)1\phi(a)$ we have that $1 = \phi(u)1 = 1\phi(a)$. Thus in particular $\phi(a) = 1$, which is a contradiction. Hence none of the strings in A^*WA^* can be mapped to 1.

Now let $x \in A^* \setminus A^*WA^*$ be a non-empty string. Then $x = a_1 \ldots a_n$ where $a_i \in A$. Suppose that $\phi(x) \neq 1$. Then for at least one i, we must have $\phi(a_i) \neq 1$. But then $a_i \in W$ and so $x \in A^*WA^*$, which is a contradiction. We have therefore proved the result. □

In order to prove our main theorem, it remains therefore to prove that the set $\phi^{-1}(m)$ is star-free for each element m of our monoid, which is not of maximum rank. Our approach will be to write the set $\phi^{-1}(m)$ as a 'combination' of sets $\phi^{-1}(n)$, where the rank of n is strictly greater than the rank of m, and where by 'combination', we mean using only Boolean operations or concatenation.

The following is a technical lemma whose significance will soon become clear.

Lemma 11.4.8 *Let $\phi\colon A^* \to M$ be a monoid homomorphism to an aperiodic monoid M. Let $m \in M \setminus \{1\}$.*

(i) *Let $a \in A$ and $n \in M$ be such that $n\phi(a)M = mM$ and $n \notin mM$. Then $\mathbf{r}(n) > \mathbf{r}(m)$.*

(ii) *Let $a \in A$ and $n \in M$ be such that $M\phi(a)n = Mm$ and $n \notin Mm$. Then $\mathbf{r}(n) > \mathbf{r}(m)$.*

(iii) *Let $a, b \in A$ and $n \in M$ be such that*

$$m \in M\phi(a)nM \cap Mn\phi(b)M \text{ and } m \notin M\phi(a)n\phi(b)M.$$

Then $\mathbf{r}(n) > \mathbf{r}(m)$.

Proof (i) Observe that $MmM = Mn\phi(a)M \subseteq MnM$, so that $\mathbf{r}(m) \leq \mathbf{r}(n)$. Suppose that $\mathbf{r}(m) = \mathbf{r}(n)$. By finiteness, $MmM = MnM$. It follows that we can write $n = umv$ for some $u, v \in M$. Since $m \in n\phi(a)M$, we also have that $m = np$ for some $p \in M$. Thus $n = umv = un(pv)$. By the cancellation property for aperiodic monoids, we therefore have that $n = un = npv$. Thus $n = mv$, which implies that $n \in mM$. But this is a contradiction. It follows that $\mathbf{r}(n) > \mathbf{r}(m)$ as required.

The proof of (ii) is simply the left-right dual of the proof of (i).

(iii) Let $m \in M\phi(a)nM \cap Mn\phi(b)M$ and $m \notin M\phi(a)n\phi(b)M$. Clearly $\mathbf{r}(m) \leq \mathbf{r}(n)$. Suppose that $\mathbf{r}(m) = \mathbf{r}(n)$. Then $n = umv$ for some $u, v \in M$. In addition, we have from our assumptions that $m = r\phi(a)ns = pn\phi(b)t$ for some $p, r, s, t \in M$. Now $n = umv = u(r\phi(a)ns)v = (ur\phi(a))n(sv)$. Thus by the cancellation property for aperiodic monoids, we have that $n = ur\phi(a)n$. But then $m = pn\phi(b)t = p(ur\phi(a)n)\phi(b)t = (pur)\phi(a)n\phi(b)t$, which contradicts our assumptions. It follows that $\mathbf{r}(n) > \mathbf{r}(m)$, as required. □

The key argument in our proof of Schützenberger's theorem is the following.

Proposition 11.4.9 *Let $\phi\colon A^* \to M$ be a monoid homomorphism to an aperiodic monoid M. Let $m \in M \setminus \{1\}$. Then*

$$\phi^{-1}(m) = (UA^* \cap A^*V) \setminus A^*WA^*,$$

where U, V and W are subsets of A^* defined as follows:

- *The set U is the union of sets of the form $\phi^{-1}(n)a$, where $n \in M$, $a \in A$, $n\phi(a)M = mM$ but $n \notin mM$.*

- *The set V is the union of sets of the form $a\phi^{-1}(n)$, where $n \in M$, $a \in A$, $M\phi(a)n = Mm$ but $n \notin Mm$.*
- *The set W is the union of the set $\{a \in A\colon m \notin M\phi(a)M\}$ together with sets of the form $a\phi^{-1}(n)b$, where $n \in M$, $a, b \in A$, $m \in M\phi(a)nM \cap Mn\phi(b)M$, but $m \notin M\phi(a)n\phi(b)M$.*

Proof We prove first that

$$\phi^{-1}(m) \subseteq (UA^* \cap A^*V) \setminus A^*WA^*.$$

Let $w \in \phi^{-1}(m)$, so that $\phi(w) = m$. We have to show that

(1) $w \in UA^*$.

(2) $w \in A^*V$.

(3) $w \notin A^*WA^*$.

(1) We prove that $w \in UA^*$. Let $u \in A^*$ be the shortest prefix of w such that $\phi(u)M = mM$. We claim that $u \neq \varepsilon$: for if it were, we would have $mM = M$ and so $mp = 1$ for some $p \in M$ which, by the cancellation property for aperiodic monoids, would give $m = 1$. Thus $u \neq \varepsilon$ as claimed. We can therefore write $u = ra$, where $a \in A$ and $r \in A^*$. Put $n = \phi(r)$. Then

$$mM = \phi(u)M = \phi(r)\phi(a)M = n\phi(a)M.$$

We claim that $n \notin mM$: for suppose we did have $n \in mM$. Then $n = mm'$ for some $m' \in M$. Thus $\phi(r)M = mm'M \subseteq mM$, and $mM = \phi(u)M = \phi(r)\phi(a)M \subseteq \phi(r)M$, which gives $\phi(r)M = mM$, which contradicts our choice of u. It follows that $w \in UA^*$, as claimed.

(2) The proof that $w \in A^*V$ is similar to the above proof.

(3) We prove that $w \notin A^*WA^*$. Suppose that $y \in W$. Then there are two possibilities. Either $y = a$ and $m \notin M\phi(a)M$ or $\phi(y) = \phi(a)n\phi(b)$ and $m \notin M\phi(a)n\phi(b)M$. In either case, $m \notin M\phi(y)M$. If $w \in A^*WA^*$ then $w = uyv$ for some $y \in W$ and $u, v \in A^*$ and so $m = \phi(w) = \phi(u)\phi(y)\phi(v) \in M\phi(y)M$, which is a contradiction.

We now have to prove the reverse inclusion:

$$(UA^* \cap A^*V) \setminus A^*WA^* \subseteq \phi^{-1}(m).$$

Let

$$w \in (UA^* \cap A^*V) \setminus A^*WA^*$$

and put $n = \phi(w)$. We shall prove that $n = m$ by using Proposition 11.4.5.

Now $w \in UA^*$ implies that $w = uv$ where $u \in U$. Thus $\phi(u) = n\phi(a)$, where $n\phi(a)M = mM$. Thus $n = \phi(w) = \phi(u)\phi(v) = n\phi(a)\phi(v) \in mM$. From the fact that $w \in A^*V$, we may similarly show that $n \in Mm$. Thus $n \in mM \cap Mm$.

It remains to be proved that $m \in MnM$. So to this end suppose to the contrary that $m \notin MnM$. This means that $m \notin M\phi(w)M$. Let z be a factor of w, so that $w = uzv$, of smallest possible length such that $m \notin M\phi(z)M$. Clearly $z \neq \varepsilon$ since $m \in M = M^2$. Suppose that $z \in A$. By assumption $m \notin M\phi(z)M$ and so $z \in W$, which implies that $w \in A^*WA^*$, a contradiction. It follows that z has length at least 2. Accordingly we can write $z = arb$, where $a, b \in A$ and $r \in A^*$. Then $m \in M\phi(a)\phi(r)M$ since z was chosen with minimum length. Similarly $m \in M\phi(r)\phi(b)M$. It follows that $z \in W$, which is a contradiction. Hence $n \notin W(m)$ and we have proved the result. □

We have now done most of the work needed to prove the hard direction of our main theorem.

Theorem 11.4.10 *Let L be a recognisable language whose syntactic monoid $M(L)$ is aperiodic. Then L is star-free.*

Proof Let $\phi\colon A^* \to M(L)$ be the natural monoid homomorphism and let P be the subset of $M(L)$ such that $L = \phi^{-1}(P)$. We can write

$$L = \sum_{m \in P} \phi^{-1}(m),$$

and so to prove that L is star-free it is enough to prove that each $\phi^{-1}(m)$ is star-free. If $m = 1$, then this is proved by Lemma 11.4.7. Suppose that for all $n \in M$ such that $\mathbf{r}(n) > \mathbf{r}(m)$ we have proved that $\phi^{-1}(n)$ is star-free. The fact that $\phi^{-1}(m)$ is star-free now follows from Proposition 11.4.9 and Lemma 11.4.8. □

Combining Theorems 11.4.1 and 11.4.10, we obtain the following.

Theorem 11.4.11 (Schützenberger) *A recognisable language is star-free if and only if its syntactic monoid is aperiodic.* □

Exercises 11.4

1. Show that the syntactic monoids of the basic languages are aperiodic.

2. Let M be a finite monoid and let $a \in M$. Prove that the $\mathcal{H}$-class containing the element a is equal to

$$(aM \cap Ma) \setminus W(a).$$

11.5 An example

The goal of this section is to prove that the language $L = (ab + ba)^*$ is star-free and to find a star-free generalised regular expression that describes it.

In Example 9.4.5, we calculated a presentation of the syntactic monoid $M = M(L)$ of L, using the minimal automaton recognising L. From the table of representatives, we see that none of them induces a non-trivial permutation on a subset of $\{0, 1, 2, 3\}$. Thus L is star-free. We wish to determine a star-free expression for L, and this involves more work. Before dealing with this specific case, let me describe the method we shall use. Let L be a recognisable language over the alphabet A whose syntactic monoid is aperiodic. Let $\phi\colon A^* \to M(L)$ be the natural homomorphism corresponding to the syntactic congruence of L, and let $P = \phi(L)$ so that $L = \phi^{-1}(P)$ by Lemma 10.2.4. The proof of Theorem 11.4.10 tells us that we should first write L as the union of the inverse images of the elements m of P, since to find a star-free expression for L, it is enough to find star-free expressions for each of these inverse images. The proof of Theorem 11.4.10 depends on ordering the elements of P in such a way that the corresponding list of ranks is non-increasing. For each element m in P, in the order described above, we shall apply Proposition 11.4.9; this involves computing the sets U, V and W. For each of these sets we need to find the n that satisfy the given conditions. By Lemma 11.4.8 we know that $\mathbf{r}(n) > \mathbf{r}(m)$. Of course, n must satisfy other conditions, but the rank condition limits our choice and therefore the amount of work we have to do. The other conditions that n must satisfy involve Green's relations. It is clear, therefore, that our first step in finding a star-free expression for our given L should be the computation of the egg-box diagram of $M(L)$. Here is the egg-box diagram of $M(L)$ where $L = (ab + ba)^*$:

ε

ba	b
a	ab

ab^2a	ba^2	ab^2
b^2a	b^2a^2	b^2
a^2b	a^2	a^2b^2

0

Denote the σ_L-class containing x by $[x]$. We have that

$$P = \{[\varepsilon], [ab], [ba], [ab^2a]\},$$

where I have listed the elements in non-increasing rank order. Thus

$$L = \phi^{-1}([\varepsilon]) + \phi^{-1}([ab]) + \phi^{-1}([ba]) + \phi^{-1}([ab^2a]).$$

To find a star-free expression for L, it is enough to find a star-free expression for each of these four languages.

(1) *The language* $\phi^{-1}([\varepsilon])$. This can be calculated using Lemma 11.4.7. We easily find in this case that

$$\phi^{-1}([\varepsilon]) = \varepsilon.$$

(2) *The language* $\phi^{-1}([ab])$. We use Proposition 11.4.9. We know that we can write

$$\phi^{-1}([ab]) = (UA^* \cap A^*V) \setminus A^*WA^*$$

Thus we have to calculate U, V, and W. It is here that we have to use the egg-box diagram. We shall use c and d to denote elements of A to avoid confusion. Put $m = [ab]$.

U: We need to find $n \in M$ and $c \in A$ such that

$$n\phi(c) \, \mathcal{R} \, m \text{ but } \textbf{not } n \, \mathcal{R} \, m.$$

Also $\mathbf{r}(n) > \mathbf{r}(m)$. The only possibility for n is therefore $n = [\varepsilon]$. Thus $c = a$. It follows that

$$U = \phi^{-1}([\varepsilon])a = a.$$

V: We need to find $n \in M$ and $c \in A$ such that

$$\phi(c)n \, \mathcal{L} \, m \text{ but } \textbf{not } n \, \mathcal{L} \, m.$$

Once again $\mathbf{r}(n) > \mathbf{r}(m)$. The only possibility for n is therefore $n = [\varepsilon]$. Thus $c = b$. It follows that

$$V = b\phi^{-1}([\varepsilon]) = b.$$

W: This breaks down into two parts, which I shall call W_1 and W_2, respectively. First we calculate

$$W_1 = \{c \in A \colon m \notin M\phi(c)M\}.$$

Now both $M[a]M$ and $M[b]M$ contain $[ab]$. It follows that $W_1 = \emptyset$. We now turn to W_2. This involves looking for the sets $c\phi^{-1}(n)d$, where

$$m \in M\phi(c)nM \cap Mn\phi(d)M \text{ but } m \notin M\phi(c)n\phi(d)M.$$

Again $\mathbf{r}(n) > \mathbf{r}(m)$. The only possibility for n is $n = [\varepsilon]$. We now need to find the values of c and d so that the triple $(c, [\varepsilon], d)$ satisfies the conditions above. There are four cases to check, and we quickly find that the only possible values of (c, n, d) are: $(a, [\varepsilon], a)$ and $(b, [\varepsilon], b)$. Thus

$$W = W_2 = a\phi^{-1}([\varepsilon])a + b\phi^{-1}([\varepsilon])b = a^2 + b^2.$$

Putting all these calculations together we obtain

$$\phi^{-1}([ab]) = (aA^* \cap A^*b) \setminus (A^*(a^2 + b^2)A^*) = (ab)^+$$

(3) *The language* $\phi^{-1}([ba])$. The calculations for this case are similar to the calculations in (2) above with the roles of a and b interchanged. Thus

$$\phi^{-1}([ba]) = (bA^* \cap A^*a) \setminus (A^*(a^2 + b^2)A^*) = (ba)^+.$$

(4) *The language* $\phi^{-1}([ab^2a])$. Once again we have to calculate the sets U, V, and W. I shall leave it as an exercise for you to check the following:

$$U = \phi^{-1}([ba])a + \phi^{-1}([ab])b$$

$$V = b\phi^{-1}([ba]) + a\phi^{-1}([ab])$$

$$W = a\phi^{-1}([a])a + b\phi^{-1}([b])b.$$

Make the following definitions:

- $R = (ab)^+$.
- $S = (ba)^+$.
- $X = a(ba)^*$.
- $Y = b(ab)^*$.

Then $U = Sa + Rb$, $V = aR + bS$, and $W = aXa + bYb$. Thus

$$\phi^{-1}([ab^2a]) = ((Sa + Rb)A^* \cap A^*(aR + bS)) \setminus (A^*(aXa + bYb)A^*).$$

11.6 Summary of Chapter 11

- *Star-free languages*: A language is star-free if it can be constructed from finite languages using the Boolean operations and product but not Kleene star.

- *Groups*: A group is a monoid in which each element has an inverse. A group with exactly one element is said to be trivial. A group in a semigroup is simply a subsemigroup that is also a group.

- *Aperiodic monoids*: A monoid is aperiodic if the only groups in the monoid are trivial. A submonoid of an aperiodic monoid is aperiodic, a homomorphic image of an aperiodic monoid is aperiodic, products of aperiodic monoids are aperiodic, and Schützenberger products of aperiodic monoids are aperiodic.

- *Schützenberger's Theorem*: A language is star-free if and only if its syntactic monoid is aperiodic.

11.7 Remarks on Chapter 11

The class of star-free languages seems to have first been studied by Trakhtenbrot in 1958 and McNaughton in 1960. Schützenberger's Theorem was first proved in [116]. The book by McNaughton and Papert [88] gives a fascinating account of the many different, but equivalent, ways that the class of star-free languages can be characterised. These results lead one to surmise that the class of star-free languages is a natural one. The proof I give of Schützenberger's Theorem is due to Perrin [100]. A short proof can be found in [62]. The proof of Proposition 11.2.15 is adapted from [54].

There is an alternative proof of Schützenberger's Theorem that uses completely different ideas. In Section 1.7, I defined what is meant by a 'semiautomaton.' This notion underlies both finite automata in the sense of this book and the automata with output such as the Moore and Mealy machines also introduced in Section 1.7. Much of automata theory can be developed as the theory of semiautomata and then adapted to deal with acceptors or automata with output. This was the approach adopted by John Rhodes in his early work and developed with the help of colleagues and students at Berkeley. The key theorem of the Rhodes School is the Krohn-Rhodes Theorem, which I shall briefly sketch. By using output functions in the sense of Moore machines, it is possible to connect semiautomata in series so that the output of one machine becomes the input of the next. Such a combination of semiautomata is called a 'cascade product' and is itself a semiautomaton. We now single out two classes of semiautomata. A *reset semiautomaton* has the property that the effect of each input letter is to act either as the identity function on the set of states, or as a constant function. A *grouplike semiautomaton* has as the set of states a group G, input alphabet G, and the action of an input letter is by right multiplication. A group G is said to be *simple* if the only normal subgroups are the trivial subgroup and G itself; normal subgroups were defined in Question 11 of Exercises 11.2. A *simple grouplike semiautomaton* is a grouplike semiautomaton in which the group involved is simple. Finally, we say that the semiautomaton **A** *covers* the semiautomaton **B** if **B** is a homomorphic image of a subsemiautomaton of **A**; I shall not define this more precisely here. We can now state the Krohn-Rhodes Theorem [76]: every semiautomaton **A** can be covered by a cascade product of semiautomata each of which is either a two-state reset semiautomaton or a simple grouplike semiautomaton. Furthermore, the groups occurring in the grouplike semiautomata divide the groups in the transition monoid of **A**. A proof of this theorem is described in [54]. The Krohn-Rhodes Theorem can be used to prove the hard direction of Schützenberger's Theorem: if a recognisable language L has an aperiodic syntactic monoid then L is star-free. The proof runs as follows. Let **A** be the minimal automaton for L. Then the transition monoid of **A** is isomorphic to the syntactic monoid of L and so is aperiodic. The Krohn-Rhodes Theorem applies to the semiautomaton underlying **A**. Because the transition monoid is aperiodic, we conclude that the semiautomaton underlying **A** can be covered by a cascade product of two-state reset semiautomata, i.e., because of aperiodicity no groups are involved. It is then reasonably easy to prove that L can be accepted by an automaton whose underlying semiautomaton is itself a cascade product of two-state reset semiautomata. The proof that L is star-free can now be completed by induction on the number of reset semiautomata involved in the cascade representation. The proof of this can be found in [54]. The Krohn-Rhodes Theorem is important in the study of sequential functions. I shall say a little more about this topic in the Remarks on Chapter 12.

Despite their lack of idempotents, groups are quite interesting. An introductory account can be found in [46]. A more advanced work is Stewart [127], who explains how groups were first introduced as a tool for studying algebraic equations.

Chapter 12

Varieties of languages

In Theorem 9.4.3, we proved that a language is recognisable precisely when its syntactic monoid is finite. This result opened the door to an algebraic approach to recognisable languages. In Chapter 11, we showed that the syntactic monoid is an important tool in studying recognisable languages when we characterised star-free languages in terms of their syntactic monoids. However, we showed in Section 10.3 that not every finite monoid is a syntactic monoid, and that different languages can have the same syntactic monoid. What these results point us toward is this: there is a correspondence, not between *individual* finite monoids and *individual* recognisable languages, but between *families* of monoids and *families* of languages. The aim of this chapter is to explain what this means and to prove the theorem on which it depends: the Variety Theorem of Eilenberg and Schützenberger. This theorem provides a framework for thinking about finite monoids and recognisable languages.

All monoids in this chapter are finite.

12.1 Pseudovarieties and varieties

Proposition 10.3.4 gave us examples of finite monoids that are not syntactic monoids. However, there is a connection between arbitrary finite monoids and syntactic monoids. To see what this is, we first need a lemma.

Lemma 12.1.1 *Let* $\{\rho_i : i \in I\}$ *be a set of congruences on a semigroup* S. *Put* $\rho = \bigcap_{i \in I} \rho_i$. *Then:*

(i) *ρ is a congruence on S.*

(ii) *S/ρ is isomorphic to a subsemigroup of* $\prod_{i \in I} S/\rho_i$.

Proof (i) This is left as an exercise.

(ii) Let $\pi_i : S \to S/\rho_i$ be the natural homomorphism $\pi_i(s) = \rho_i(s)$. Define $\pi : S \to \prod_{i \in I} S/\rho_i$ by $\pi(s) = (\pi_i(s))_{i \in I}$. This is clearly a homomorphism.

We now calculate the kernel of π. Then $(s,t) \in \ker(\pi)$ iff $\pi(s) = \pi(t)$ iff $\pi_i(s) = \pi_i(t)$ for all $i \in I$ iff $(s,t) \in \bigcap_{i\in I} \rho_i = \rho$. Thus $\ker(\pi) = \rho$. By Theorem 9.2.11, there is an injective homomorphism from S/ρ into $\prod_{i\in I} S/\rho_i$. □

Our next result assuages the disappointment of Proposition 10.3.4 and points us in the right direction.

Theorem 12.1.2 *Let M be a finite monoid. Then there is a finite alphabet A and finitely many languages $L_i \subseteq A^*$, where $1 \leq i \leq n$, such that each language L_i is recognised by M, and M is isomorphic to a submonoid of*

$$M(L_1) \times \ldots \times M(L_n).$$

Proof Let M be a finite monoid, in particular M is finitely generated. Thus there exists a finite alphabet A and a surjective monoid homomorphism $\alpha: A^* \to M$ by Proposition 10.2.2. For each $m \in M$, define $L_m = \alpha^{-1}(m) \subseteq A^*$, which, by construction, is a language recognised by M. Let σ_m be the syntactic congruence of L_m. Then $A^*/\sigma_m = M(L_m)$, the syntactic monoid of L_m. Put $\sigma = \bigcap_{m\in M} \sigma_m$, a congruence by Lemma 12.1.1(i). By Lemma 12.1.1(ii), there is an injective homomorphism from A^*/σ to $\prod_{m\in M} M(L_m)$. To conclude the proof, we show that $\sigma = \ker(\alpha)$ which, by Theorem 9.2.11, will prove that A^*/σ is isomorphic to M, giving us the result. Let $(x,y) \in \sigma$. Then $(x,y) \in \sigma_m$ for all $m \in M$. In particular, $(x,y) \in \sigma_{\alpha(x)}$. Now $\varepsilon x \varepsilon \in L_{\alpha(x)} = \alpha^{-1}(\alpha(x))$. Thus $\varepsilon y \varepsilon \in L_{\alpha(x)}$ and so $\alpha(y) = \alpha(x)$, giving $(x,y) \in \ker(\alpha)$ as required. Now let $(x,y) \in \ker(\alpha)$. It is easy to check that $uxv \in L_m \Leftrightarrow uyv \in L_m$. It follows that $(x,y) \in \sigma_m$ for each $m \in M$ and so $(x,y) \in \sigma$. □

The theorem above highlights three operations: taking submonoids of monoids, taking homomorphic images of monoids, and taking finite products of monoids. This leads us to the following definition. A *pseudovariety of monoids* M is a non-empty collection of monoids satisfying the following three conditions:

(VM1) If $S \in \mathsf{M}$ and T is a submonoid of S, then $T \in \mathsf{M}$.

(VM2) If $S \in \mathsf{M}$ and T is a homomorphic image of S, then $T \in \mathsf{M}$.

(VM3) If $S_1, \ldots, S_n \in \mathsf{M}$, then $S_1 \times \ldots \times S_n \in \mathsf{M}$.

Examples 12.1.3 Here are some examples of pseudovarieties of monoids.

(1) *Finite monoids.* This is clear.

(2) *Finite aperiodic monoids.* We proved this in Theorem 11.3.4.

(3) *Finite commutative monoids.* This is left as an exercise.

(4) *Finite groups.* This needs a little work. The proof of (VM2) follows from Lemma 11.2.13, and the proof of (VM3) follows from Question 2 of Exercises 11.2. The proof of (VM1) is more involved. Let H be a submonoid of the finite group G. We need to show that H is a group. Notice that we are not assuming that H is a subgroup of G. We therefore have to prove that for each $g \in H$ the inverse of g also belongs to H. Because G is finite, H is finite and so by Theorem 10.1.2, there is an integer $n \geq 1$ such that g^n is an idempotent. Now G is a group and so has exactly one idempotent, the identity, and H is a submonoid of G and so it has exactly one idempotent, the identity. It follows that $g^n = 1$. Thus $g^{n-1}g = 1 = gg^{n-1}$, and so $g^{-1} = g^{n-1} \in H$, as required.

□

We shall now describe a way of constructing further examples of pseudovarieties of monoids. Let C be any collection of monoids. I shall define four operations that may be applied to C to yield a potentially new class of monoids.

- Define $\mathbf{S}(\mathsf{C})$ to be the collection of all submonoids of elements of C.
- Define $\mathbf{H}(\mathsf{C})$ to be the collection of all homomorphic images of elements of C.
- Define $\mathbf{P}(\mathsf{C})$ to be the collection of all finite products of elements of C.
- Define $\mathbf{I}(\mathsf{C})$ to be the collection of all monoids that are isomorphic to elements of C.

In the lemma below, **IP** is the operation **P** followed by the operation **I**. We shall explain after the proof why we use this and not **P** on its own.

Lemma 12.1.4 *Let* **O** *be any one of the operations* **H**, **S**, *or* **IP**. *Let* C, C' *be collections of monoids. Then the following three conditions hold:*

(i) $\mathsf{C} \subseteq \mathbf{O}(\mathsf{C})$.

(ii) *If* $\mathsf{C} \subseteq \mathsf{C}'$ *then* $\mathbf{O}(\mathsf{C}) \subseteq \mathbf{O}(\mathsf{C}')$.

(iii) $\mathbf{O}(\mathbf{O}(\mathsf{C})) = \mathbf{O}(\mathsf{C})$.

Proof We shall prove (i), (ii), and (iii) for the operation **H** and leave the proofs of the two remaining cases as exercises.

(i) This follows from the fact that the identity function is a homomorphism.

(ii) If T is a homomorphic image of a monoid in C, then it is trivially a homomorphic image of a monoid in C'.

(iii) We have to show that $\mathbf{H}(\mathbf{H}(\mathsf{C})) \subseteq \mathbf{H}(\mathsf{C})$. Let $S \in \mathbf{H}(\mathbf{H}(\mathsf{C}))$. Then S is a homomorphic image of a monoid $T \in \mathbf{H}(\mathsf{C})$, and T is a homomorphic image of a monoid $U \in \mathsf{C}$. Thus S is also a homomorphic image of U since the composition of two monoid homomorphisms is a monoid homomorphism by Proposition 9.1.8(iii). Thus $S \in \mathbf{H}(\mathsf{C})$, as required. □

Remark We now explain why we considered the operation **IP** above and not the operation **P**. It is true that **P** satisfies both properties (i) and (ii). The problem is (iii). If $S_1, S_2, S_3, S_4 \in \mathsf{C}$ then $S_1 \times S_2, S_3 \times S_4 \in \mathbf{P}(\mathsf{C})$. Thus $(S_1 \times S_2) \times (S_3 \times S_4) \in \mathbf{PP}(\mathsf{C})$. However, this monoid is isomorphic to the monoid $S_1 \times S_2 \times S_3 \times S_4$ in $\mathbf{P}(\mathsf{C})$ not equal to it.

The next lemma describes some relations amongst our operations.

Lemma 12.1.5 *Let* C *be a collection of monoids.*

(i) $\mathbf{SH}(\mathsf{C}) \subseteq \mathbf{HS}(\mathsf{C})$.

(ii) $\mathbf{PS}(\mathsf{C}) \subseteq \mathbf{SP}(\mathsf{C})$.

(iii) $\mathbf{PH}(\mathsf{C}) \subseteq \mathbf{HP}(\mathsf{C})$.

Proof (i) Let $S \in \mathbf{SH}(\mathsf{C})$. Then S is a submonoid of a monoid T that is a homomorphic image of a monoid $U \in \mathsf{C}$. By Proposition 9.1.8, S is a homomorphic image of a submonoid of U, which proves the inclusion.

(ii) Let $S \in \mathbf{PS}(\mathsf{C})$. Then $S = S_1 \times \ldots \times S_n$ where each S_i is a submonoid of a monoid $T_i \in \mathsf{C}$. However, $S_1 \times \ldots \times S_n$ is a submonoid of $T_1 \times \ldots \times T_n$ by Lemma 10.1.10. Thus S is a submonoid of a finite product of monoids belonging to C, as required.

(iii) Let $S \in \mathbf{PH}(\mathsf{C})$. Then $S = S_1 \times \ldots \times S_n$ where each S_i is a homomorphic image of a monoid $T_i \in \mathsf{C}$. But $S_1 \times \ldots \times S_n$ is a homomorphic image of $T_1 \times \ldots \times T_n$ by Lemma 10.1.10. Thus S is a homomorphic image of a finite product of monoids from C, as required. □

Theorem 12.1.6 *Let* C *be a collection of monoids. Then* $\mathbf{HSP}(\mathsf{C})$ *is a pseudovariety of monoids containing* C*, and is in fact the smallest pseudovariety of monoids containing* C*.*

Proof By Lemma 12.1.4 and the remark following, we have that

$$\mathsf{C} \subseteq \mathbf{P}(\mathsf{C}) \subseteq \mathbf{SP}(\mathsf{C}) \subseteq \mathbf{HSP}(\mathsf{C}).$$

Thus

$$\mathsf{C} \subseteq \mathbf{HSP}(\mathsf{C}),$$

as claimed. Next we prove that $\mathbf{HSP}(\mathsf{C})$ is closed under the operations **H**, **S**, and **P**, which will show that it is a pseudovariety. Closure under **H** follows from Lemma 12.1.4(iii) since $\mathbf{H}^2 = \mathbf{H}$. Closure under **S** follows from Lemma 12.1.5(i) and Lemma 12.1.4(iii) because

$$\mathbf{SHSP}(\mathsf{C}) \subseteq \mathbf{HSSP}(\mathsf{C}) = \mathbf{HSP}(\mathsf{C}).$$

Finally, we prove closure under $\mathbf{P}$. We have that

$$\mathbf{PHSP}(\mathsf{C}) \subseteq \mathbf{HPSP}(\mathsf{C}) \subseteq \mathbf{HSPP}(\mathsf{C}),$$

using Lemma 12.1.5(iii) and (ii). It is clear that for any collection of monoids D we have that $\mathbf{P}(\mathsf{D}) \subseteq \mathbf{IP}(\mathsf{D})$. Thus

$$\mathbf{HSPP}(\mathsf{C}) \subseteq \mathbf{HSIPIP}(\mathsf{C}).$$

But by Lemma 12.1.4(iii), we have that

$$\mathbf{HSIPIP}(\mathsf{C}) = \mathbf{HSIP}(\mathsf{C}).$$

It is clear that for any collection of monoids D we have that

$$\mathbf{I}(\mathsf{D}) \subseteq \mathbf{H}(\mathsf{D}).$$

Thus

$$\mathbf{HSIP}(\mathsf{C}) \subseteq \mathbf{HSHP}(\mathsf{C}).$$

But then by Lemma 12.1.5(i) and Lemma 12.1.4(iii) we have that

$$\mathbf{HSHP}(\mathsf{C}) \subseteq \mathbf{HHSP}(\mathsf{C}) = \mathbf{HSP}(\mathsf{C}).$$

Thus

$$\mathbf{PHSP}(\mathsf{C}) \subseteq \mathbf{HSP}(\mathsf{C}),$$

as required. To show that $\mathbf{HSP}(\mathsf{C})$ is the smallest pseudovariety containing C, let M be any pseudovariety containing C. Then $\mathbf{HSP}(\mathsf{C}) \subseteq \mathbf{HSP}(\mathsf{M}) = \mathsf{M}$.□

The pseudovariety $\mathbf{HSP}(\mathsf{C})$ is called the *pseudovariety of monoids generated by* C. The following provides a slightly more succint way of describing the pseudovariety generated by C, and is left as an exercise.

Proposition 12.1.7 *A monoid belongs to the pseudovariety generated by* C *if and only if it divides a finite product of monoids belonging to* C. □

What we have done for monoids, we shall do for languages. Before that we prove a technical lemma, which will partially motivate the definition that follows. On a point of notation: it is easy to check that $(u^{-1}L)v^{-1} = u^{-1}(Lv^{-1})$ so that the expression $u^{-1}Lv^{-1}$ is unambiguous.

Lemma 12.1.8 *Let* $L \subseteq A^*$ *be a recognisable language with syntactic monoid* $M(L)$ *and where* $\alpha\colon A^* \to M(L)$ *is the natural monoid homomorphism associated with the syntactic congruence* σ_L. *Then for each* $m \in M(L)$ *the set*

$$\alpha^{-1}(m) = \sigma_L(x),$$

for some x, *can be written as a finite Boolean combination of the languages* $u^{-1}Lv^{-1}$, *where* $u, v \in A^*$.

Proof For $x \in A^*$ define

$$C(x) = \{(u, v) \in A^* \times A^* \colon uxv \in L\},$$

the set of contexts of x in L. Thus $(x, y) \in \sigma_L \Leftrightarrow C(x) = C(y)$. We prove that

$$\sigma_L(x) = (\bigcap_{(u,v)\in C(x)} u^{-1}Lv^{-1}) \setminus (\bigcup_{(u,v)\notin C(x)} u^{-1}Lv^{-1}).$$

We prove first that the left-hand side is contained in the right-hand side. Let $y \in \sigma_L(x)$. If $(u, v) \in C(x)$, then $uxv \in L$ and so $uyv \in L$. Hence $y \in u^{-1}Lv^{-1}$. If $(u, v) \notin C(x)$ then $uxv \notin L$ and so $uyv \notin L$, which implies $y \notin u^{-1}Lv^{-1}$. It follows that the left-hand side is contained in the right-hand side.

We now prove that the right-hand side is contained in the left-hand side. Let y be an element of the right-hand side. Suppose $uxv \in L$. Then $(u, v) \in C(x)$ and so $y \in u^{-1}Lv^{-1}$ from which we get $uyv \in L$. Now suppose that $uyv \in L$. Then $y \in u^{-1}Lv^{-1}$. Thus $(u, v) \in C(x)$ and so $uxv \in L$. It follows that $(x, y) \in \sigma_L$.

The *finiteness* of this Boolean combination follows from the fact that by Proposition 7.5.5(i) the left or right quotient of a recognisable language is recognisable and by Theorem 7.5.5(ii) a recognisable language has only a finite number of left and right quotients. □

The definition below is not an obvious one, although it is partially motivated by Lemma 12.1.7. We shall see, however, why it is correct when we prove the Variety Theorem. A ***variety of languages*** L is defined to be a family of languages L_A, where A ranges over all finite alphabets A, such that the following conditions hold:

(VL1) For each finite alphabet A, the set of languages L_A is a subset of $\mathsf{P}(A^*)$, which is closed under the Boolean operations.

(VL2) For each finite alphabet A, each language $L \in \mathsf{L}_A$ and each letter $a \in A$, both $a^{-1}L$ and La^{-1} belong to L_A.

(VL3) If $\alpha \colon A^* \to B^*$ is a monoid homomorphism and $L \in \mathsf{L}_B$ then $\alpha^{-1}(L) \in \mathsf{L}_A$.

Thus a variety of languages is a collection of sets of languages over different alphabets that fit together under taking inverse images. Our goal is to show that there is a bijection between the collection of pseudovarieties of monoids and the collection of varieties of languages. We now define two functions.

- Let M be a pseudovariety of monoids. Define $\mathcal{L}(\mathsf{M})$ as follows: for each finite alphabet A put

$$\mathcal{L}(\mathsf{M})_A = \{L \subseteq A^* \colon M(L) \in \mathsf{M}\}.$$

- Let L be a variety of languages. Define $\mathcal{M}(\mathsf{L})$ to be the pseudovariety of monoids generated by the collection of all syntactic monoids of languages belonging to L.

The following result ensures that $\mathcal{L}(\mathsf{M})$ is a variety of languages.

Proposition 12.1.9 *Let* M *be a pseudovariety of monoids. Then* $\mathcal{L}(\mathsf{M})$ *is a variety of languages.*

Proof We prove first that $\mathcal{L}(\mathsf{M})_A$ is the set of all languages over A, which are recognised by monoids in M. If $L \in \mathcal{L}(\mathsf{M})_A$, then $M(L) \in \mathsf{M}$ and $M(L)$ recognises L by Lemma 10.2.4. Now let $L \subseteq A^*$ be any language recognised by a monoid M in M. Then $M(L)$ divides M by Theorem 10.2.6. But $M \in \mathsf{M}$ and so $M(L) \in \mathsf{M}$ because pseudovarieties are closed under division by Proposition 12.1.7. Thus $L \in \mathcal{L}(\mathsf{M})_A$. We now have to check that (VL1), (VL2), and (VL3) hold. These verifications follow from Proposition 10.2.8. □

We can now prove our main result due to Eilenberg and Schützenberger.

Theorem 12.1.10 (The Variety Theorem) *The functions* $\mathcal{L}$ *and* $\mathcal{M}$ *are mutually inverse. There is therefore a bijection between the collection of all monoid pseudovarieties and the collection of all language varieties.*

Proof We show first that $\mathcal{L}$ is injective. Let M and M' be two pseudovarieties of monoids such that $\mathcal{L}(\mathsf{M}) = \mathcal{L}(\mathsf{M}')$. Let $M \in \mathsf{M}$. Then by Theorem 12.1.2, M divides a finite product $M(L_1) \times \ldots \times M(L_n)$, where each language $L_i \subseteq A^*$ is recognised by M for some alphabet A. It follows from the proof of Proposition 12.1.9 that $L_i \in \mathcal{L}(\mathsf{M})_A$. By assumption $\mathcal{L}(\mathsf{M})_A = \mathcal{L}(\mathsf{M}')_A$. Thus $L_i \in \mathcal{L}(\mathsf{M}')_A$. By definition $M(L_i) \in \mathsf{M}'$, and so $M(L_1) \times \ldots \times M(L_n) \in \mathsf{M}'$. But M divides $M(L_1) \times \ldots \times M(L_n)$ and so $M \in \mathsf{M}'$ because pseudovarieties are closed under division. We have shown that $\mathsf{M} \subseteq \mathsf{M}'$. By symmetry, the reverse inclusion holds, and we have proved the claim.

Let L be a variety of languages. We show that $\mathcal{L}\mathcal{M}(\mathsf{L}) = \mathsf{L}$, which implies that $\mathcal{L}$ is surjective. It is immediate from the definitions that $\mathsf{L}_A \subseteq \mathcal{L}(\mathcal{M}(\mathsf{L}))_A$ for each alphabet A. We prove the reverse inclusion. Let $L \in \mathcal{L}(\mathcal{M}(\mathsf{L}))_A$. Then $M(L)$ divides a product of the form $M = M(L_1) \times \ldots \times M(L_n)$, where $L_i \in \mathsf{L}_{A_i}$ for some finite alphabets A_i. We shall prove that $L \in \mathsf{L}_A$. Because $M(L)$ is the syntactic monoid of L and divides M, we know by Theorem 10.2.6 that M recognises L. Thus there is a monoid homomorphism ϕ: $A^* \to M$ and a subset $P \subseteq M$ such that $L = \phi^{-1}(P)$. Let π_i: $M \to M(L_i)$ be the projection homomorphism, and let η_i: $A_i^* \to M(L_i)$ be the natural homomorphism corresponding to the syntactic congruence σ_{L_i}. Define $\phi_i = \pi_i\phi$. We therefore have the following diagram of monoid homomorphisms where the unbroken lines are the known functions:

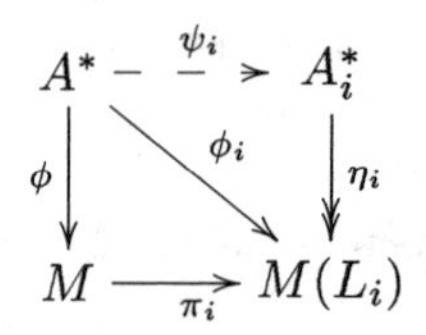

By Proposition 10.2.3, there is a monoid homomorphism $\psi_i\colon A^* \to A_i^*$ such that $\eta_i \psi_i = \phi_i$. Let $m \in M$. Then $m = (m_1, \ldots, m_n)$ where each $m_i \in M(L_i)$. The proof that $L \in \mathsf{L}_A$ occurs in stages:

(1) $\eta_i^{-1}(m_i) \in \mathsf{L}_{A_i}$ by Lemma 12.1.8 and (VL1) and (VL2).

(2) $\phi_i^{-1}(m_i) = \psi_i^{-1}(\eta_i^{-1}(m_i)) \in \mathsf{L}_A$ by (VL3) and (1).

(3) $\phi^{-1}(m) = \bigcap_{i=1}^{n} \phi_i^{-1}(m_i) \in \mathsf{L}_A$ by (VL1) and (2). We leave the proof of this as an exercise.

(4) $L = \phi^{-1}(P) = \bigcup_{m \in P} \phi^{-1}(m) \in \mathsf{L}_A$ by (VL1) and (3).

We have therefore proved that $L \in \mathsf{L}_A$, as required.

We have proved that $\mathcal{L}$ is a bijection, and that $\mathcal{L}\mathcal{M}(\mathsf{L}) = \mathsf{L}$ for each variety of languages L. Let M be a pseudovariety of monoids and suppose that $\mathcal{M}\mathcal{L}(\mathsf{M}) = \mathsf{M}'$. Then $\mathcal{L}\mathcal{M}\mathcal{L}(\mathsf{M}) = \mathcal{L}(\mathsf{M}')$. Thus $\mathcal{L}(\mathsf{M}) = \mathcal{L}(\mathsf{M}')$ and so $\mathsf{M} = \mathsf{M}'$. It follows that $\mathcal{L}$ and $\mathcal{M}$ are mutually inverse operations. □

Exercises 12.1

1. Prove Lemma 12.1.1(i): if $\{\rho_i\colon i \in I\}$ is a set of congruences on a semigroup S, then $\rho = \bigcap_{i \in I} \rho_i$ is a congruence on S.

2. Show that finite commutative monoids form a pseudovariety.

3. Complete the proof of Lemma 12.1.4 for **S** and **IP**.

4. Show that **P** satisfies conditions (i) and (ii) of Lemma 12.1.4.

5. Prove Proposition 12.1.7: a monoid belongs to the pseudovariety generated by C iff it divides a finite product of monoids belonging to C.

6. Show that the intersection of any family of pseudovarieties of finite monoids is a pseudovariety of finite monoids. Show that the intersection of all pseudovarieties of finite monoids containing a class C of finite monoids is precisely the pseudovariety generated by C.

7. Prove (3) in Theorem 12.1.10.

8. Let G be a pseudovariety of finite groups. Let $\mathsf{M}(\mathsf{G})$ be the collection of all finite monoids whose maximal groups belong to G. Show that $\mathsf{M}(\mathsf{G})$ is a pseudovariety of finite monoids. You will find Exercises 11.2, Questions 8 and 9 helpful.

12.2 Equations for pseudovarieties

A pseudovariety of monoids is a collection of monoids closed under homomorphic images, finite products, and submonoids. In this section, we shall derive an alternative characterisation of pseudovarieties which is of practical importance.

If ρ is a relation on a monoid M, then we can define the intersection $\rho^\sharp$ of all congruences on M that contain ρ. Since the intersection of congruences is a congruence by Lemma 12.1, the relation $\rho^\sharp$ is a congruence. It is clear that $\rho^\sharp$ is the smallest congruence on M containing ρ. We say that ρ *generates* the congruence $\rho^\sharp$. If σ is a congruence such that $\sigma = \rho^\sharp$ where ρ is finite, then we say that σ is *finitely generated.*

Proposition 12.2.1 *Let σ be a congruence on A^* such that A^*/σ is finite. Then σ is finitely generated.*

Proof Denote the σ-congruence class of x by $[x]$. Let $l[x]$ stand for the length of the shortest strings in $[x]$. Put $k = 1 + \max\{l[x]\colon x \in A^*\}$. Since there are only finitely many congruence classes, this number is well-defined and finite. By construction, each congruence class contains a string of length strictly less than k. Put

$$\rho = \{(x, y) \in A^* \times A^*\colon x\sigma y,\ \text{and } |x| \leq k, |y| < k\}.$$

For each $x \in A^*$ such that $|x| \leq k$, there exists $y \in A^*$ such that $|y| < k$ and $x\sigma y$. We prove that $\rho^\sharp = \sigma$. Since $\rho \subseteq \sigma$ it follows that $\rho^\sharp \subseteq \sigma$. Suppose that that the reverse inclusion did not hold; we shall obtain a contradiction. Assume therefore that for some u and v we have that $u\sigma v$ but $(u, v) \notin \rho^\sharp$. Choose such a pair (u, v) with $|u| + |v|$ minimal. Suppose that $|u| \geq k$. Then $u = xu'$ for some u' such that $|u'| = k$. By definition, there exists $v' \in A^*$ such that $(u', v') \in \rho$ and $|v'| < k$. From $(u', v') \in \rho$ we get that $(xu', xv') \in \rho^\sharp$ and so $u = xu'\sigma xv'$. It follows that $xv'\sigma v$. Observe that $(xv', v) \notin \rho^\sharp$, for if it were then $(u, v) \in \rho^\sharp$, which is a contradiction. In addition, $|xv'| < |u|$ since $u = xu'$ and $|v'| < k$ and $|u'| = k$. However, the pair (xv', v) now contradicts the minimality of the pair (u, v). It follows that $|u| < k$. Now repeat the argument above with v. It follows that $u\sigma v$ and $|u|, |v| < k$. Hence $(u, v) \in \rho$ and so $(u, v) \in \rho^\sharp$, which is a contradiction. □

We shall now describe a method for describing pseudovarieties of semigroups. We have already met examples of this method. For instance, a semigroup S is commutative if $ab = ba$ for all $a, b \in S$. This can be expressed more formally as follows. Let x and y be two variables. Then a semigroup is commutative if the equation $xy = yx$ holds whenever we substitute a for x and b for y, where a and b are any elements of S. We now generalise this idea.

Let A be a countably infinite alphabet; this is the only place where we use an infinite alphabet. Let $u, v \in A^*$. We say that a monoid M *satisfies the equation* $u = v$ if $\alpha(u) = \alpha(v)$ for every monoid homomorphism $\alpha\colon A^* \to M$. Just as in the case of A finite, a monoid homomorphism from A^* is determined by its values on the elements of the set A. It follows that this formal definition says precisely that $u = v$ is satisfied by M if, whenever we substitute the symbols in u and in v by elements of M, the resulting two elements of M are always equal.

It is easy to show that the collection of all finite monoids $\mathsf{M}(u, v)$ satisfying the equation $u = v$ is a pseudovariety. The intersection of any family of pseudovarieties is a pseudovariety, so if we have a sequence of equations $(u_n = v_n)_{n \geq 1}$ whether finite or infinite, then $\mathsf{M}' = \bigcap_{n \geq 1} \mathsf{M}(u_n, v_n)$ is also a pseudovariety. We say that M' is *defined* by the equations $(u_n = v_n)_{n \geq 1}$. In other words, a monoid belongs to M' if it satisfies all of the equations $u_n = v_n$. I shall write $\mathsf{M}(u_n = v_n)_{n \in I}$ for the pseudovariety of monoids satisfying all the equations $u_n = v_n$.

Example 12.2.2 The pseudovariety defined by $\mathsf{M}(xy = yx)$ is the pseudovariety of commutative monoids. The pseudovariety $\mathsf{M}(x^2 = x)$ is the pseudovariety of all monoids in which every element is idempotent: the *band monoids.* A commutative band is called a *semilattice.* Thus $\mathsf{M}(xy = yx, x^2 = x)$ is the pseudovariety of semilattice monoids. □

The notion of a pseudovariety being defined by a set of equations turns out not to be sufficient to describe all pseudovarieties. To do this, we need a more general notion.

Suppose we have an infinite sequence of equations $(u_n = v_n)_{n \geq 1}$. We say that a monoid M *ultimately satisfies* these equations if there is a k such that M satisfies $u_n = v_n$ for all $n \geq k$. The collection of monoids M'' ultimately satisfying the equations $(u_n = v_n)_{n \geq 1}$ is also a pseudovariety.

Example 12.2.3 By Theorem 10.3.2, a finite monoid S is aperiodic iff $a^n = a^{n+1}$ for some $n > 0$ and for all $a \in S$. Thus a finite monoid is aperiodic iff it ultimately satisfies the equations $(x^n = x^{n+1})_{n>0}$. □

We now have the following alternative characterisation of pseudovarieties of finite monoid.

Theorem 12.2.4 *Every pseudovariety of finite monoids is ultimately defined by a sequence of equations.*

Proof Let M be a pseudovariety of finite monoids. Let $A = \{a_1, a_2, \ldots\}$ be a countably infinite alphabet. Put $A_n = \{a_1, \ldots a_n\}$. Clearly A_n^* is a submonoid of A^*. All the monoids in M are finite. Each finite monoid on n elements is isomorphic to one defined on the set $X_n = \{1, 2, \ldots, n\}$. Thus each monoid in M of size n is isomorphic to a monoid defined on X_n. We can clearly list all the monoids defined on X_n that belong to M. There is therefore a list of monoids,

$$M_1, M_2, \ldots,$$

of non-decreasing size such that each of them belongs to M and each monoid in M is isomorphic to exactly one of them.

For each $n \geq 1$, put $S_n = M_1 \times \ldots \times M_n$. There are only finitely many monoid homomorphisms $\phi: A_n^* \to S_n$ because A_n and S_n are both finite and each such homomorphism is determined by its values on the set A_n. Let ρ_n be the congruence defined on A_n^* as the intersection of all congruences $\ker(\phi)$ where ϕ is a monoid homomorphism. By the above there are only finitely many such congruences. By Lemma 12.1.1, A_n^*/ρ_n is isomorphic to a submonoid of the finite product of the finite monoids $A^*/\ker(\phi)$. Each of these monoids is isomorphic to a submonoid of S_n. It follows that A_n^*/ρ_n is isomorphic to a submonoid of a product of finitely many copies of S_n with itself. In particular, A^*/ρ_n is finite. By Proposition 12.2.1, the congruence ρ_n is finitely generated. Thus there is a finite set $E_n \subseteq A_n^* \times A_n^*$ generating ρ_n. Put $E = \bigcup_{n \geq 1} E_n$. We claim that M is ultimately defined by the equations E.

Let $M \in$ M. Then there exists k such that M is isomorphic to a submonoid of S_n for every $n \geq k$. By construction, S_n satisfies the equations E_n for each $n \geq k$. It follows that M satisfies the equations E_n for each $n \geq k$.

To prove the converse, let M be a finite monoid that satisfies E_n for each $n \geq k$ for some k. Choose $n \geq \max(k, |M|)$. Let ϕ: $A_n^* \to M$ be a surjective monoid homomorphism. By definition $\phi(u) = \phi(v)$ for all $(u, v) \in E_n$. It follows that $(u, v) \in \rho_n$ implies that $\phi(u) = \phi(v)$. There is therefore a surjective monoid homomorphism from A_n^*/ρ_n to M by Proposition 9.2.12. But A_n^*/ρ_n is isomorphic to a submonoid of a product of finitely many copies of S_n. Thus M divides a product of finitely many copies of S_n. Hence $M \in$ M. □

Exercises 12.2

1. Show that $\mathsf{M}(u = v)$ is a pseudovariety.

2. Show that the collection of monoids ultimately satisfying the set of equations $(u_n = v_n)_{n \geq 1}$ is a pseudovariety.

3. Prove that if M is a pseudovariety generated by a single monoid, then M is defined by a sequence of equations (rather than merely being ultimately defined).

4. Let $\mathcal{K}$ be one of Green's relations. We say that a finite monoid is *$\mathcal{K}$-trivial*

if $\mathcal{K}$ is the equality relation. Thus by Theorem 11.3.2, the finite aperiodic monoids are just the $\mathcal{H}$-trivial ones. Prove the following:

(i) A finite monoid is $\mathcal{R}$-trivial iff it satisfies $(xy)^n x = (xy)^n$ for some $n > 0$.

(ii) A finite monoid is $\mathcal{L}$-trivial iff it satisfies $y(xy)^n = (xy)^n$ for some $n > 0$.

(iii) A finite monoid is $\mathcal{J}$-trivial iff it satisfies $(xy)^n x = (xy)^n = y(xy)^n$ for some $n > 0$.

Show also that a finite monoid is $\mathcal{J}$-trivial iff it satisfies $(xy)^n = (yx)^n$ and $x^n = x^{n+1}$ for some $n > 0$.

12.3 Summary of Chapter 12

- *Pseudovariety of monoids*: A non-empty collection of finite monoids closed under taking submonoids, homomorphic images, and finite products.

- *Variety of languages*: A collection of languages over each finite alphabet, where the set of languages over a fixed alphabet is closed under the Boolean operations and the taking of left and right quotients. The whole ensemble is closed under inverse images of monoid homomorphisms between free monoids.

- *The Variety Theorem*: Each pseudovariety is paired with a unique variety of languages and vice versa.

- *Pseudovarieties and equations*: Each pseudovariety is ultimately defined by an infinite sequence of equations. This means that each monoid in the pseudovariety satisfies all the equations from some point on.

12.4 Remarks on Chapter 12

In writing this chapter, I have consulted both Lallement [77] and Pin [103] constantly. The proof of Theorem 12.1.6 is adapted from [70]. Theorem 12.1.10 is proved in [40] and Theorem 12.2.4 in [41]. Both results are due to the combined efforts of two outstanding mathematicians: Eilenberg and Schützenberger.

I indicated in Section 12.1, that the definition of a variety of languages is perhaps not the most obvious, although all three conditions are certainly plausible: condition (VL1) is natural from the point of view of the results we proved back in Section 2.6; condition (VL2) highlights the importance of taking left and right quotients, something we first met in Chapter 7; Lemma 12.1.8 is further evidence for the importance of the Boolean operations taken in tandem with left and right quotients; and, finally, condition (VL3) is reasonable, particularly given what we said in the second paragraph of the Remarks on Chapter 4. However more mathematical motivation for the definition can be found in Lallement [77] (pp 165–169).

The Variety Theorem provides the framework for talking about recognisable languages and finite monoids. It says that if you have a pseudovariety of finite monoids then there will be an associated variety of languages, although it will not tell you what these languages look like; that involves extra work and depends on the properties of the pseudovariety of monoids in question. Likewise, if you have a variety of languages, the Theorem tells us that there is an associated pseudovariety of finite monoids, but again it will not tell us what they look like; you have to do extra work. For example, Schützenberger's Theorem can be interpreted in terms of the Variety Theorem: it identifies the pseudovariety of monoids that correspond to the variety of star-free languages as being the pseudovariety of aperiodic monoids.

In this chapter, we set up a correspondence between pseudovarieties of finite monoids and varieties of languages. This correspondence can also be extended to deal with semigroups rather than monoids. A *pseudovariety of semigroups* is a collection of semigroups closed under subsemigroups, homomorphic images, and finite products of semigroups. In this context, a language L is a subset of A^+ not A^*. The *syntactic semigroup* of L, denoted $S(L)$, is defined to be A^+/σ_L where σ_L is the congruence defined on A^+ as follows: $(x, y) \in \sigma_L$ iff for all $u, v \in A^*$ (sic) we have that $uxy \in L \Leftrightarrow uyv \in L$. The results we proved about syntactic monoids generalise to syntactic semigroups. *Varieties of languages* in this context are defined as before except that (VL3) is replaced by homomorphisms between free semigroups. These varieties are sometimes called *+-varieties* to distinguish them from the varieties of languages I defined in the text that are sometimes called *$*$-varieties.* The Variety Theorem can then be extended to give a correspondence between pseudovarieties of finite semigroups and +-varieties of languages.

The Variety Theorem is, in effect, a research program, and as such it has been enormously fruitful. We have seen that the finite aperiodic monoids are just the $\mathcal{H}$-trivial ones and, of course, Schützenberger's Theorem tells us that the associated languages are the star-free ones. Similar analyses have also been carried out on the $\mathcal{J}$-trivial, $\mathcal{R}$-trivial, and $\mathcal{L}$-trivial finite monoids. The $\mathcal{J}$-trivial monoids correspond to the class of 'piecewise testable languages.' These are the languages in the Boolean algebra generated by languages of the form $A^*a_1A^*a_2 \dots A^*a_nA^*$ where the a_i are letters; this means that every piecewise testable language can be described by Boolean combinations of such languages. This is a deep result first proved by Simon [121], and described in Pin [103]. A more recent account can be found in [61]. The $\mathcal{R}$-trivial and $\mathcal{L}$-trivial monoids are much easier to handle. See Pin, again, for a nice account [103].

An important class of languages was introduced in Section 2.4. A language $L \subseteq A^+$ is said to be *locally testable* iff it is in the Boolean algebra generated by languages of the form uA^*, A^*v, and A^*wA^*. This variety of languages was first identified as promising by McNaughton and Papert [88]. The characterisation of the corresponding pseudovariety of monoids is difficult and is due to McNaughton [89], Zalcstein [137], and Brzozowski and Simon [21]: a language L is locally testable iff its syntactic semigroup $S = S(L)$ has the following property: for each idempotent $e \in S$ the local monoid eSe is a semilattice. A sketch of the proof of this result is given by Perrin [100]. The ramifications of this result were unexpected. Through work of Straubing [128] and Tilson [133], it became clear that the study of certain kinds of pseudovarieties depended on 'category theory,' where, roughly, this theory is a generalisation of both graph theory and monoid theory. A recent paper on locally testable languages is [134].

Another strand in the relationship between semigroups and languages, which I have only been able to touch on in this book, concerns the sequential functions defined at the end of Chapter 1. It is interesting to try to decompose such functions into series or parallel arrangements of simpler sequential functions. The theory that tells us whether and how this is possible is intimately connected with the theory of pseudovarieties of monoids and semigroups. This theory is described in Chapter VI of [40].

If you want to find out more about research on finite monoids and regular languages, there are a number of places to start. Howie's book [67] covers roughly the same material as mine. Pin's book [103] is probably the obvious next text to read, since it overlaps with my book, but then goes considerably further. Useful surveys of the field are [105] and [15]. The collections [45] and [56], and the paper [125] will give you some idea of the range of work being carried out in this area.

The description of pseudovarieties in terms of equations being ultimately satisfied is replaced in more advanced work by a description in terms of 'pseudoidentities' due to Reiterman [110]. This approach is described in Jorge Almeida's advanced book [3].

And to make an end is to make a beginning.

T. S. Eliot

Appendix A

Discrete mathematics

The aim of this appendix is to sketch out the background you need from discrete mathematics to read this book. It is not intended to be a replacement for having attended a course or worked through a book, but it can serve as a detailed synopsis or an *aide-memoire.* There are many good introductions to discrete mathematics. The book by Johnsonbaugh [71], for example, contains more than enough.

A.1 Logic and proofs

Mathematics is first and foremost about proofs. Human beings are already endowed with reasoning abilities, even if they are a bit rough-and-ready, so my philosophy on proofs is that the best way to learn what they are is to try to do them. If you have not met proofs before, then it will take time before you begin to feel comfortable with them. But the effort is worth it in the depth of understanding gained as a result. In this section, I will just give you the bare essentials on proofs, and encourage you, even if you are not familiar with them, to dive in anyway.

It is important to understand at the outset, that in mathematics we only deal in things that are either true or false: one or other but not both. Sentences such as: 'To be or not to be?', 'Out damn spot!' and 'Aaah!' ask questions, make demands, express emotions, and so on, but are neither true nor false. All these things play a big role in trying to do mathematics, but are rigorously expunged from the writing of mathematics. A *statement* is a sentence that is either true or false. Mathematics, once all the human factors have been wrung from it, is about statements. The goal of mathematics is to determine which statements are true and which false. As human beings we add the word 'interesting' in front of the word 'statements.' To show that a statement is false it is enough to find one single example, called a *counterexample*, where the statement is false. To show that a statement is true you have to prove it.

Most statements in mathematics take the form: 'if P, then Q' meaning 'if P is true, then Q is true.' We also say 'P implies Q', or 'P is a sufficient condition for Q,' or that 'Q is a necessary condition for P.' We often write '$P \Rightarrow Q$.' The letter P stands for the assumptions, and the letter Q stands for the conclusions. The statement '$Q \Rightarrow P$' is called the *converse* of the statement '$P \Rightarrow Q$.' A statement and its converse say very different things, and the truth or falsity of one does not imply the truth or falsity of the other. For example, the statement 'if n is a positive number, then n^2 is a positive number' is true, but the converse, 'if n^2 is a positive number, then n is a positive number,' is false, as the counterexample $n = -1$ shows. If '$P \Rightarrow Q$' and '$Q \Rightarrow P$' are both true, then we write '$P \Leftrightarrow Q$,' which is read 'P if and only if Q' often abbreviated to 'P iff Q.' We also say that 'P is a necessary and sufficient condition for Q.'

There are two ways to try to prove a statement of the form 'if P, then Q.' The simplest is called a *direct proof*, and takes the following form: we write down our assumption P; we carry out an argument; we deduce Q. An *argument* consists of things we know to be true, perhaps because they were proved earlier, or because of a definition, or because of what we are allowed to assume in P. In addition, we can write down statements that are deductions from previous statements in our argument. For example, if 'R or S' is true and I know that R is false then I know that S must be true. Little arguments such as this are often described at the beginning of books on discrete mathematics, but are usually easy to pick up from working through specific proofs. Unfortunately, direct proofs do not always work and then we have recourse to what are known as *indirect proofs*. These take the following form: assume that P is true; then for the sake of argument assume that Q is *false*, i.e., the reverse of what we want. Then carry out an argument that leads to an assertion contradicting something we know to be true. Because assuming Q false leads to a contradiction, and because statements are either true or false but not both, it follows that Q must really be true. Most arguments in this book are direct, and many mathematicians have a bias in favour of direct arguments.

There is one type of direct proof that we often use in this book: *proof by induction*. This works for those statements P whose truth is determined by the truth of a sequence of special cases: $P(0)$, $P(1)$, $P(2)$, $\ldots$, that is, to prove P is true we have to prove that the statements $P(n)$ are true for all natural numbers n. The general form a proof by induction is as follows:

- *Base case:* Check first that $P(0)$ is true.

- *Induction hypothesis:* Assume that $P(k)$ is true.

- *Deduction:* Use the truth of $P(k)$ to establish the truth of $P(k+1)$; it is at this stage we usually have to do some work.

- *Conclusion:* Conclude that $P(n)$ is true for all n, from which it follows that P is true, as required.

There are a number of variations on this basic pattern. First, $P(n)$ may only be true from some number n_0 onward, in which case the base case above will be 'check $P(n_0)$ is true.' Second, we often have to assume that $P(n_0), \ldots, P(k)$ are *all* true to prove that $P(k+1)$ is true.

Mathematicians have developed some special terms concerned with proofs. Proved statements are called *propositions* or *theorems.* There is no real agreement on how to use these terms, but a rule of thumb is that theorems are more important than propositions; that is how I distinguish between them in this book. A *lemma* is a proved statement that is more useful in proving other results than of particular interest in itself. A *corollary* is a proved statement that follows almost immediately from a proposition or theorem. A *conjecture* is a statement that we think is true but do not yet have a proof for (and so could turn out to be false). The conclusion to a proof is nowadays marked by drawing a box. Older books mark the ends of proofs by writing 'q.e.d.,' which is an abbreviation of the Latin 'quod erat demonstrandum,' meaning 'that which was to be proved.' The notion of proving things was due to the Greeks, which is why the terminology has a classical ring to it.

We prove a wide variety of propositions and theorems in this book, but an important class concern algorithms. Informally, an *algorithm* is a recipe for doing something. In school, we learn algorithms for adding, subtracting, multiplying, and dividing numbers. Mathematics is full of algorithms. They are useful because they tell us how to actually do something rather than just telling us that something exists. Most often they enable us to write a program to get a computer to do something. Every program is an implementation of an algorithm, and an algorithm is only an algorithm if it could be implemented by some computer. However, if I give you an algorithm, how do you know that it does what it says on the wrapper? This is another role for proofs: proving or verifying that an algorithm is correct. Many of the proofs in this book do exactly that. I should add that the word 'algorithm,' like the word 'algebra,' is of Arabic origin: what the Greeks started the Arabs continued.

Everything I have said applies to writing down proofs, and how to read proofs, it does not tell you how to find proofs. This is an exercise in creativity as much as logic. The book by Polya [108] contains useful hints.

A.2 Set theory

The most basic mathematical notion is that of a *set.* This is simply a collection of objects that we wish to regard as a whole. The members of a set are called its *elements.* We usually, but not invariably, use capital letters to name sets: such as A, B, or C or fancy capital letters such as $\mathbb{N}$ and $\mathbb{Z}$. The elements of a set are usually, but not invariably, denoted by lower case letters. If x *is an element of* the set A then we write

$$x \in A,$$

and if x *is not an element of* the set A then we write

$$x \notin A.$$

A set is completely defined once we have said what its elements are; there is nothing more to them than that. If a set has only finitely many elements, then we might be able to list them: this is done by putting them in 'curly brackets' $\{$ and $\}$. It is important to remember that order and repetition do not matter when describing the elements of a set.

Examples A.2.1 Here are two examples of sets.

(1) The set of binary digits (or bits) is

$$\{0, 1\}.$$

Both 0 and 1 are elements of this set, so we can write $0 \in \{0, 1\}$ and $1 \in \{0, 1\}$. However, 2 is not an element of this set and so we can write $2 \notin \{0, 1\}$.

(2) The set of all letters in the English alphabet is

$$\{a, b, c, \ldots, x, y, z\}.$$

We have that $a \in \{a, b, c, \ldots, x, y, z\}$ but $\alpha \notin \{a, b, c, \ldots, x, y, z\}$.

$\square$

We often describe a set by saying what properties an element must have to belong to the set:

$$\{x \colon P(x)\}$$

means 'the set of all things x that satisfy the condition P.' For example,

$$\{x \colon x \text{ is a whole number divisible by 2 }\}$$

is the set

$$\{0, 2, 4, 6, 8, 10, \ldots\}.$$

It is quite possible for a set to contain elements that are themselves sets. For example, $A = \{\{a\}, \{a, b\}\}$ is a set whose elements are $\{a\}$ and $\{a, b\}$, which both happen to be sets. This is important, because it means that the notion of set, which at first appears rather simple-minded, is capable of great elaboration.

Examples A.2.2 Some more examples of sets.

(1) The empty set $\{\}$, which is the unique set with no elements. We usually denote this set by $\emptyset$, a letter from the Danish alphabet being the first one of the Danish word for 'empty.'

(2) The set $\mathbb{N} = \{0, 1, 2, 3, \ldots\}$ of all *natural numbers*.

(3) The set $\mathbb{Z} = \{\ldots, -3, -2, -1, 0, 1, 2, 3, \ldots\}$ of all *integers*. The reason $\mathbb{Z}$ is used to designate this set is because 'Z' is the first letter of the word 'Zahl' the German for number.

(4) The set $\mathbb{Q}$ of all *rational numbers*; i.e., all fractions both positive and negative. Here $\mathbb{Q}$ is used to designate this set because 'Q' is the first letter of the word 'quotient'.

(5) The set $\mathbb{R}$ of all *real numbers*; i.e., all numbers that can be represented by decimals with potentially infinitely many digits after the decimal point.

$\square$

Notice that in some of the above definitions of sets, I have used '...' to mean 'and so on in the obvious way': if you are going to define a set in this way, you should be pretty certain that the remaining elements really are obvious; a definition of a set shouldn't be an intelligence test.

Sometimes it is easy to decide what the elements of a set are and sometimes we need to do a lot of work: for example, we might need to prove a theorem or apply an algorithm.

If you have a set A, then you can form a new set B by *choosing* elements from A to put in B. We say that B is a *subset* of A, which is written $B \subseteq A$. For example, if $A = \{a, b, c, d\}$. The set $\{a, d\}$ is a subset of A. If A is not a subset of B, then we write $A \nsubseteq B$. So $\{a, \alpha\} \nsubseteq A$. If $A \subseteq B$ and $A \neq B$, then we say that A is a *proper* subset of B.

Examples A.2.3 Some inclusions among sets.

(1) $\emptyset \subseteq A$ for every set A. This often surprises people and needs proving. To do this, we need to argue indirectly. Suppose that $\emptyset$ were not a subset of A. Then there would have to be some element of $\emptyset$ that was not an element of A. But this is impossible, because $\emptyset$ is empty.

(2) $A \subseteq A$ for every set A.

(3) $\mathbb{N} \subseteq \mathbb{Z} \subseteq \mathbb{Q} \subseteq \mathbb{R}$.

$\square$

Other familiar subsets of $\mathbb{N}$ are the set of even numbers, the set of odd numbers, and the set of primes.

All the subsets of a set A together form a new set called the *powerset* of A, denoted $\mathsf{P}(A)$. For example, if $A = \{a, b\}$ then $\mathsf{P}(A) = \{\emptyset, \{a\}, \{b\}, \{a, b\}\}$.

There are three basic operations on sets. To define them, let A and B be sets:

- $A \cap B$, the *intersection* of A and B, is the set of all elements that belong both to A and to B. If A and B are a pair of sets such that $A \cap B = \emptyset$ then they are said to be *disjoint*.

- $A \cup B$, the *union* of A and B, is the set of all elements that belong to A or to B or to both. The word 'or' in mathematics always means 'or both.' In everyday use, we tend to use the word 'or' in an exclusive sense: 'Would you like this bar of delicious Belgian chocolate, or would you like this slice of apple strudel with whipped cream?'; I'm expecting you to make a decision, not to eat both.

- $A \setminus B$, read '*A minus B*,' also called *relative complement*, the set of all elements that belong to A but not to B.

These operations are called *Boolean operations* after George Boole.

Notation Throughout this book, union is also denoted by + when refering to unions of languages.

Set operations have a number of important properties. Let A, B, and C be any sets:

- *Empty set laws:* $A \cap \emptyset = \emptyset$ and $A \cup \emptyset = A$.

- *Idempotency laws:* $A \cap A = A$ and $A \cup A = A$.

- *Associative laws:* $A \cap (B \cap C) = (A \cap B) \cap C$ and $A \cup (B \cup C) = (A \cup B) \cup C$.

- *Distributive laws:* $A \cap (B \cup C) = (A \cap B) \cup (A \cap C)$ and $A \cup (B \cap C) = (A \cup B) \cap (A \cup C)$.

- *De Morgan's laws:* $A \setminus (B \cap C) = (A \setminus B) \cup (A \setminus C)$, and $A \setminus (B \cup C) = (A \setminus B) \cap (A \setminus C)$.

To prove these laws hold, we often use the following standard method. For two sets A and B, we have that:

$$A = B \Leftrightarrow A \subseteq B \text{ and } B \subseteq A.$$

Thus to prove that $A = B$, it is enough to prove separately that $A \subseteq B$ and $B \subseteq A$.

If there are several sets $A_1, \ldots, A_n$, then the intersection $A_1 \cap \ldots \cap A_n$ is unambiguous because of associativity. This intersection can also be written $\bigcap_{i=1}^{n} A_i$. If $\{A_i \colon i \in I\}$ is a set of sets, then their intersection can be written $\bigcap_{i \in I} A_i$. Similar comments apply to union. Observe that forming infinite intersections and infinite unions is mathematically unproblematical.

Notation In the case of languages L_i, I write $\sum_{i=1}^{n} L_i$ rather than $\bigcup_{i=1}^{n} L_i$ in this book.

It often happens that we are interested in a fixed set U and its subsets. In this context, U is called (rather grandly) a *universe*. If $A \in \mathsf{P}(U)$ then $A' = U \setminus A$ is called the *complement* of A. For example, if we take as our universe the set $\mathbb{N}$, and let A be the set of even numbers, then A' is the set of odd numbers.

I mentioned above that we can form sets whose elements are themselves sets. This is where the real power of set theory manifests itself. For example, consider the two sets

$$A = \{a, \{a, b\}\} \text{ and } B = \{c, \{c, d\}\}.$$

When is $A = B$? It is not hard to check that we need $a = c$ and $b = d$. We define $(a, b) = \{a, \{a, b\}\}$, called an *ordered pair*. Whereas $\{a, b\}$ contains the elements a and b in no particular order, (a, b) contains the elements a and b in the given order. Thus set notation, which is essentially un-orderly, can be used to define the orderly. The element a is called the *first component* and the element b is called the *second component*. We can also define *ordered triples*, which look like (a, b, c), and more generally *ordered n-tuples*, which look like $(a_1, \ldots, a_n)$. Ordered n-tuples are also called *lists* or *finite sequences*.

If A and B are sets, then the set $A \times B$, called the *product of A by B*, is defined to be the following set:

$$A \times B = \{(a, b) \colon a \in A \text{ and } b \in B\},$$

the set of all ordered pairs where the first component comes from A and the second component comes from B. For example, if $A = \{1, 2\}$ and $B = \{a, b, c\}$, then

$$A \times B = \{(1, a), (1, b), (1, c), (2, a), (2, b), (2, c)\}.$$

More generally we can define $A \times B \times C$ to consist of all ordered triples where the first component comes from A, the second from B, and the third from C. Yet more generally we can define $A_1 \times \ldots \times A_n$ to consist of all n-tuples where the ith component comes from A_i.

If you want to find out more about set theory, then I recommend Paul Halmos' book [59].

A.3 Numbers and matrices

I do not use numbers very much in this book, but there are some properties we shall need. If a and b are integers then we say that '*a divides b*,' written $a \mid b$, if $a = bq$ for some integer q. The symbol '$|$' is just a replacement for the word 'divides,' it does not mean the same as a/b, which is either the fraction $\frac{a}{b}$ or means 'a divided by b.' If a and b are natural numbers and d is a number that divides them both then we call d a *common divisor* of a and b. The largest common divisor of a and b is called their *greatest common divisor*. If the greatest common divisor of a and b is 1, then a and b are said to be *coprime*.

If $a, b \in \mathbb{Z}$ and $a > 0$, then we can find integers r, q such that $b = aq + r$ where $0 \leq r < a$. This result is called the *Remainder Theorem*. An *even number* is an integer n that can be written $n = 2m$ for some integer m. An *odd number* is an integer n that can be written $n = 2m + 1$ for some integer m.

The Remainder Theorem is the basis for a fast algorithm, called *Euclid's algorithm*, for computing the greatest common divisor of two natural numbers. This algorithm can be used to prove the following theorem: a and b are coprime if and only if there are integers x and y such that $1 = xa + yb$.

A natural number p is said to be *prime* if it satisfies *two* conditions: first, $p \geq 2$ and second, the only divisors of p are 1 and p itself. Notice that 1 is *not* prime. This exception is for technical reasons. The only even prime number is 2. There are infinitely many primes, a theorem first proved by Euclid over 2000 years ago. The proof goes as follows: let $p_1, \ldots, p_n$ be the first n prime numbers. Put $N = (p_1 \ldots p_n) + 1$. Then there are two possibilities. Either N is itself prime or if not it must be divisible by a prime. If p is a prime dividing N, then it must be bigger than p_n, because dividing N by any of the primes $p_1, \ldots, p_n$ leaves remainder one. In either event, we have found a prime bigger than p_n. It follows that there can be no largest prime, and so there are infinitely many primes.

I use matrices in one or two places for convenience. A *matrix* is just a rectangular array consisting of m rows and n columns of *elements* or *entries*. We speak of an $m \times n$-matrix. If $m = n$ we say the matrix is *square*. If $m = 1$ we say the matrix is a *row matrix*, if $n = 1$ we say the matrix is a *column matrix*. Matrices are usually denoted by upper case letters $A, B, \ldots$, sometimes in bold. If A is an $m \times n$-matrix then the element in the ith row and jth column, where $1 \leq i \leq m$ and $1 \leq j \leq n$, is denoted A_{ij}. If A is a square matrix, then the entries A_{ii} are said to form the *leading diagonal* of A. Let A be an $m \times n$-matrix. The *transpose* A^T of A is the $n \times m$-matrix defined by interchanging rows and columns of A. Thus $(A^T)_{ji} = A_{ij}$. A matrix A is *symmetric* if $A^T = A$.

Let A be an $m \times n$-matrix and let B be an $n \times p$-matrix. Suppose that the entries of A and B are of the same type and can be added and multiplied, and assume further that at least addition is associative. Then we can define the *product matrix* AB to be the matrix, where

$$(AB)_{ik} = \sum_{j=1}^{n} A_{ij} B_{jk}.$$

The matrix AB is an $m \times p$-matrix. This operation is called *matrix multiplication*. For reasonable choices of entries for our matrices, say numbers or, as we shall use in this book, languages, it turns out that $(AB)C = A(BC)$. Thus matrix multiplication is associative. If A and B are both $m \times n$-matrices with entries that can be added, then we can add A and B to form a new matrix $A + B$ whose entries are defined as follows:

$$(A + B)_{ij} = A_{ij} + B_{ij}.$$

The set of all $n \times n$-matrices with entries from some set R is denoted by $M_n(R)$.

A.4 Graphs

I do not use any graph theory as such, but I do use graphs to represent information. A *graph* is a diagram consisting of points, called *vertices*, joined together by lines, called *edges*. Each edge joins exactly two vertices. Each graph can be drawn in many ways; think of the edges as being elastic so that they can be stretched and deformed as much as you like. We usually try to draw a graph in the simplest possible way. When we draw graphs we are sometimes forced to make edges cross, such points are not vertices. A pair of vertices joined by an edge are said to be *adjacent*. Two or more edges joining the same pair of vertices are called *multiple edges*. An edge joining a vertex to itself is called a *loop*. A graph with no loops or multiple edges is called a *simple graph*.

Be warned that the terminology of graph theory is not completely standardised. Furthermore, the word 'graph' is used with another, and quite different, meaning in mathematics: we talk about 'graphs of functions.'

The *degree* of a vertex is the number of edges that emerge from it. A loop at a vertex counts 2 to the degree because we count each end of the loop. A vertex of degree zero is called an *isolated vertex*. If G is a graph, then a *subgraph* of G is a graph G', which is obtained from G by erasing some of the edges and some of the vertices (where 'some' can mean anything between 'none' and 'all'). A graph that is all in one piece is said to be *connected*, whereas one that splits into two or more pieces is said to be *disconnected* and the connected pieces that make up a graph are called its *connected components*. A *path* in a graph is a succession of edges with the property that the second vertex of each edge is the first vertex of the next one. The *length* of a path is just the number of edges involved in the path. A path is said to be *closed* if it begins and ends at the same vertex. A *cycle* in a graph is a closed path with distinct vertices. Observe that a graph is connected precisely when any two vertices can be joined by a path. A *directed graph* or *digraph* is a graph in which edges are replaced by arrows. A *labelled* graph whether directed or not is simply a graph of either sort with symbols labelling the edges or arrows. Let G be a graph with set of vertices V. Then the *adjacency matrix* A of G is the $|V| \times |V|$-matrix with rows and columns labelled by the vertices defined as follows: A_{ij} is the number of edges in G joining i and j. If G is a directed graph, then the entry A_{ij} of its adjacency matrix is defined to be the number of edges *from i to j*. A connected graph without cycles is called a *tree*. A tree with a distinguished vertex is called a *rooted tree*. In a rooted tree, the vertices of degree one (excluding if necessary the root) are called the *leaves*. A vertex in a tree which is not a leaf is said to be *interior*.

A.5 Functions

Probably the single most important definition in mathematics is that of a function. It took hundreds of years to evolve into its current form, and at first sight may seem 'overprecise.' However, experience has shown that it is the right one.

Let X and Y be sets. A *function* α *from* X *to* Y, written $\alpha\colon X \to Y$, is a rule that associates with each element $x \in X$ exactly one element $y \in Y$. We write $\alpha(x) = y$ or $x \mapsto y$. The set X is called the *domain of* α and the set Y is called the *codomain of* α. The element $\alpha(x)$ is called the *value of* α *at* x. The domain of alpha is denoted $\mathrm{dom}(\alpha)$, and the codomain of α is denoted $\mathrm{cod}(\alpha)$.

It is important to note that a function is determined by *three* pieces of information: the domain, the codomain, and the rule. Change *any one* of these three pieces of information and you change the function. It is for this reason that I said above that the definition may look 'overprecise.'

Examples A.5.1 Some examples of functions.

(1) The function that takes as input an ordered pair of integers (a, b) and produces as output their sum. The domain of this function is $\mathbb{Z} \times \mathbb{Z}$, the codomain is $\mathbb{Z}$ and the rule is $(a, b) \mapsto a + b$. We can similarly define the subtraction function and the multiplication function with the same domains and codomains. Notice that we write $a + b$ rather than $+(a, b)$. The former is known as *infix notation* and the latter as *prefix notation*.

(2) The function that takes as input an ordered pair of rational numbers (a, b) where $b \neq 0$ and produces as output a/b. The domain of this function is $\mathbb{Q} \times (\mathbb{Q} \setminus \{0\})$ and the codomain is $\mathbb{Q}$. Observe that we have used set notation to make clear what the domain of this function is. If I had defined the domain to be $\mathbb{Q} \times \mathbb{Q}$, then I would have been obliged to assign a meaning to '$a/0$.'

(3) Let X be any set. Then the function $1_X\colon X \to X$ is defined by $1_X(x) = x$ for each $x \in X$. It is called the *identity function on* X.

(4) Let $y \in Y$ be a fixed element and let $\alpha\colon X \to Y$ be the function defined by $\alpha(x) = y$ for each $x \in X$. Then α is called a *constant function*.

$\square$

Notation If $\alpha\colon X \to Y$ is a function, then the value of α at the input x is written $\alpha(x)$. We say that the *argument* x *is written on the right*. However, it is just as valid to write $(x)\alpha$. In this case, we say that the *argument* x *is* *written on the left*. Now we have to choose which convention to follow. It is not an important choice, but we should all agree. In this book, all my arguments will be written on the right **except** when I deal with the full transformation monoid $T(X)$ in Chapter 8 and thereafter.

Functions were traditionally defined by means of formulae; most functions in this book are defined by means of tables: for each element of the domain x, I tell you the corresponding element of the codomain $\alpha(x)$. Another method that is occasionally used is *recursion*. I only use this method when I define 'extended transition functions.'

Let $\alpha: X \to Y$ and $\beta: W \to Z$ be two functions. Then we can define a new function

$$\alpha \times \beta: X \times W \to Y \times Z$$

by $(\alpha \times \beta)(x, w) = (\alpha(x), \beta(w))$.

A function $\alpha: X \to X$ is sometimes called a *unary operation* on X. A function $\alpha: X \times X \to X$ is sometimes called a *binary operation* on X. Let X be a set, $\alpha: X \to X$ and $\beta: X \times X \to X$ be functions and let $X' \subseteq X$. We say that X' is *closed under* α if $\alpha(x') \in X'$ for each $x \in X'$, and we say that X' is *closed under* β if $\beta(x', x'') \in X'$ for each $(x', x'') \in X' \times X'$.

Let $\alpha: X \to Y$ and $\beta: Y \to Z$ be functions. Then we can combine these two functions to form a new function $\beta \circ \alpha$ called the *composite of* β *and* α: the domain of $\beta \circ \alpha$ is X, the codomain of $\beta \circ \alpha$ is Z, and the rule is given by: $(\beta \circ \alpha)(x) = \beta(\alpha(x))$. This operation on functions is called *composition*.

Functions are classified according to their properties. Particularly important are the following. Let $\alpha: X \to Y$ be a function. Then α is said to be:

- *injective (one-to-one)* if $\alpha(x) = \alpha(x')$ implies $x = x'$.
- *surjective (onto)* if for each $y \in Y$, there exists an $x \in X$ such that $\beta(x) = y$.
- *bijective* if it is both injective and surjective.

In words: a function is injective if two different arrows cannot converge to the same element of the codomain; a function is surjective if every element of the codomain is at the receiving end of an arrow. It can be checked that the composition of injective functions is injective, and the composition of surjective functions is surjective. In general, if $\alpha: X \to Y$ the set

$$\{\alpha(x): a \in X\}$$

is a subset of Y but need not equal Y. We call it the *image of* α, denoted $\text{im}(\alpha)$. A function is surjective precisely when its codomain is equal to its image.

Let $\alpha: X \to Y$ be a *bijective* function. Because α is bijective it is in particular surjective. This means that for each $y \in Y$, there is *at least one*

$x \in X$ such that $\alpha(x) = y$. Because α is bijective it is in particular injective. This means that if $\alpha(x) = \alpha(x')$ then $x = x'$. It follows that for each $y \in Y$ there is *exactly one* $x \in X$ such that $\alpha(x) = y$. We may therefore define a function $\alpha^{-1}: Y \to X$ by

$$\alpha^{-1}(y) = x \text{ iff } \alpha(x) = y.$$

The function α^{-1} is called the *inverse of* α. This function has the effect of 'undoing' the effect of α. By this, I mean the following. Let $x \in X$. Calculate $(\alpha^{-1} \circ \alpha)(x)$. This is just $\alpha^{-1}(\alpha(x))$. But α maps x to $\alpha(x)$ and so $\alpha^{-1}(\alpha(x)) = x$. Hence $(\alpha^{-1} \circ \alpha)(x) = x$. In fact, it is not hard to show that α^{-1} is itself bijective and that its inverse is α. That is $(\alpha^{-1})^{-1} = \alpha$. It follows that $(\alpha \circ \alpha^{-1})(y) = y$ for each $y \in Y$.

Let $\alpha: X \to Y$ be a function. If $X' \subseteq X$ then define

$$\alpha(X') = \{\alpha(x): x \in X'\}.$$

If $Y' \subseteq Y$ then define

$$\alpha^{-1}(Y') = \{x \in X: \alpha(x) \in Y'\},$$

called the *inverse image of* α. There is a clash of notations here: we can calculate the inverse image of a function even when it is not bijective.

The number of elements in a set X is called the *cardinality of* X, denoted by $|X|$. Two sets have the same cardinality iff there is a bijection between them. A set X is said to be *countably infinite* if there is a bijection from X to $\mathbb{N}$.

There is a slight generalisation of the notion of a function that we shall need. A *partial function* $\alpha: X \to Y$ from X to Y is defined by the condition that for each $x \in X$ there is *at most one* element $y \in Y$. This means that $\alpha(x)$ is either not defined *or* if it is defined then it has only one value. The set of elements x of X for which $\alpha(x)$ is defined is called the *domain of definition* of α. I shall denote the domain of definition of the partial function α by $\text{dom}(\alpha)$. If α is a partial funtion such that $\alpha(x) = \alpha(x')$ implies $x = x'$ whenever $\alpha(x)$ and $\alpha(x')$ are defined, then we say that α is a *partial injection*.

Let $\alpha, \beta: X \to Y$ be two partial functions. We say that β is an *extension* of α or that α is a *restriction* of β if $\text{dom}(\alpha) \subseteq \text{dom}(\beta)$ and for all $x \in \text{dom}(\alpha)$ we have that $\alpha(x) = \beta(x)$. We write $\alpha \subseteq \beta$, if β is an extension of α. Let α_i, where $i \in \mathbb{N}$, be a collection of functions such that $\alpha_i \subseteq \alpha_{i+1}$. Let $x \in \text{dom}(\alpha_i) \cap \text{dom}(\alpha_j)$. Then either $i \leq j$ or $j \leq i$. In either case, $\alpha_i(x) = \alpha_j(x)$. It follows that if we define $\alpha: X \to Y$ by $\alpha(x) = \alpha_i(x)$ if $x \in \text{dom}(\alpha_i)$, then α is a well-defined partial function. We say that α is defined by taking the 'union' of the functions α_i.

A.6 Relations

Let A and B be sets. A *relation* ρ *from* A *to* B is simply a subset $\rho \subseteq A \times B$. If $A = B$ we say that ρ is a *relation on* A. We usually write $x \,\rho\, y$ instead of

$(x, y) \in \rho$. If ρ is a relation from A to B and $a \in A$ then we define

$$\rho(a) = \{b \in B : (a, b) \in \rho\}.$$

To describe a relation ρ we have to say which ordered pairs belong to ρ. If ρ is finite then we can simply list them. Alternatively, we can draw a *directed graph* or *arrow diagram*: the vertices of the graph are $A \cup B$ and we draw an arrow from a to b if $a \in A$, $b \in B$ and $(a, b) \in \rho$. If $A = B$ then we have one set of vertices and draw arrows between them.

Examples A.6.1 Some relations.

(1) Let A be the set of all people now alive. Let μ be the relation on A defined by: $a \, \mu \, b$ if a is the mother of b. Family relations of this type are the motivation for the terminology.

(2) Let $A = \{a, b, c, d, e, f\}$ and $B = \{\diamondsuit, \heartsuit, \spadesuit, \clubsuit\}$ and let

$$\rho = \{(a, \heartsuit), (a, \spadesuit), (c, \clubsuit), (f, \heartsuit), (f, \spadesuit)\}.$$

This is a random example. A relation is defined as soon as the sets A and B are given and the set of ordered pairs. A relation need not have any particular meaning, although in practice we are mainly interested in ones that do.

(3) Let $A = \{1, 2, 3, 4, 5\}$ and $B = \{1, 10, 11, 100, 101\}$ and

$$\sigma = \{(1, 1), (2, 10), (3, 11), (4, 100), (5, 101)\}.$$

This is a relation that tells us the binary representations of the first few integers.

(4) Let $A = \mathbb{N}$. Define $a \mid b$ iff a divides b.

□

I will now concentrate on relations defined on a set X so that the relation in question is a subset of $X \times X$. Relations are classified according to their properties. The following are the most important. Let ρ be a relation on a set X. Then ρ is said to be:

- *reflexive* if $(x, x) \in \rho$ for every $x \in X$.
- *symmetric* if $(x, y) \in \rho$ implies $(y, x) \in \rho$.
- *antisymmetric* if $(x, y) \in \rho$ and $(y, x) \in \rho$ imply $x = y$.
- *transitive* if $(x, y) \in \rho$ and $(y, z) \in \rho$ imply $(x, z) \in \rho$.

A relation that is reflexive, symmetric, and transitive is called an *equivalence relation.* Equivalence relations are discussed in more detail in Chapter 7, and play an important role thereafter.

A relation $\leq$ on a set X is called a *partial order (relation)* if it is reflexive, *antisymmetric*, and transitive. Finite partial orders can be described by means of *Hasse diagrams*: these are graphs in which the vertices are the elements of the partially ordered set, and if $x \leq y$ and there is no element z such that $x \leq z \leq y$, then we draw x lower down the page than y and connect x and y by an edge. The whole partial order can be recovered from this diagram.

A partial order is called a *linear order* if for each pair of elements $x, y \in X$ we have that either $x \leq y$ or $y \leq x$. This is true of the usual order on the integers, for example. The natural numbers have an important order property: each non-empty subset of $\mathbb{N}$ has a smallest element. The set of integers $\mathbb{Z}$ with its usual order is linearly ordered, but it does not have the property that every non-empty subset has a smallest element.

Bibliography

[1] A. V. Aho, R. Sethi, J. D. Ullman, *Compilers: principles, techniques, and tools*, Addison-Wesley, 1986.

[2] A. V. Aho, Algorithms for finding patterns in strings, in *Algorithms and complexity, Volume A* (editor J. van Leeuwen), Elsevier, 1992, 257–300.

[3] J. Almeida, *Finite semigroups and universal algebra*, World Scientific, 1994.

[4] D. Angluin, Inference of reversible languages, *Journal of the Association for Computing Machinery* **29** (1982), 741–765.

[5] M. A. Arbib (editor), *Algebraic theory of machines, languages, and semigroups*, Academic Press, 1968.

[6] M. A. Arbib, *Brains, machines and mathematics*, Springer-Verlag, 1987.

[7] D. N. Arden, Delayed-logic and finite-state machines, in *Theory of computing machine design*, University of Michigan Press, Ann Arbor, Michigan, 1960, 1–35.

[8] R. A. Baeza-Yates, Searching subsequences, *Theoretical Computer Science* **78** (1991), 363–376.

[9] V. Bar-Hillel, M. Perles, E. Shamir, On formal properties of simple phrase structure grammars, *Zeitschrift für Phonetik, Sprachwissenschaft, und Kommunicationsforschung* **14** (1961), 143–172.

[10] M. P. Béal, D. Perrin, Symbolic dynamics and finite automata, in *Handbook of formal languages, Volume 2* (editors G. Rozenberg, A. Salomaa), Springer, 1997, 463–506.

[11] G. Berry, R. Sethi, From regular expressions to deterministic automata, *Theoretical Computer Science* **48** (1986), 117–126.

[12] J. Berstel, *Transductions and context-free languages*, B.G. Teubner Stuttgart, 1979.

[13] J. Berstel, J.-E. Pin, Local languages and the Berry-Sethi algorithm, *Theoretical Computer Science* **155** (1996), 439–446.

[14] J. Berstel, C. Reutenauer, *Rational series and their languages*, Springer-Verlag, 1988.

[15] M. J. J. Branco, Varieties of languages, in *Semigroups, algorithms, automata and languages* (editors G. M. S. Gomes, J.-E. Pin, P. V. Silva), World Scientific, 2002, 91–132.

[16] W. Brauer, *Automatentheorie*, B.G. Teubner, Stuttgart, 1984.

[17] A. Brüggemann-Klein, Regular expressions into finite automata, *Lecture Notes in Computer Science* **583** (1992), 87–98.

[18] J. A. Brzozowski, Canonical regular expressions and minimal state graphs for definite events, in *Mathematical theory of automata*, Polytechnic Press of the Polytechnic Institute of Brooklyn, Brooklyn, New York, 1962, 529–561.

[19] J. A. Brzozowski, Derivatives of regular expressions, *Journal of the Association for Computing Machinery* **11** (1964), 481–494.

[20] J. A. Brzozowski, E. J. McCluskey, Signal flow graph techniques for sequential circuit state diagrams, *IEEE Transactions on Electronic Computers*, **12** (1963), 67–76.

[21] J. A. Brzozowski, I. Simon, Characterization of locally testable events, *Discrete Mathematics* **4** (1973), 243–271.

[22] J. R. Büchi, *Finite automata, their algebras and grammars* (editor D. Siefkes), Springer-Verlag, 1989.

[23] J. Carroll, D. Long, *Theory of finite automata*, Prentice-Hall International, 1989.

[24] J.-M. Champarnaud, J.-E. Pin, A maxmin problem on finite automata, *Discrete Applied Mathematics* **23** (1989), 91–96.

[25] C. Choffrut, Minimizing subsequential transducers: a survey, *Theoretical Computer Science* **292** (2003), 131–143.

[26] S. Choi, D. T. Huynh, Finite automaton aperiodicity is PSPACE-complete, *Theoretical Computer Science* **88** (1991), 99–116.

[27] N. Chomsky, Three models for the description of languages, *IRE Transactions of Information Theory* **2** (1956), 113–124.

[28] N. Chomsky, *Syntactic structures*, Mouton, 1975.

[29] N. Chomsky, G. A. Miller, Finite state languages, *Information and Control* **1** (1958), 91–112.

[30] N. Chomsky, M. P. Schützenberger, The algebraic theory of context-free languages, in *Computer programming and formal systems* (editors P. Brafford, E. Hirschberg), North-Holland, Amsterdam, 1963, 118–161.

[31] P. S. Churchland, *Neurophilosophy*, The MIT Press, 1990.

[32] A. H. Clifford, G. B. Preston, *The algebraic theory of semigroups*, vol. I, American Mathematical Society, Providence, R.I., 1961.

[33] A. H. Clifford, G. B. Preston, *The algebraic theory of semigroups*, vol. II, American Mathematical Society, Providence, R.I., 1967.

[34] D. I. A. Cohen, *Introduction to computer theory*, Second Edition, John Wiley and Sons, 1997.

[35] J. H. Conway, *Regular algebra and finite machines*, Chapman and Hall, 1971.

[36] T. H. Cormen, C. E. Leiserson, R. L. Rivest, *Introduction to algorithms*, The MIT Press, 1990.

[37] M. Chrochemore, C. Hancart, Automata for matching patterns, in *Handbook of formal languages, Volume 2* (editors G. Rozenberg, A. Salomaa), Springer, 1997, 399–462.

[38] K. Culik II, J. Kari, Digital images and formal languages, in *Handbook of formal languages, Volume 3* (editors G. Rozenberg, A. Salomaa), Springer, 1997, 599–616.

[39] S. Eilenberg, *Automata, languages and machines, Volume A*, Academic Press, 1974.

[40] S. Eilenberg, *Automata, languages and machines, Volume B*, Academic Press, 1976.

[41] S. Eilenberg, M. P. Schützenberger, On pseudovarieties, *Advances in Mathematics* **19** (1976), 413-418.

[42] T. S. Eliot, *Four quartets*, Faber Paperbacks, 1976.

[43] D. B. A. Epstein, J. W. Cannon, D. F. Holt, S. V. F. Levy, M. S. Paterson, W. P. Thurston, *Word processing in groups*, Jones and Bartlett, 1992.

[44] R. W. Floyd, R. Beigel, *The language of machines*, Computer Science Press, 1994.

[45] J. Fountain (editor), *Semigroups, formal languages and groups*, Kluwer Academic Publishers, 1995.

[46] J. B. Fraleigh, *A first course in abstract algebra*, Fifth Edition, Addison-Wesley, 1994.

[47] J. E. F. Friedl, *Mastering regular expressions*, Second Edition, O'Reilly, 2002.

[48] R. Fröberg, C. Gottlieb, R. Häggkvist, On numerical semigroups, *Semigroup Forum* **35** (1987), 63–83.

[49] W. H. Gardner, *Poems and prose of Gerard Manley Hopkins*, Penguin Books, 1978.

[50] F. Gécseg, I. Peák, *Algebraic theory of automata*, Akadémiai Kiadó, Budapest, 1972.

[51] R. H. Gilman, Formal languages and infinite groups, *DIMACS Series in Discrete Mathematics and Theoretical Computer Science* **25** (1996), 27–51.

[52] D. Giammarresi, A. Restivo, Two-dimensional languages, in *Handbook of formal languages, Volume 3* (editors G. Rozenberg, A. Salomaa), Springer, 1997, 215–267.

[53] S. Ginsburg, G. F. Rose, Operations which preserve definability in languages, *Journal of the Association for Computing Machinery* **10** (1963), 175–195.

[54] A. Ginzburg, *Algebraic theory of automata*, Academic Press, 1968.

[55] V. M. Glushkov, The abstract theory of automata, *Russian Mathematical Surveys* **16** (1961), 1–53.

[56] G. M. S. Gomes, J.-E. Pin, P. V. Silva, *Semigroups, algorithms, automata and languages*, World Scientific, 2002.

[57] R. I. Grogorchuk, V. V. Nekrashevich, V. I. Sushchanskii, Automata, dynamical systems, and groups, *Proceedings of the Steklov Institute of Mathematics* **231** (2000), 128–203.

[58] F. von Haeseler, *Automatic sequences*, Walter de Gruyter, 2003.

[59] P. R. Halmos, *Naive set theory*, Springer-Verlag, 2001.

[60] J. Hartmanis, R. E. Stearns, Sets of numbers defined by finite automata, *American Mathematical Monthly* **74** (1967), 539–542.

[61] P. M. Higgins, A proof of Simon's theorem on piecewise testable languages, *Theoretical Computer Science* **178** (1997), 257-264.

[62] P. M. Higgins, A new proof of Schützenberger's theorem, *International Journal of Algebra and Computation* **10** (2000), 217–220.

[63] A. Hodges, *Alan Turing: the enigma*, Vintage, 1992.

[64] J. E. Hopcroft, An $n \log n$ algorithm for minimising the states in a finite automaton, in *Theory of machines and computations* (editor Z. Kohavi), New York, Academic Press, 1971, 189–196.

[65] J. E. Hopcroft, J. D. Ullman, *Introduction to automata theory, languages and computation*, Addison-Wesley, 1979.

[66] J. E. Hopcroft, R. Motwani, J. D. Ullman, *Introduction to automata theory, languages and computation*, Second Edition, Addison Wesley, 2001.

[67] J. M. Howie, *Automata and languages*, Clarendon Press, Oxford, 1991.

[68] J. M. Howie, *Fundamentals of semigroup theory*, Clarendon Press, Oxford, 1995.

[69] D. A. Huffman, The synthesis of sequential switching circuits, *Journal of the Franklin Institute* **257** (1954), 161–190, 275–303.

[70] Th. Ihringer, *Allgemeine Algebra*, B. G. Teubner, Stuttgart, 1988.

[71] R. Johnsonbaugh, *Discrete mathematics*, Fifth Edition, Prentice Hall, 2001.

[72] D. Jurafsky, J. H. Martin, *Speech and language processing*, Prentice Hall, 2000.

[73] A. J. Kfoury, R. N. Moll, M. A. Arbib, *A programming approach to computability*, Springer-Verlag, 1982.

[74] S. C. Kleene, Representation of events in nerve nets and finite automata, in *Automata studies* (editors C. E. Shannon, J. McCarthy), Princeton University Press, 1956, 3–42.

[75] D. C. Kozen, *Automata and computability*, Springer-Verlag, 1997.

[76] K. Krohn, J. Rhodes, The algebraic theory of machines I, *Transactions of the American Mathematical Society* **116** (1965), 450–464.

[77] G. Lallement, *Semigroups and combinatorial applications*, John Wiley and Sons, 1979.

[78] M. V. Lawson, *Inverse semigroups*, World Scientific, 1998.

[79] D. Lewin, D. Protheroe, *Design of logic systems*, Chapman and Hall, 1994.

[80] H. R. Lewis, C. H. Papadimitriou, *Elements of the theory of computation*, Second Edition, Addison Wesley Longman, 1998.

[81] D. Lind, B. Marcus, *Symbolic dynamics and coding*, Cambridge University Press, 1995.

[82] M. Lothaire, *Combinatorics on words*, Cambridge University Press, 1997.

[83] S. W. Margolis, J.-E. Pin, Expansions, free inverse semigroups and Schützenberger product, *Journal of Algebra* **110** (1987), 298–305.

[84] S. W. Margolis, J. C. Meakin, Free inverse monoids and graph immersions, *International Journal of Algebra and Computation* **3** (1993), 79–99.

[85] W. S. McCulloch, W. Pitts, A logical calculus of the ideas immanent in nervous activity, *Bulletin of Mathematical Biophysics* **5** (1943), 115–133.

[86] R. McNaughton, H. Yamada, Regular expressions and state graphs for automata, *IRE Transactions on Electronic Computers* **9** (1960), 39–47.

[87] R. McNaughton, S. Papert, The syntactic monoid of a regular event, in *Algebraic theory of machines, languages and semigroups* (editor M. A. Arbib), Academic Press, 1968, 297–312.

[88] R. McNaughton, S. Papert, *Counter-free automata*, MIT Press, Cambridge, 1971.

[89] R. McNaughton, Algebraic decision procedures for local testability, *Mathematical Systems Theory* **8** (1974), 60–76.

[90] G. H. Mealy, A method for synthesizing sequential circuits, *Bell System Technical Journal* **34** (1955), 1045–1079.

[91] Yu. T. Medvedev, On the class of events representable in a finite automaton, 1956, in Russian, reprinted in English in [96].

[92] B. Mikolajczak (editor), *Algebraic and structural automata theory*, North-Holland, 1991.

[93] M. Minsky, *Computation: finite and infinite machines*, New York, Prentice-Hall, 1967.

[94] R. Moll, M. A. Arbib, A. J. Kfoury, *An introduction to formal language theory*, Springer-Verlag, 1988.

[95] E. F. Moore, Gedanken-Experiments on sequential machines, in *Automata studies* (editors C. E. Shannon, J. McCarthy), Princeton University Press, 1956, 129–153.

[96] E. F. Moore (editor), *Sequential machines: selected papers*, Addison-Wesley, 1964.

[97] J. Myhill, Finite automata amd the representation of events, *Wright Air Development Command Technical Report* **57–624**, (1957), 112–137.

[98] A. Nerode, Linear automaton transformations, *Proceedings of the American Mathematical Society* **9** (1958), 541–544.

[99] M. Perles, M. O. Rabin, E. Shamir, The theory of definite automata, *IEEE Transactions of Electronic Computing* **12** (1963), 233–243.

[100] D. Perrin, Finite automata, in *Handbook of theoretical computer science* (editor J. van Leeuwen), Elsevier Science Publishers B.V., 1990, 3–57.

[101] D. Perrin, Les débuts de la theorie des automates, *Technique et Science Informatique* **14** (1995), 409–433.

[102] C. Petzold, *Codes*, Microsoft Press, 1999.

[103] J.-E. Pin, *Varieties of formal languages*, North Oxford Academic, 1986.

[104] J.-E. Pin, On reversible automata, in *Lecture Notes in Computer Science* **583**, Springer, 401–416.

[105] J.-E. Pin, Finite semigroups and recognisable languages: an introduction, in *Semigroups, formal languages and groups* (editor J. Fountain) Kluwer Academic Publishers, 1995, 1–32.

[106] J.-E. Pin, Syntactic semigroups, in *Handbook of formal languages, Volume 1* (editors G. Rozenberg, A. Salomaa), Springer, 1997, 679–746.

[107] J.-E. Pin, Tropical semirings, in *Idempotency* (editor J. Gunawardena), CUP, 1998, 50–69.

[108] G. Polya, *How to solve it*, Penguin Books, 1990.

[109] M. O. Rabin, D. Scott, Finite automata and their decision problems, *IBM Journal of Research and Development* **3** (1959), 114–125. Reprinted in *Sequential machines* (editor E. F. Moore), Addison-Wesley, Reading, Massachusetts, 1964, 63–91.

[110] J. Reiterman, The Birkhoff theorem for finite algebras, *Algebra Universalis* **14** (1982), 1–10.

[111] J. Rhodes, *Applications of automata theory and algebra*, Department of Mathematics, University of California, Berkeley, California.

[112] E. Roche, Y. Schabes (editors), *Finite-state language processing*, The MIT Press, 1997.

[113] A. Salomaa, *Theory of automata*, Pergamon Press, 1969.

[114] M. P. Schützenberger, Une théorie algébrique du codage, in *Séminaire Dubreil-Pisot* (1955/56), exposé no. 15.

[115] M. P. Schützenberger, Une théorie algébrique du codage, *Comptes Rendus des Séances de l'Académie des Sciences Paris* **242** (1956), 862–864.

[116] M. P. Schützenberger, On finite monoids having only trivial subgroups, *Information and control* **8** (1965), 190–194.

[117] M. P. Schützenberger, Sur certaines variétés de monoids finis, in *Automata theory* (editor E. R. Cainiello), Academic Press, New York, 1966.

[118] C. E. Shannon, J. McCarthy (editors), *Automata studies*, Princeton University Press, Princeton, New Jersey, 1956.

[119] D. Shasha, C. Lazere, *Out of their minds*, Copernicus, 1995.

[120] J. C. Shepherdson, The reduction of two-way automata to one-way automata, *IBM Journal of Research and Development* **3** (1959), 198–200.

[121] I. Simon, Piecewise testable events, *Lecture Notes in Computer Science* **33**, Springer-Verlag, 1975, 214–222.

[122] C. C. Sims, *Computation with finitely presented groups*, Cambridge University Press, 1994.

[123] M. Sipser, *Introduction to the theory of computation*, PWS Publishing Company, 1997.

[124] M. Smith, *Station X*, Channel 4 Books, 2000.

[125] B. Steinberg, Finite state automata: a geometric approach, *Transactions of the American Mathematical Society* **353** (2001), 3409–3464.

[126] G. A. Stephen, *String searching algorithms*, World Scientific, 1994.

[127] I. Stewart, *Galois theory*, Second Edition, Chapman and Hall, 1998.

[128] H. Straubing, Finite semigroup varieties of the form $V * D$, *Journal of Pure and Applied Algebra* **36** (1985), 53–94.

[129] H. Straubing, *Finite automata, formal logic, and circuit complexity*, Birkhäuser, 1994.

[130] W. Thomas, Languages, automata, and logic, in *Handbook of formal languages, Volume 3* (editors G. Rozenberg, A. Salomaa), Springer, 1997, 389–455.

[131] K. Thompson, Regular expression search algorithm, *Communications of the ACM* **11** (1968), 419–422.

[132] W. P. Thurston, Groups, tilings and finite state automata, *Summer 1989 AMS Colloquium Lectures*, National Science Foundation, University of Minnesota.

[133] B. Tilson, Categories of algebra: an essential ingredient in the theory of monoids, *Journal of Pure and Applied Algebra* **48** (1987), 83–198.

[134] A. N. Trahtman, A polynomial time algorithm for local testability and its level, *International Journal of Algebra and Computation* **9** (1998), 31–39.

[135] A. M. Turing, On computable numbers with an application to the Entscheidungsproblem, *Proceedings of the London Mathematical Society* **2** (1936), 230–265. Erratum: Ibid **43** (1937), 544–546.

[136] S. Yu, Regular languages, in *Handbook of formal languages, Volume 1* (editors G. Rozenberg, A. Salomaa), Springer, 1997, 41–110.

[137] Y. Zalcstein, Locally testable languages, *Journal of Computing System Science* **6** (1972), 151–167.

Index